FAMILIES MENTAL HEALTH AND CHALLENGES IN THE 21ST CENTURY

Families Mental Health and Challenges in the 21st Century contains the papers presented at the 1st International Conference of Applied Psychology on Humanity 2022 (ICAPH 2022, Malang, Indonesia, 27 August 2022). The contributions focus on the challenges in micro-family environments that are faced with rapid developments of technology and information in the 21st century. The issues addressed in the book include:

- Family Strengthening
- Principles and Practices
- Children and Woman Protection
- Family Resilience
- Crisis and Challenge

Families Mental Health and Challenges in the 21st Century will be of interest to professionals and academics involved or interested in psychology, the field of mental health and related disciplines.

PROCEEDINGS OF THE 1ST INTERNATIONAL CONFERENCE OF APPLIED PSYCHOLOGY ON HUMANITY (ICAPH 2022), MALANG, INDONESIA, 27 AUGUST 2022

Families Mental Health and Challenges in the 21st Century

Edited by

Syamsul Arifin
Department of Islamic Education, University of Muhammadiyah Malang, Indonesia

Salis Yuniardi, Diah Karmiyati, Iswinarti, Nida Hasanati, Tulus Winarsunu and Cahyaning Suryaningrum
Department of Psychology, University of Muhammadiyah Malang, Indonesia

Routledge
Taylor & Francis Group
LONDON AND NEW YORK

First published 2023
by Routledge
4 Park Square, Milton Park, Abingdon, Oxon OX14 4RN

and by Routledge
605 Third Avenue, New York, NY 10158
e-mail: enquiries@taylorandfrancis.com
www.routledge.com – www.taylorandfrancis.com

Routledge is an imprint of the Taylor & Francis Group, an informa business

ISBN: 978-1-032-51442-0 (hbk)
ISBN: 978-1-032-51467-3 (pbk)
ISBN: 978-1-003-40238-1 (ebk)

DOI: 10.1201/9781003402381

Typeset in Times New Roman
by MPS Limited, Chennai, India

Families Mental Health and Challenges in the 21st Century – Arifin et al. (eds)
© 2024 The Editor(s), ISBN: 978-1-032-51442-0

Table of Contents

Families Mental Health and Challenges in the 21st Century – Arifin et al. (eds)
© 2024 The Editor(s), ISBN: 978-1-032-51442-0

Preface

The International Conference of Applied Psychology on Humanity is the first international conference organized by the Faculty of Psychology, University of Muhammadiyah Malang. The 1st ICAP on Humanity is a part of the international conference series held by *Lembaga Pengembangan Publikasi Ilmiah* (Institute of Scientific Publication Development) UMM.

By looking at the rapid growing of technology in the 21st century, many things can be done easily with technology. Humans want all things effective and efficient so they tend to adapt easily, but new problems arise behind this technological revolution. Technology has a big impact on human life, both physically and psychologically. In several studies, this impact spreads across all ages, both in the macro- and micro-environment. In the micro-family environment, many challenges are faced with the development of technology and information, such as communication problems, fatigue at work, balance in work/school life and family life, roles in the family, financial, parenting, and many other problems that are increasingly developing for the family looking for a solution.

Therefore, the conference aims are means for researchers, academia, professionals, and stakeholders to share information that can be used as input to solve the current families' mental health issues or problems. It also works as a means for researchers to publish their findings and opens up great opportunities to build collaborative networks between national and international scientists.

The conference was held on 27 August 2022 both online and offline from the University of Muhammadiyah Malang. We received an overwhelming response with more than 50 submissions from Indonesia and also overseas. Unfortunately, many papers could not be accepted due to the reviewing outcomes and our capacity limits. However, we would like to sincerely thank all the presenters for the great enthusiasm, also for the reviewers, editors, faculty members, who contributed so impressively to the success of the conference. We hope you will find the proceeding beneficial, exciting and encouraging for the future development in the related studies.

Acknowledgements

This endeavour would not have been possible without the support by Faculty of Psychology, University of Muhammadiyah Malang, Faculty of Psychology, University of Pendidikan Sulan Idris, Malaysia, as well *Lembaga Pengembangan Publikasi Ilmiah* (Institute of Scientific Publication Development) UMM.

The ICAP-H organizing committee wishes to acknowledge the scientist committee for the valuable contribution in managing the peer review process by providing scientific advice and information to the paper submissions.

Many thanks to the steering committee and organizing committee of ICAP-H 2022 for their substantial role in succeessing ICAP-H at Malang, Indonesia. Furthermore, the deepest gratitude to all presenters and delegates for their great enthusiasm and active participation in the conference events.

Committee Members

Scientific Committee

1. Name: Assoc. Prof. Dr. Iswinarti, M.Si
   Affiliation: *University of Muhammadiyah Malang, Indonesia*
   University E-mail address: iswinarti@umm.ac.id
2. Name: Assoc. Prof. Dr. Diah Karmiyati, M.Si
   Affiliation: *University of Muhammadiyah Malang, Indonesia*
   University E-mail address: diah@umm.ac.id
3. Name: Assoc. Prof. Dr. Latipun, M.Kes
   Affiliation: *University of Muhammadiyah Malang, Indonesia*
   University E-mail address: latipun@umm.ac.id
4. Name: Assoc. Prof. Dr. Tulus Winarsunu, M.Si
   Affiliation: *University of Muhammadiyah Malang, Indonesia*
   University E-mail address: Tulus@umm.ac.id
5. Name: Assoc. Prof. Dr. Datin Roslinda Binti Mustapha
   Affiliation: *University of Pendidikan Sultan Idris, Malaysia*
   University E-mail address: roslinda.m@fpm.upsi.edu.my
6. Name: Assoc. Prof. Dr. rer. nat. Nurul Ain Hidayah binti Abas
   Affiliation: *University of Pendidikan Sultan Idris, Malaysia*
   University E-mail address: nurulain@fpm.upsi.edu.my
7. Name: M. Salis Yuniardi, M.Psi., Ph.D
   Affiliation: *University of Muhammadiyah Malang, Indonesia*
   University E-mail address: salis@umm.ac.id
8. Name: Dr. Cahyaning Suryaningrum
   Affiliation: *University of Muhammadiyah Malang, Indonesia*
   University E-mail address: Cahyaning@umm.ac.id
9. Name: Dr. Dini Permanasari, M.M., Psikolog
   Affiliation: *University of Muhammadiyah Malang, Indonesia*
10. Name: Dr. Nida Hasanati, M.Si
    Affiliation: *University of Muhammadiyah Malang, Indonesia*
    University E-mail address: nida@umm.ac.id
11. Name: Dr. Siti Suminarti, M.Si
    Affiliation: *University of Muhammadiyah Malang*
    University E-mail address: suminarti@umm.ac.id
12. Name: Udi Rosida Hijrianti, S.Psi., M.Psi., Psikolog
    Affiliation: *University of Muhammadiyah Malang*
    University E-mail address: udi@umm.ac.id
13. Name: Nandy Agustin Syakarofath, S.Psi., M.A
    Affiliation: *University of Muhammadiyah Malang*
    University E-mail address: nandysyakarofath@umm.ac.id

 ISBN: 978-1-032-51442-0

Organizing Committee

Steering COMMITTEE

Patrons	• Prof. Dr. Syamsul Arifin, M.Si.
Advisors	• M. Salis Yuniardi, M.Psi, Ph.D.
Person in charge	• Dr. Fardini Sabilah, M.Pd.
	• Thomy Eko Saputro, Ph.D.
	• Dr. Siti Suminarti F., M.Si.
	• Dr. Nida Hasanati, M.Si.
	• Dr. Ari Firmanto, M.Si.
General Chair	• Uun Zulfiana, M.Psi.
Treasurer	• Retno Firdiyanti, M.Psi.
Publicity and Sponsorship Chair	• Adhyatman Prabowo, M.Psi. (Co)
	• Devina Andriany, M.Psi.
	• Ifatul Masbahah, S.Psi.
	• Siti Maimunah, S.Psi., M.M., M.A.
Program Chair	• Nandy Agustin Syakarofath, S.Psi., M.A. (Co)
	• Atika Permata Sari, S.Psi., M.Psi.
	• Sulivan Fitriani, S.Psi, M.A.
	• Sadia Mewar, S.Pd., M.Si.
	• Ni'matuzzahroh, S.Psi., M.Si, Ph.D
Secretary Chair	• Udi Rosida Hijrianti, S.Psi., M.Psi. (Co)
	• Diana Savitri Hidayati, S.Psi., M.Psi.
	• Istiqomah, S.Psi., M.Si.
	• Navy Tri Indah Sari, S.Psi., M.Si.
Facilities Chair	• Eli Kamila, S.T. (Co)
	• Tri Hartanto, S.Pd
	• Moh. Afif Setiawan, S.T.
Accommodation and Transportation Chair	• Farisha Prasidya Putri, S.Psi. (Co)
	• Siti Fatimah, S.Ag., M.Si.
	• Yasin Kusumo Pringgodigdo, M.H.I.
	• Hufron Rosidin

Families Mental Health and Challenges in the 21st Century – Arifin et al. (eds)
© 2024 The Author(s), ISBN 978-1-032-51442-0

Relationship between sedentary behavior and quality of life among youth with disabilities during the COVID-19 pandemic in Malaysia

Nor Aida Ahmad & Nor Firdous Mohamed*
Department of Psychology and Counselling, Faculty of Human Development, Sultan Idris Education University, Malaysia

ABSTRACT: Research suggests that physical activity plays an important role to improve mental health and quality of life. However, there were limited studies done among youth with disabilities, especially during the COVID-19 pandemic in Malaysia. This study aims: (1) to determine the status of sedentary behavior, mental health, and quality of life; (2) to determine the relationship between sedentary behavior and mental health; (3) to investigate the relationship between sedentary behavior and quality of life. A cross-sectional (online survey) design comprising youth (N = 113) aged 15 to 35 years (mean age = 27.36) was used in this study. The Godin Leisure Time-Exercise Questionnaire (GLTEQ) and the World Health Organization Quality of Life-Disability (WHOQOL-DIS) were used. Then, Spearman's rank correlation coefficient was conducted to determine the relationship between sedentary behavior and quality of life. Results showed a positive relationship between sedentary behavior and quality of life (r (110) = .0214, $p<0.05$). In conclusion, the importance of an active lifestyle among youth with disabilities in Malaysia was presented in this study.

Keywords: Sedentary behavior, mental health, quality of life, youth with disabilities

1 INTRODUCTION

Physical activity during a pandemic can be very limited because of the Movement Restriction Order (MCO) in Malaysia. The research found that a sedentary lifestyle for a long time can increase the problems such as hypertension, obesity, muscle atrophy, postural disorder, and inadequate cardiovascular system (Çetin *et al.* 2019). Without adequate physical activity at home and stress caused by the lockdown, people will have a higher risk of developing anxiety and depression.

Indeed, several studies have already raised concerns about the impact of COVID-19 on the health and well-being of people with a disability as well as access to resources and funding (Kwok 2020; Smith & Judd 2020). The current study needs to be done because there are more than 80% of adolescents insufficiently physically active globally (World Health Organization 2018).

Furthermore, a study found that youth with disabilities are not achieving the target recommended amount of daily physical activity for youth and they are significantly less active and obese compared to their non-disabled peers (Rimmer & Rowland 2008). Moreover, COVID-19-related restrictions likely exacerbate the current public health problems of low levels of physical activity and high prevalence of sedentary behaviors in children and adolescents in all communities—disabled or non-disabled youth (King *et al.* 2020).

*Corresponding Author: norfirdous@fpm.upsi.edu.my

DOI: 10.1201/9781003402381-1

Despite the impact, as reported by those with a disability, it is currently limited, especially in the Malaysian local context. This study aims to help us to understand the importance of being active and the effect of sedentary behavior on mental health and quality of life among youth with disabilities by doing a survey using the Godin Leisure-Time Exercise Questionnaire (GLTEQ) developed by Godin in 1985 and World Health Organization Quality of Life Disabilities (WHOQOL-DIS) developed by Power and Green in 2010.

1.1 *Research questions*

1. What is the status of sedentary behavior, and quality of life among youth with disabilities during the pandemic COVID-19 restrictions?
2. What is the relationship between sedentary behavior and quality of life among youth with disabilities during the pandemic COVID-19 restrictions?

1.2 *Research objectives*

1. To determine the status of sedentary behavior, and quality of life of youth with disabilities during the pandemic COVID-19 restrictions.
2. To seek the relationship between sedentary behavior and quality of life among youth with disabilities during the pandemic COVID-19 restrictions.

2 METHOD

2.1 *Research design*

This study consists of a quantitative cross-sectional online survey design. The cross-sectional design is a study to collect data from the intended population at one specific period. The study based on this design can help us to understand the characteristics and allow us to compare certain variables in a population (Olsen & St George 2004).

2.2 *Sample*

For this quantitative cross-sectional online survey design study, the criteria of the participants consist of youth with disabilities aged 15 to 35 years with a fairly distribution of male and female participants. The inclusion criteria were youth with disabilities having their gadget (handphones, tablets) and independence. Independence here is defined as a youth who can use a smartphone and understand the questions because we want to make sure that respondents can answer the question without help and influence from others. The exclusion criteria included the following: youth with disabilities who were younger than 15 years, cognitive impairment, being pregnant, and not providing informed consent. This is because people in this category were not suitable to answer the questions related to our research. The purposive sampling technique was used to recruit participants who were most accessible to the researcher in this study.

2.3 *Procedure*

An anonymous online survey (google form) was assembled by a steering group of multidisciplinary co-researchers and reviewed by health service providers and parents of children with disabilities. On February 1, 2022, the online survey was opened for two months, and it closed on March 30, 2022. The link to the online survey was disseminated through health service providers, associations, and charities throughout Malaysia, as well as through social networks such as Facebook, WhatsApp, and Twitter. The survey included online questions with Likert scales and free-text answers. These participants were recruited from several associations and organizations such as the Department of Social Welfare Malaysia, the Society of Muscular Atrophy Malaysia, the Malaysian Rare Disorders Society, the Development Organization for Blind Malaysia, and the Malaysian Deaf Muslim Association and Social Security Organization (SOCSO). The participants were stratified based on their types of disabilities, namely, hearing and speech impairment, visual impairment, and physical disability, as well as those with intellectual disabilities (such as learning disabilities and dyslexia). The questionnaire consists of demographic questions (age, gender, disabilities), the GLTEQ, and the WHOQOL-DIS. The suggested time spent for them to answer the questionnaire was around 20 minutes. Then, the collected data were analyzed for discussion and interpretation.

Flowchart
The following flow chart shows the study procedure

Month 1–2: Ethics Approval Granted (2021-0415-01) and Participants Recruitment

↓

Month 2–4: Cross-sectional online survey

↓

Month 5–6: Statistical Analysis and Report Writing

3 RESULTS

3.1 *Sample's profile*

For the participants' gender, it was reported that 61.9% (n = 70) of respondents are male and 38.1% (n = 43) of respondents are female. There were 4.4% respondents in the age range of 18 to 20 years, 17.7% respondents in the age range of 21 to 23 years, 21.2% respondents in the age range of 24 to 26 years, 21.2% respondents in the age range of 27 to 29 years, 20.4% respondents at the age range of 30 to 32 years and 15% of respondents at the age range of 33 to 35 years.

For the participants' types of disability, most of the respondents have a physical disability with a total of 85% respondents while there are 2.7% respondents with hearing disability, 3.5% respondents with a learning disability, 1.8% respondents with multiple disabilities, 2.7% respondents with a speech disability, and 4.4% of the respondents with visual disability.

3.2 *Descriptive analysis of GLTEQ*

The analysis of the sedentary lifestyles of people with disability was performed by using the GLTEQ. Based on Figure 4.1, the results showed that 69.9% (n = 79) respondents were having active lifestyles during the pandemic, 9.7% (n = 11) respondents were living moderately active lifestyles and 20.4% (n = 23) of the respondents were living a sedentary lifestyle

Table 1. Sample's Socio-demographic profile (N = 113).

Variable		n	Percentage
Age			Mean = 27.36
Gender	Male	70	61.9
	Female	43	38.1
Types of disability	Physical disability	96	85.0
	Visual disability	5	4.4
	Learning disability	4	3.5
	Hearing disability	3	2.7
	Speech disability	3	2.7
	Multiple disabilities	2	1.8

Notes: *N = 113*

during the pandemic. The statistical results showed that more than half of the represented sample of youth with disability in Malaysia had an active lifestyle during the COVID-19 pandemic.

3.3 *Descriptive analysis of world health organization quality of life—disability (WHOQOL-DIS)*

Based on the data presented in Table 2, the Total QoL was 14.36 (SD = 2.1), control/autonomy has the highest value mean score (14.79, SD = 2.94) among other domains of quality of life, and the lowest mean score was discrimination domain with a mean score of 13.19 and a standard deviation of 3.44. The physical domain has a mean score of 13.25 (SD = 2.93), while the psychological and environmental domains have a mean score of 14.6 (SD = 2.79) and 14.45 (SD = 3.12), respectively. Domain social has a mean score of 14.23 (SD = 3.82), while the inclusion domain which measures directly the effect of disability on the respondents has a mean score of 14.75 (SD = 3.95).

Table 2. Mean and standard deviation of youth with disability on world health organization quality of life—disability (N = 113).

Domain	Mean	Std. Deviation
Physical	13.25	2.93
Psychological	14.60	2.79
Social	14.23	3.82
Environment	14.45	3.12
Discrimination	13.19	3.44
Control/Autonomy	14.79	2.94
Inclusion	14.75	3.95
Total QoL	14.36	2.10

3.4 *The relationship between sedentary behavior and quality of life*

The relationship between sedentary behavior and quality of life among youth with disabilities is presented in Table 3 based on the GLTAQ and WHOQOL-Disability results. A positive relationship was found between sedentary behavior and quality of life ($r(110) = .0214$, $p<0.05$). Following the result, we rejected the null hypothesis, and the researched hypothesis was accepted.

Table 3. Relationship between sedentary behavior and quality of life among youth with disabilities (N = 110).

Variable	Sedentary behavior	*p-value*
Quality of Life	0.214*	0.025

*Correlation is significant at the 0.05 level (2-tailed)

4 DISCUSSION

This study included 113 samples with the majority of the sample being from the male population 61.9% compared to the female population which covers the other 38.1% of the sample. The mean age of the samples was 27.36 and the majority of it comes from youth with a physical disability which covers 85% of the sample and the least comes from youth with multiple disabilities with only 1.8%.

The objective of this research was to determine the status of sedentary behavior, mental health, and quality of life among youth with disability in Malaysia during the COVID-19 pandemic. Although there are other studies done to identify the status of sedentary behavior, mental health, and quality of life, there is no research done on youth with disability, especially during the COVID-19 pandemic. The status of the quality of life was found by using the seven domains from the WHOQOL-DIS. The Control/Autonomy domain has the highest average value followed by the Inclusion domain and the Psychological domain. These also indicated that the questions related to Control/Autonomy domain were the least quality of life obtained by youth with disability. The same result was found by past research in which physical deterioration and increased difficulties in managing everyday activities lead to a lack of control in deciding which leads to decreased quality of life (Durstime *et al.* 2000; Fernhall & Unnithan 2002; Lobenius-Palmér *et al.* 2018). Interestingly, from the current finding, there was a positive significant relationship found between sedentary behavior and quality of life ($p = 0.025$).

The positive relationship indicates that when the sedentary behavior was decreased among the samples, their quality of life would increase. The research hypothesis of this study was accepted, following the previous study that also reported the same result as the current study. The study suggested that sedentary behavior in youth with a disability would decrease overall physical health, and decreased social interaction with peers which contribute to a lower quality of life (Diaz *et al.* 2019). Besides, a previous study done on children and adolescents with physical disabilities found that those who participate at least twice a week in sports had more beneficial scores on quality of life when compared to those who did not participate (Te Velde *et al.* 2018). Furthermore, a higher level of physical health was found in a previous study as an important predictor of quality of life among older people with a disability (Yeung & Breheny 2019). This is because higher levels of physical activity are associated with greater meaningful participation in society, improve quality of relationships, and better well-being in general which further reduce the potential loneliness, and increase overall satisfaction with life (La Grow *et al.* 2013; Yeung & Breheny 2016). Lastly, a previous study on Korean people with physical disabilities also found significant direct effects of leisure-time physical activity on emotional and informational support and positive social interaction (Kim *et al.* 2021). This previous finding was consistent with our current findings which support the significant relationship between sedentary behavior and quality of life among youth with disabilities during the pandemic.

Based on the findings and discussion, this section discusses the implication of the current study. The current study managed to determine the relationship between sedentary behavior and quality of life. From this finding, the study managed to understand the importance of physical activity towards the quality of life among youth with disabilities. First, the finding suggested that youth with disabilities were active during the COVID-19 pandemic as they get a lot of time to do other physical activities at home and they did not need to spend time taking public transport to go to work or school every day. Furthermore, with the use of smartphones, various types of physical activity can be learned from home and they would feel more comfortable trying new things at home compared to doing physical exercise in public places because of their disabilities which can make them feel uncomfortable and fear of judgment from others.

4.1 *Limitation of study*

The limitation of this current research was the data were not collected properly as most of the samples come from youth with a physical disability while only several samples come from youth with speech, hearing, learning, vision, and multiple disabilities. Besides, because of the MCO in Malaysia caused by the COVID-19 pandemic, it is hard to find suitable samples, especially for the minority group like youth with disabilities that are being studied in this research. Furthermore, the sample size was inadequate to represent the whole population result, and the data were not normally distributed for sedentary behavior and quality of life. Lastly, the measurement used to calculate the status of sedentary behavior, Godin Leisure Time Activity Questionnaire, might not be suitable to correctly assess sedentary behavior among youth with disabilities. The related questions might be too subjective and not suitable for respondents to self-report, especially when doing online surveys where bias and other environmental conditions can influence their answers.

4.2 *Conclusion and implication*

To help increase data related to youth with disability in Malaysia, the current study has examined the status of sedentary behavior, mental health, and quality of life among youth with disability in Malaysia. Sedentary behavior shows a positively significant relationship with quality of life. This result suggested that youth with disabilities that have an active lifestyle would have a better quality of life. This indicated the importance of physical activity even for disabled people as physical activity helps to improve their physical health and confidence, and increases social interaction which overall increased their quality of life. From past research, the results related to youth with disabilities in Malaysia were not adequate. It is hoped that in the future, there will be a lot more research conducted to continue this study to help improve our findings related to sedentary behavior, mental health, and quality of life among youth with disabilities in Malaysia.

REFERENCES

Çetin, S., Ece, C., Şen, M., Çetin, H. N., & Aydoğan, A. (2019). The Effects of Pilates and Aerobic Exercise on Blood Pressure, Heart Rates, and Blood Serum Lipids in Sedentary Females. *Journal of Education and Training Studies*, *7*(4), 229–235. doi:10.11114/jets.v7i4.4077

Costello, A. B., & Osborne, J. (2005). Best Practices in Exploratory Factor Analysis: Four Recommendations for Getting the Most From Your Analysis. *Practical Assessment, Research, and Evaluation*, *10*(1), 7. DOI: https://doi.org/10.7275/jyj1-4868

DeVellis, R.F. (2017). *Scale Development: Theory and Applications* (4th ed.). Thousand Oaks, CA: Sage. [Paper reference 9] DOI: 10.4236/psych.2018.911145

Diaz, R., Miller, E. K., Kraus, E., & Fredericson, M. (2019). Impact of Adaptive Sports Participation on Quality of Life. *Sports Medicine and Arthroscopy Review*, *27*(2), 73–82.

Durstine, J.L., Painter, P., Franklin, B.A., Morgan, D., Pitetti, K.H., & Roberts, S.O. (2000). Physical Activity for the Chronically ill and Disabled. *Sports Medicine*, 30, 207–219. PubMed doi:10.2165/00007256-200030030-00005

Fernhall, B., & Unnithan, V.B. (2002). Physical Activity, Metabolic Issues, and Assessment. *Physical Medicine and Rehabilitation Clinics of North America*, 13, 925–947. PubMed, doi:10.1016/S1047-9651(02)00031-1

Kim, J., Kim, J., Kim, Y., Han, A., & Nguyen, M. C. (2021). The Contribution of Physical and Social Activity Participation to Social Support and Happiness Among People with Physical Disabilities. *Disability and Health Journal, 14*(1), 100974, https://doi.org/10.1016/j.dhjo.2020.100974

King, A.J., Burke, L.M., Halson, S.L., & Hawley, J.A. (2020). The Challenge of Maintaining Metabolic Health During a Global Pandemic. *Sport. Med.*, 50, 1233–1241. https://doi.org/10.1007/s40279-020-01295-8

Kwok N. (2020). Adapted Physical Activity Through COVID-19. *Eur J Adapt Phys Activ.*, 13. https://doi.org/10.5507/euj.2020.003.

Kyriazos, T. (2018). Applied Psychometrics: Sample Size and Sample Power Considerations in Factor Analysis (EFA, CFA) and SEM in General. *Psychology, 9*(8), 2207–2230. https://doi.org/10.4236/psych.2018.98126

La Grow, S., Yeung, P., Towers, A., Alpass, F., & Stephens, C. (2013). The Impact of Mobility on Quality of Life Among Older Persons. *Journal of Aging and Health, 25*(5), 723–736. https://doi.org/10.1177/0898264313490198

Laferrier, J., Teodorski, E., & Cooper, R. (2015). Investigation of the Impact of Sports, Exercise, and Recreation Participation on Psychosocial Outcomes in a Population of Veterans with Disabilities. *American Journal of Physical Medicine & Rehabilitation, 94*(12), 1026–1034. https://doi.org/10.1097/phm.0000000000000263

Lobenius-Palmér, K., Sjöqvist, B., Hurtig-Wennlöf, A., & Lundqvist, L. O. (2018). Accelerometer-assessed Physical Activity and Sedentary Time in Youth with Disabilities. *Adapted Physical Activity Quarterly, 35* (1), 1–19. DOI: https://doi.org/10.1123/apaq.2015-0065

Olsen, C., & St George, D. M. M. (2004). Cross-sectional Study Design and Data Analysis. *College Entrance Examination Board*, 26(3), 2006.

Rimmer, J. A., & Rowland, J. L. (2008). Physical Activity for Youth with Disabilities: a Critical Need in an Underserved Population. *Developmental Neurorehabilitation, 11*(2), 141–148. DOI: 10.1080/17518420701688649

Sagar-Ouriaghli, I., Godfrey, E., Graham, S., & Brown, J. (2020). Improving Mental Health Help-Seeking Behaviours for Male Students: A Framework for Developing a Complex Intervention. *International Journal of Environmental Research and Public Health, 17*(14), 4965. doi: 10.3390/ijerph17144965

Smith, J. A., & Judd, J. (2020). COVID-19: Vulnerability and the Power of Privilege in a Pandemic. *Health Promotion Journal of Australia, 31*(2), 158. doi: 10.1002/hpja.333

Souter, G., Lewis, R., & Serrant, L. (2018). Men, Mental Health and Elite Sport: A Narrative Review. *Sports Medicine – Open, 4*(1). https://doi.org/10.1186/s40798-018-0175-7

Te Velde, S., Lankhorst, K., Zwinkels, M., Verschuren, O., Takken, T., & de Groot, J. (2018). Associations of Sport Participation with Self-perception, Exercise Self-efficacy and Quality of Life Among Children and Adolescents with a Physical Disability or Chronic Disease—a Cross-sectional Study. *Sports Medicine – Open, 4*(1). https://doi.org/10.1186/s40798-018-0152-1

World Health Organization. Physical activity. Geneva: World Health Organization; 2018 [cited 2021 August 2]. Available from: https://www.who.int/news-room/fact-sheets/detail/physical-activity

Yeung, P., & Breheny, M. (2016). Using the Capability Approach to Understand the Determinants of Subjective Well-being Among Community-dwelling Older People in New Zealand. *Age And Ageing, 45*(2), 292–298. https://doi.org/10.1093/ageing/afw002

Yeung, P., & Breheny, M. (2019). Quality of Life Among Older People with a Disability: The Role of Purpose in Life and Capabilities. *Disability and Rehabilitation, 43*(2), 181–191. https://doi.org/10.1080/09638288.2019.162087

Families Mental Health and Challenges in the 21st Century – Arifin et al. (eds)
© 2024 The Author(s), ISBN 978-1-032-51442-0

The role of coping with stress and self-esteem in the academic burnout of high school students in the post-pandemic era

Hana Aulia* & Udi Rosida Hijrianti
Universitas Muhammadiyah Malang, Malang, Indonesia

ABSTRACT: Academic burnout is emotional, physical, and mental exhaustion due to excessive long-term academic demands. Students who struggle to cope with stress and self-esteem can experience academic burnout. The research aimed to determine how coping with stress and self-esteem affected high school students' academic burnout during the pandemic era. The accidental sampling method was used for 100 high school students in Malang. The Maslach Burnout Inventory-Student Survey (MBI-SS) scale, the Coping Questionnaire for Children and Adolescents (CQCA), and the Self-Esteem Scale (RSES) were used as data collection methods. For the test analysis, multiple linear regression was used. The results showed F count=(19.282>3.089) and sig value (0.000<0.05) which indicates that coping with stress and self-esteem has an impact on academic burnout, with a 28.4% contribution.

Keywords: Avoidant coping, blended learning, cynicism, low self-esteem, worthiness

1 INTRODUCTION

The COVID-19 pandemic has given rise to two new educational learning systems. Students benefit greatly from the availability of online and blended learning, in particular. According to the results of a survey given to one of the senior high schools in Malang on Thursday, October 21, 2021, students reported feeling unmotivated, which interfered with learning activities, procrastinating on homework, feeling hopeless, and losing motivation, resulting in a decline in academic achievement. According to Ilahi *et al.* (2022), the offline learning system that was re-implemented in schools today causes students to feel physically unprepared, including sleepiness and lethargy, as a result of students' being accustomed to online learning. According to field observation, high school students in Malang frequently lose focus during class activities and appear fatigued. Guidance and Counseling teacher said, "Students' willingness to learn is low because they have grown accustomed to comfortable settings while studying at home for years, such as leaning and reclining. Because schools cannot reprimand students for low motivation, students develop negative character traits. Students are becoming more passive and simply want to be entertained."

As a result of this occurrence, high school students in Malang are more likely to experience academic burnout. Academic burnout is a stressful condition brought on by long-term, excessive academic pressure that results in emotional exhaustion, an inclination to depersonalize, and a low sense of accomplishment (Yang 2004). Supporting findings made by Teuber *et al.* (2021), Burnout was positively related to student workload and academic demands in Chinese secondary schools. This is consistent with the occurrences found in the field.

*Corresponding Author: hanaaauliaaa26@webmail.umm.ac.id

DOI: 10.1201/9781003402381-2

As per the results of a field survey done on October 21, 2021, many students plan on having good academic results, but on the other hand, they also struggle with their procrastinating habits, laziness, and choosing to do other things instead of studying, such as playing games and using social media, which, of course, impacts their stress levels. Additionally, the Guidance and Counseling teacher said "During the pandemic period, playing games was exciting, but lessons were tedious. It's truly the case. It's how the students in X High School feel. We try to avoid burdening students whenever possible."

Based on this occurrence, some students' responses to pressure are less effective. Palupi and Findyartini (2019) discovered that academic burnout is negatively related to adaptive coping and positively related to maladaptive coping. Furthermore, research by Vizoso *et al.* (2019) shows that academic burnout has a negative relationship with adaptive coping but a positive relationship with maladaptive coping. These findings demonstrate the link between coping with stress and academic burnout. Coping with stress is a technique that people use to manage potentially dangerous situations or control their behavior based on the type of stressor they perceive (Folkman & Lazarus 1980). As a result, in this pandemic era, high school students in Malang require an effective method of coping with stress in order to avoid academic burnout.

Furthermore, identity conflicts are prevalent among high school students, according to the theory of adolescent development. Students who are successful in resolving identity conflicts will find it easy to fit in with their peers; however, students who are unsuccessful in resolving identity conflicts will experience identity confusion, which can lead to withdrawal and isolation from friends and family (Santrock 2012).

This phenomenon is consistent with the results of the self-understanding questionnaire given to High School students in Malang. Many students are unconfident in their abilities. For instance, "always feel like the worst one among other friends," "insecure," "feel lazy," "do not recognize their strengths," and a number of other issues that can lead to academic burnout. According to Maharani (2019), there is a negative relationship between academic burnout and self-esteem. Moreover, Mun and Kim (2022) discovered a significant negative relationship between academic burnout and self-esteem in nursing students. According to Tanzil *et al.* (2022), the low social interaction during online learning contributed to the increased levels of loneliness among students throughout the pandemic era. Students with high self-esteem will have their social needs met. In accordance with the research, students at one of the senior high schools in Malang typically have low self-assessment.

Self-esteem is an individual's positive or negative opinion of themselves, as well as how they evaluate the entirety of their thoughts and feelings about themselves (Rosenberg 1965). Accurate assessments of an individual's worth, as well as achievements and successes, can assist in developing self-esteem. Those with high self-esteem, in contrast to narcissists, are better able to recognize their own strengths and weaknesses, as well as to evaluate, value, and accept both their own strengths and weaknesses. People with low self-esteem are less capable of respecting themselves, are more likely to point out flaws, engage in risky activities, and feel inferior (Santrock 2012).

In contrast to previous research, researchers wanted to try to find the effect of coping with stress and self-esteem in academic burnout of High School students in Malang during learning in the current pandemic era. In this regard, the purpose of this research was to determine the effect of coping with stress and self-esteem on academic burnout in high school students during learning activities throughout the pandemic era. The findings of this research were expected to make a significant contribution, particularly to the field of educational psychology.

2 METHODS

This research used a quantitative approach, which means that the data were collected in the form of quantitative data and then processed using statistical techniques. The research design used in this research was multiple linear regression analysis, which was used to determine whether or not there was a relationship between the variables. The accidental sampling method was used in this research's sampling, which involves selecting respondents who just happen to be present and available somewhere in accordance with the research context (Etikan & Bala 2017). The subject criteria included: (1) high school students from Malang; and (2) participation in limited face-to-face learning (blended learning) or distance learning.

In this research, three variables were used. There were two independent variables, (X1) coping with stress and (X2) self-esteem, as well as the dependent variable (Y) academic burnout. The coping with stress in this research was the coping methods used by students, including (1) Seeking social support, (2) Problem-solving, (3) Avoidant coping, (4) Palliative emotion regulation, (5) Anger-related emotion regulation. Coping with stress was measured using the Coping Questionnaire for Children and Adolescents (CQCA) scale developed by Eschenbeck *et al.* (2012) and adapted by Novrizal *et al.* (2018). The CQCA consisted of 23 Likert model items with the answer choices of 1 = never, 2 = rarely, 3 = often, and 4 = always on the seeking social support and problem-solving subscale. Meanwhile, for avoidant coping, palliative emotion regulation, and anger-related emotion regulation, the answer choices were 1 = always, 2 = often, 3 = rarely, and 4 = never. The CQCA reliability index is 0.834.

The following condition was used to define burnout in this research: (1) Exhaustion, fatigue caused by academic demands; (2) Cynicism, a pessimistic attitude toward learning activities; and (3) Reduced academic efficacy, a decrease in academic confidence. The scale instrument used the Maslach Burnout Inventory-Student Survey (MBI-SS) developed by Schaufeli *et al.* (2002) and adapted by Maharani (2019), consisting of 15 Likert model items with five answer choices, namely: Very Appropriate (VA), Appropriate (A), Less Appropriate (LA), Not Appropriate (NA), and Very Inappropriate (VI). The correlation range for the valid statement items is 0.478 to 0.870, with an r table of 0.361.

The self-esteem referred to in this research was (1) Competence related to cognitive, physical, and individual motivation in facing demands and challenges; and (2) worthiness, or the feeling of achieving results. Self-esteem was measured using the Self Esteem Scale (RSES) developed by Rosenberg (1965) and adopted by Maharani (2019) with 10 Likert model items with five answer choices, namely: Very Appropriate (VA), Appropriate (A), Less Appropriate (LA), Not Appropriate (NA), and Very Inappropriate (VI). The correlation range for the valid statement items is 0.412 to 0.777, with an r table of 0.361.

The following were the three main procedures used in the research: Preparation was the stage at which the researcher began to learn about the theoretical study of the variables to be studied. At this point, the researcher started looking for measuring instruments with relevant features based on the theoretical studies that had been examined. Before being used on the population sample, the measuring instrument was subjected to validity and reliability tests. The measuring instrument was distributed to students who met the predetermined subject criteria during the implementation stage. Furthermore, it also gave each student a scale of coping with stress, self-esteem, and academic burnout. The data analysis stage involved reviewing the findings of the information gathered from high school students who completed and returned the assessment. The collected data were then processed and inputted into the SPSS statistical calculation program's non-parametric analysis module. Using multiple linear regression analysis.

Before doing the multiple linear regression test, it has passed the assumption test first. The researcher initially performed a classical assumption test, which included a normality test and a linearity test, before testing the hypothesis (Hayes *et al.* 2014). Based on the Kolmogorov–Smirnov test with a significance level of 5%, the result (Asymp sig) of ($0.200>0.05$) meant that the data was normally distributed. Furthermore, based on the Durbin–Watson test with a significance level of 0.05 on the coping with stress variable, the deviation from linearity ($0.092>0.05$) and the self-esteem variable was ($0.131>0.05$), which meant that there was a linear relationship between the independent variables (coping with stress and low self-esteem) and the dependent variable (academic burnout).

3 RESULT

As suggested by Choiriah and Liana (2019), the sample was obtained by multiplying the total number of independent and dependent variables by ten to obtain the fewest number of samples required. In accordance with the preceding statement, the researcher used a sample of 100 students with gender differences, i.e., up to 36 male participants and up to 64 female participants. Researchers also distributed online questionnaires to 18 senior high schools in Malang using Google Forms.

Table 1. Subject categorization based on the type of stress coping on the Coping Questionnaire for Children (CQCA) scale.

	Category	Frequency	Percentage	Mean
Seeking social support	High	47 students	47	14.21
	Low	53 students	53	
Problem-solving	High	43 students	43	18.44
	Low	57 students	57	
Avoidant coping	High	51 students	51	18.57
	Low	49 students	49	
Palliative emotion regulation	High	53 students	53	14.72
	Low	47 students	47	
Anger-related emotion regulation	High	46 students	46	12.49
	Low	54 students	54	

Empirical statistical calculations were used to obtain subject categorization based on the CQCA scale. It was discovered that 47% of High School students in Malang had high coping with stress in the form of seeking social support, while 43% of them had high problem-solving from a total of 100 students as participants. In addition, 51% had high avoidant coping, 53% had high palliative emotion regulation, and 46% had high anger-related emotion regulation.

Based on the results of categorization through empirical statistical values, out of a total of 100 students as participants, 52 students of High School in Malang were in the high self-esteem category and 48 students were in the low self-esteem category.

Based on the results of categorization through empirical statistical values, out of a total of 100 students as participants, 45 students of High School in Malang were in the high academic burnout category and 55 students were in the low academic burnout category.

The researchers used multiple linear regression models to test whether there was a relationship between coping with stress and self-esteem and academic burnout. The following are the findings of the analysis:

Table 2. Subject categorization based on RSES and MBI-SS.

	Category	Frequency	Percentage	Mean
Self Esteem	*High Self Esteem*	52 students	52	28.16
	Low Self Esteem	48 students	48	
Academic burnout	*High*	45 students	45	52.35
	Low	55 students	55	

Based on Table 3, the value of the F table with n = 100 at $\alpha = 0.05$ is 3.089. F count is 19.282. F count = (19.282>3.089) and sig value (0.000<0.05) which indicates that coping with stress and self-esteem, together, have an influence on academic burnout. Based on calculations using multiple linear regression models on the CQCA scale, there is an impact on avoidant coping with a sig value (0.028<0.05) and a calculated T value (2.236>1.986). This indicates that there is a positive relationship of avoidant coping on academic burnout. For anger-related emotion regulation type, the sig value (0.000<0.05) and T count value (-4.789>1.986). This indicates that there is a negative relationship of anger-related emotion regulation on academic burnout. The r square value is 0.284. This indicates that the impact or contribution of coping with stress and self-esteem to academic burnout is 28.4%. In addition, the value of the effective contribution (EC) on the coping with stress variable is 29.89% and the self-esteem variable is 28.14%. The relative contribution (RC) of each independent variable to the dependent variable, namely, coping with stress and self-esteem is 1.05% and 99%, respectively.

Table 3. ANOVA.

Model		Sum of Squares	Df	Mean Square	F	Sig
H1	Regression	2350.553	2	1175.276	19.282	.000b
	Residual	5912.197	97	60.950		
	Total	8262.750	0.99			

Dependent variable: *academic burnout*
Note. R^2 = 0,284

4 DISCUSSION

The purpose of this research is to determine whether coping with stress and self-esteem affects academic burnout in high school students during the pandemic period. The results reveal that coping with stress and self-esteem, when combined, have an impact on academic burnout among High School students in Malang. Academic burnout is affected by two types of coping with stress including stress-avoidant coping and anger-related emotion regulation. 29.89% of academic burnout was attributed to stress management. Meanwhile, self-esteem affected academic burnout by 28.14%.

Academic burnout was only affected by avoidant coping and anger-related emotion regulation, so coping with stress had a minor effect. Avoidant coping had a positive effect on academic burnout. This indicated that the higher the avoidant coping, the greater the academic burnout among high school students in Malang. This is consistent with the findings of Macintyre *et al.* (2020) who stressed that using avoidant coping strategies can be harmful and should be avoided. Academic burnout was also negatively impacted by aspects of anger-related emotion regulation. This suggests that the less adept students are at controlling their anger, the greater their proclivity for academic burnout. This is in line with the research of Szczygiel and Mikolajczak (2018) who discovered a significant relationship between burnout and anger-focused coping strategies. Emotional intelligence is thus required. According to research conducted on High School students in Malang, students typically struggle to control their anger and experience moderate academic burnout. This can happen for a variety of reasons, including personality, social support, academic burden, and perceptions of personal stressors, Jacobs and Dodd (2003).

The presence of self-esteem benefits students in dealing with academic burnout. According to the findings, self-esteem had a negative effect on academic burnout. Students who have low self-esteem are more likely to become academically burned out. According to Kurtović *et al.* (2018), people who have a positive self-perception respond to stress by focusing on problems and finding solutions rather than fighting pressures passively. On the other hand, people with a stress coping strategy that emphasizes emotions are more likely to experience mental health issues. According to recent research, students with low self-esteem and a negative self-view will always deal with emotional exhaustion. Academic burnout is a risk when emotional-focused stress coping (avoidant coping) is overemphasized.

It is critical to consider how students' self-esteem affects their ability to manage stress and prevent academic burnout. Students with high self-esteem, according to Orth and Robins (2022), can manage academic demands while also receiving adequate emotional support. Students with low self-esteem, on the other hand, procrastinate on assignments and avoid task demands in order to avoid negative feedback from others. This is consistent with findings from a research on high school students in Malang. Students who frequently use avoidance coping strategies, such as procrastinating on assignments, experience academic burnout. It is also critical to assist students who have low self-esteem.

Academic burnout is more affected by self-esteem than coping with stress. This is due to the fact that some high school students in Malang still fall under the category of low self-esteem, which in the pandemic period leads to academic burnout. Coping with stress has little effect on academic burnout. This is because academic burnout, which is still in the moderate category, is only affected by avoidance coping and anger-related emotion regulation methods. Additionally, using a stress-coping scale that is less focused on students' academic problems has a smaller impact.

The current research had limitations due to the small sample size and limited focus of the participants, which prevented generalization to other pandemic-affected areas. Researchers believed that other factors were causing academic burnout among high school students in Malang because there was only a slight correlation between the variables of coping with stress and self-esteem. Furthermore, because the coping stress scale is not focused on students' academic problems, it has little impact on academic burnout.

5 CONCLUSION AND IMPLICATION

Coping with stress and self-esteem has an impact on academic burnout among High School students in Malang. Avoidant coping and anger-related emotion regulation are two types of coping stress that have an impact on academic burnout. The remaining factors do not affect academic burnout. Students who have low self-esteem must work to improve their self-esteem in order to avoid academic burnout during the pandemic.

Further research is expected to put the inclusion of additional variables that can support a greater influence on the factors contributing to academic burnout. Choosing a sample that is not solely focused on one topic will allow for a more comprehensive generalization of the research results, which will aid high school students in avoiding academic burnout. A stress-coping scale that is more appropriate for students' academic problems at school can also be used.

REFERENCES

Choiriah, E. N., & Liana, L. (2019). *Pengaruh Kualitas Produk, Citra Merek, Dan Kualitas Layanan Terhadap Loyalitas Pelanggan Dimediasi Kepuasan Pelanggan* (Studi Pada Pelanggan Sepeda Motor Honda di Kota Semarang. *MADIC*, 236–244.

Eschenbeck, H., Heim-Dreger, U., Tasdaban, E., Lohaus, A., & Kohlmann, C.-W. (2012). A Turkish Adaptation of the Coping Scales from the German Stress and Coping Questionnaire for Children and Adolescents. *European Journal of Psyhological Assessment*, *28*(1), 32–40. https://doi.org/https://psycnet.apa.org/doi/10.1027/1015-5759/a000088

Etikan, I., & Bala, K. (2017). Sampling and Sampling Methods. *Biometrics & Biostatisitcs International Journal*, *5*(6), 215–217. https://doi.org/10.15406/bbij.2017.05.00149

Folkman, S., & Lazarus, R. S. (1980). An Analysis of Coping in a Middle-aged Community Sample. *Journal of Health and Social Behavior*, *21*(3), 219–239. https://doi.org/https://doi.org/10.2307/2136617

Hayes, B., Douglas, C., & Bonner, A. (2014). Predicting Emotional Exhaustion Among Haemodialysis Nurses: A Structural Equation Model using Kanter's Structural Empowerment Theory. *Journal of Advanced Nursing*, 70(12), 2897–2909 https://doi.org/10.1111/jan.12452

Ilahi, T. A., Setiawan, M. E., & Sastria, E. (2022). Analysis of Students' Learning Conditions of a Pandemic to Post Pandemic at SMA Negeri 5 Sungai Penuh. *Journal of Biology Education Research*, *3*(1), 27–37. https://doi.org/https://e-journal.metrouniv.ac.id/index.php/Al-Jahiz/article/view/4706

Jacobs, S. R., & Dodd, D. K. (2003). Student Burnout as a Function of Personality, Social Support, and Workload. *Journal of College Student Development*, *44*(3), 291–303. https://doi.org/10.1353/csd.2003.0028

John W Santrock. (2012). *Life-Span Development* (N. I. Sallama (ed.); 13 Jilid 1). Penerbit Erlangga.

Kurtović, A., Vuković, I., & Gajić, M. (2018). The Effect of Locus of Control on University Students' Mental Health: Possible Mediation Through Self-esteem and Coping Health: Possible Mediation Through Self-esteem and Coping. *The Journal of Psychology*, *152*(6), 341–357. https://doi.org/10.1080/00223980.2018.1463962

Macintyre, P. D., Gregersen, T., & Mercer, S. (2020). Language Teachers' Coping Strategies During the Covid-19 Conversion to Online Teaching: Correlations with Stress, Wellbeing and Negative Emotions. *System*, *94*, 102–352. https://doi.org/10.1016/j.system.2020.102352

Maharani, D. M. (2019). *Hubungan antara self-esteem dengan academic burnout pada siswa kelas XI SMA Negeri 1 Semarang tahun ajaran 2018/2019*. Universitas Negeri Semarang.

Mun, M., & Kim, M. (2022). Effect of Academic Burnout on self Esteem in Nursing Students: Mediating Role of Resilience. *Medico-Legal Update*, *22*(2). https://doi.org/10.37506/mlu.v22i2.3236

Novrizal, I. A., Program, D., Jana, P., Muha, U., & Iyah, M. (2018). *Pengaruh Diskriminasi Etnik Terhadap Penyesuaian Akademik Dimoderasi Strategi Coping (studi pada siswa Suku Bajo di Kabupaten Wakatobi)*. Thesis Masters, Universitas Muhammadiyah Malang.

Orth, U., & Robins, R. W. (2022). Is High Self-esteem Beneficial? Revisiting a Classic Question. *American Psychologist*, *77*(1), 5–17. https://doi.org/https://doi.org/10.1037/amp0000922

Palupi, R., & Findyartini, A. (2019). The Relationship Between Gender and Coping Mechanisms with Burnout Events in First-year Medical Students. *Korean Journal of Medical Education*, *31*(4), 331–342. https://doi.org/10.3946/kjme.2019.143

Rosenberg, M. (1965). Society and the Adolescent Self-Image. *Science*, *148*(3671). https://doi.org/https://doi.org/10.1126/science.148.3671.804

Schaufeli, W. B., Martinez, I. M., Pinto, A. M., Salanova, M., & Bakker, A. B. (2002). Burnout and Engagement in University Students: A cross-national study. *SAGE Journals*, *33*(5), 464–481. https://doi.org/https://doi.org/10.1177%2F0022022102033005003

Szczygiel, D. D., & Mikolajczak, M. (2018). Emotional Intelligence Buffers the Effects of Negative Emotions on Job Burnout in Nursing. *Frontiers in Psychology*, *9*, 2649. https://doi.org/10.3389/fpsyg.2018.02649

Tanzil, A. T., Satiadarma, M. P., & Roswiyani. (2022). Kesepian, Kecerdasan Sosial Dan Harga Diri Pada Remaja Di Masa Pandemi Covid-19. *Jurnal Muara Ilmu Sosial, Humaniora, Dan Seni*, *6*(1), 80–90. https://doi.org/https://doi.org/10.24912/jmishumsen.v6i1.15819

Teuber, Z., Nussbeck, F., & Wild, E. (2021). School Burnout Among Chinese High School Students: The Role of Teacher–Student Relationships and Personal Resources. *Educational Psychology*, *41*(8), 985–1002. https://doi.org/10.1080/01443410.2021.1917521

Vizoso, C., Arias-gundín, O., & Rodríguez, C. (2019). Exploring Coping and Optimism as Predictors of Academic Burnout and Performance Among University Students. *Educational Psychology*, *39*(19), 1–16. https://doi.org/10.1080/01443410.2018.1545996

Yang, H. J. (2004). Factors Affecting Student Burnout and Academic Achievement in Multiple Enrollment Programs in Taiwan's Technical-vocational Colleges. *International Journal of Educational Development*, *24* (3), 283–301. https://doi.org/10.1016/j.ijedudev.2003.12.001

Families Mental Health and Challenges in the 21st Century – Arifin et al. (eds)
© 2024 The Author(s), ISBN 978-1-032-51442-0

Resilience and empty nest syndrome

Diana Savitri Hidayati* & Rachma Julianti Suparno*
University of Muhammadiyah Malang, Malang, Indonesia

ABSTRACT: Empty nest syndrome is a transitional phase experienced by parents when the child is left to wander which is characterized by the presence of feelings of emptiness such as sadness, anxiety, depression, and feelings of loss. One of the supporting factors in facing this phase is the existence of good resilience in each individual who experiences it. Resilience is the ability to survive and adapt that functions competently in the face of various difficult situations. The purpose of this study was to determine the relationship between resilience and empty nest syndrome. The subjects involved were as many as 100 people using snowball sampling techniques. The subject criteria are parents who have wandering children, aged 40–60 years, living with their partners, and left by children wandering for at least 2 years. The data collection method uses the empty nest syndrome scale and the CD-RISC scale. The data analysis process used *product moment* correlation analysis. The results of the study show (r = -0.329) and (p = 0.00), which means that the resilience is inversely proportional to empty nest syndrome.

Keywords: Empty Nest Syndrome, Resilience

1 INTRODUCTION

Events that occur in the early stages of an individual's development can exert an influence on later stages. As is the case when parents encounter a phase of change in parental roles, that is, when the child begins to grow up and is required to leave the house either to pursue a career or choose to follow his partner, in this case, marriage, where the change of role becomes the only stressor for the parents (Hurlock 2011). When the child leaves home, parents face a new problem of adjustment in the pattern of family life. The problem is referred to as empty nest syndrome.

Empty nest syndrome is a clinical condition, like depression, having a feeling of emptiness when a child grows up and begins to leave home and begin to make decisions about their future. The absence of a child in the house brings a feeling of emptiness to parents as parents receive a lot of satisfaction from the child before the child leaves the house (Santrock 2012). When the child leaves the house, the parents feel a sense of loss, which can result in the syndrome of an empty nest. Many factors influence the empty nest syndrome in mothers and fathers, including the gender of the parents, the level of education, the style of self-carried out upbringing, aging, and significant cultural variations (Mansoor & Salma Hasan 2019).

Parents with empty nest syndrome have to struggle with psychosocial problems including loneliness, anxiety, and depression. This phase provides an experience of a very meaningful sense of loss experienced by parents, of course, will have a negative impact. This is in line with research conducted by Wang et al (2017) that parents have empty nest syndrome with

*Corresponding Authors: dianasavitri@umm.ac.id and rachxys@gmail.com

DOI: 10.1201/9781003402381-3

significantly higher rates of loneliness and depression, putting a negative impact on the mental health of parents when children get married and leave home.

Abraham (2012) said that several factors influence the emergence of empty nest syndrome, namely having a protective relationship and being carried into the lives of children. The second factor is the feeling of lack of need for a return to his role in the family. The third factor is the loss of the main role of parents towards the child.

The syndrome of an empty nest that occurs in parents can cause changes, these changes can have an impact both positively and negatively. These emerging repercussions are like a form of complex emotion, although sometimes what is caused can be inconsistent. Positively, the impact that arises, such as parents becoming more optimistic as evidenced by a higher level of *well-being* where communication with children is maintained properly. But negative impact is that it can cause depression and loneliness (Bouchard & McNair 2016).

Parents who experience empty nest syndrome when left by the child to wander will experience a feeling of loss due to the parents' habit of caring for the child so when the child leaves the parents' home, they experience a change of role and readjustment. Parents' perception of how to adjust to this period of empty nest syndrome can affect the onset of empty nest syndrome.

Based on the facts around the researchers it was found that parents who had children wandering to continue their education outside the city felt lonely and long-lasting sadness. This is stated by parents who have three children and all their children leave houses either for education, work, or marriage. However, the parent said the most felt empty thing when he was left by the third child to wander. The reason for this is that after being abandoned by the first and second children, parents still get power and entertainment from the third child, but the atmosphere of the house changes and the difference feels when the third child goes wandering. Even though they still do work activities as usual, it still makes parents feel a sense of emptiness and remember their children. The age factor is also one that favors the emergence of empty nests, because according to him, at an early age, parents need to get attention from their children directly and hope that at the current age, they spend more time with family and children because they are a source of comfort for the cause of happiness. However, this condition is still appreciated because they, as parents, always support every good decision made by their children. In the face of this phase, parents still try to do actions that can adjust themselves even though sometimes there are in certain conditions experiencing sadness because they remember their three children.

The fact suggests that each parent is aware of a phase of role change or a process of self-adjustment in the family development stage. The arrival of this phase makes parents understand the situation and prepare what they should do, so they can be able to carry out the process of adjusting to this stage. These actions refer to an attitude of resilience that parents who are experiencing a phase of empty nest syndrome need to carry out resilient measures that can help them in the process of self-adjustment. That is, not all parents who are left by wandering children will experience empty nest syndrome.

Empty nest syndrome is an adjustment problem that parents have to face when their children leave home to wander, this adjustment problem is termed as part of a transition phase where there is an emotional response felt by parents when starting a new life characterized by a general feeling of loneliness and sadness (Hurlock 2011; Shakya 2014; Singh *et al.* 2017). This empty nest syndrome refers to a sense of oppression, sadness, depression, and anxiety. Anxiety which is one of the symptoms of empty nest syndrome can be caused by several factors, one of which is resilience.

Hurlock (2011) revealed that resilience becomes one of the contributing factors to the appearance of anxiety. Resilience is a learning process of an individual in understanding, understanding, and striving to do what he wants so that the individual can adjust to the changes that occur in his environment both the family environment and the surrounding environment Handono *et al.* (2013). Martínez-Martí & Ruch (2017) explained that emotional strength is a significant force and has a high predictive value in a person's resilience, whereas this emotional power is part of the mental response. When the individual faces difficulties, this emotional strength is positively necessary for overcoming difficulties so as to become resilient.

Hendriani (2018) described resilience as an individual's ability to respond to difficulties or trauma experienced in healthy and productive ways. Research conducted by (Darmayanthi & Lestari 2019) shows that individuals who keep themselves busy with various activities or work, in this case, carry out productive activities to restrict their over-focus on thinking about wandered children can help face the empty nest phase situation. In general, resilience is characterized by various characteristics, such as having the ability to face difficulties, being resilient in the face of stress, or being able to rise from the trauma experienced. This resilience is one of the important things that every parent should have when faced with a period of changing roles and phases of an empty nest caused by the departure of his children to leave home for a more independent life.

Resilience is the ability to maintain existence or survival to survive and to achieve well-being, both physically and mentally. The resilient person, when facing the time of the empty nest, tends to prepare and plan well so that when that phase arrives the person can overcome or at least reduce the concerns that arise and be more realistic and objective with any conditions he is experiencing. Good resilience is resilience that includes individual initiative, willingness to communicate, and paying attention to mutual satisfaction with needs.

Resilience and empty nest syndrome are related to the health of individuals. The relationship between resilience and empty nest syndrome with health can also be explained through the role of positive emotions in it. Research conducted by Tugade & Fredrickson (2004) analyzed that individuals or a group of resilient people will do a lot of emotional regulation by using their positive emotions to replace negative emotions that often arise when they are facing difficult situations or stressful conditions. They also noted that resilient individuals have psychologically healthier characteristics, such as being optimistic, dynamic, enthusiastic about the things they encounter in life, open to new experiences, and having a positive emotionality.

Resilience is a set of adaptive characteristics that an individual has to overcome, recover, or develop after a stressful state and even trauma. Resilience is the ability to adapt according to experience and knowledge and adjust to changes both externally and internally because this resilience is a human effort to achieve harmony in oneself and in the environment so as not to cause negative emotions that are not appropriate and less effective (Loviello & Charney 2019).

Based on the foregoing, it can be said that when a parent leaves his child determines his future whether it is starting a career or getting married and chooses to live with his partner, then the parent is likely to feel empty, this is the so-called empty nest syndrome. This is because parents get satisfaction from their children when they are at home. The emptiness felt by parents when children leave home can have a negative impact on their activities, but it can also have a positive impact when parents have good resilience. Therefore, this study was conducted to see if there was a relationship between resilience and empty nest syndrome.

2 METHOD

This study uses a quantitative approach that is correlational in order to see the relationship between two variables and the extent to which the two variables are interconnected.

The subject of the study was a father or mother who lived with his partner, had a wandering child, and had been wandering for at least two years. The number of study subjects was 100 people aged 40–60 years, while the research sampling technique used snowball sampling.

The bound variable in the study is the empty nest syndrome which is defined as a state that describes the condition of parents when left by their offsprings, feelings of sadness or loneliness, have feelings of emptiness or feel emptiness, feel useless, feel guilty, and inhibition of activity. The research instrument uses Oktaria's Empty Nest Syndrome Scale (2019) which refers to aspects of Hellweig. The reliability test of this type of Likert scale using the Alpha Cronbach technique obtained a reliability coefficient of 0.831 with 19 valid items that are favorable and unfavorable.

The free variable of this research is resilience, namely the capacity to maintain abilities and function competently in the face of various life stressors by having good personal competence, being strong in facing various pressures, being able to accept changes positively, being able to exercise self-control, and having good spirituality. The research instrument uses *the Connor Davidson Resilience Scale* (CD-RISC), refers to the Theory of Resilience by Connor & Davidson (2003) from Nabila (2021). The reliability test using the Alpha Cronbach technique obtained a reliability coefficient of 0.931 with 25 valid items being favorable and Likert scale types.

Analysis of research data using a *product moment* correlation test by first conducting a data normality test due to abnormal research data.

3 RESULT

In Table 1, in the empty nest syndrome variables the low classification is shown as 60%, the medium classification as 38%, and the high classification as 2%. The resilience variable has only a moderate classification category of 10% and a high classification of 90%.

Table 1. Categorization of resilience data and empty nest syndrome.

	Resilience			Empty Nest Syndrome		
Category	Interval	F	%	Interval	F	%
Low	x < 50	0	0	x<36	60	60
Keep	50–74.5	10	10	36–53.5	38	38
High	75≥x	90	90	54≥x	2	2

Normality test results using the *kolmogorov-smirnov test* obtained a sig value on the resilience scale and an empty nest syndrome scale of 0.053 ($\geq$ 0.05), which means that the data are normally distributed.

Based on Table 2, it is known that the value of the correlation coefficient (r) is−0.329 with a significance value of 0.00 and an error rate of 5%. This shows about the nature of the relationship between the two variables is a negative relationship which means that the higher

Table 2. Results of correlation analysis of resilience and empty nest syndrome.

	Index Analysis
Correlation Coefficiency (r)	–0.329
Coefficient of Determination (r^2)	0.108
The degree of possible errors	5%
ρ (Significance Value)	0.00
N	100

the resilience, the lower the empty nest syndrome and vice versa. Other results showed that the coefficient of determination (r^2) for the resilience variable was 0.108, meaning that resilience had an effective contribution of 10.8% in empty nest syndrome, of which 89.2% was influenced by other factors.

4 DISCUSSION

Based on the analysis of research data, results were obtained that showed that there was a relationship between resilience and empty nest syndrome with a value of $r = -0.329$ and a value of $\rho = 0.00$. This condition indicates the presence of a negative relationship between the two variables. Thus, it can be concluded that the higher the resilience, the lower the empty nest syndrome and vice versa which means the hypothesis is accepted.

The results of this study show that parents with high resilience can overcome psychosocial problems, including loneliness, anxiety, and depression, as well as be more optimistic in facing future situations, so that they are more adaptable in the face of empty nest syndrome with all the risks.

This resilience can be a supporting factor that has the power to cope with stress due to events that occur in life. Emotional support from a partner will go a long way to adapting to the empty nest syndrome. One of the factors influencing the empty nest syndrome in parents is marital status. Parents who have a partner are easier to adjust to compared to parents who do not have a partner. Living with a partner is an action that can help achieve a process of resilience in the phase of empty nest syndrome. The action that can be done by parents is to do activities together with their partners such as *refreshing*, taking advantage of the time alone so that parents do not feel loneliness and sadness. Mutually express the problems that are being experienced or complaints that are felt either due to the departure of the child or because of other things (Darmayanthi & Lestari 2019). The existence of the couple is very influential in the resilience of parents who experience empty nest syndrome since the orientation of roles in life will return to the couple (Ghafur & Hidayah 2014).

Another factor that can influence the phase of empty nest syndrome is the relationship with the child. Resilience in the empty nest phase can occur when parents perform actions such as having intense communication with the child, this is in line with the research conducted by Dharmawati (2016) that one of the efforts made to prevent empty nest syndrome in middle adulthood is to maintain good relations with the child.

The next action that can be taken to obtain good resilience is to meet and carry out positive activities with family and friends to reduce the feelings of loneliness experienced by parents. Shakya (2014) said that visiting the home or family of a married child and living with his partner, is one of the things that can reduce the effects of empty nest syndrome. In addition, it is important to cultivate positive relationships with others as a way to adapt to the phase of empty nest syndrome which includes tenacity, pleasure, and joy as well as the familiarity and love gained when establishing a positive relationship with another person (Wardani 2012).

Individuals who have good resilience are individuals who take actions that can overcome their problems well. To overcome this, it is necessary to take actions that can achieve resilience, such as drawing closer to God. Having spirituality in a person is one aspect of the resilience proposed by Connor and Davidson, he believes that a difficulty experienced is a decree that comes from God and there must be a way out. This is in line with research conducted by Akmalah (2014) in addition to social support from the family, drawing closer to God is an effort for coping strategies to feel calmer in the face of changes that occur in the empty nest phase. Other studies have also shown that spirituality greatly affects a person's life such as research conducted by Mabruroh (2018) that congregational prayer can reduce loneliness in the elderly due to experiencing empty nest syndrome. His research shows that doing congregational prayers can foster positive relationships with others and get emotional support from those closest to him.

This phase of empty nest syndrome in middle adulthood does not always have a negative impact on the person who experiences it, but this phase of empty nest syndrome can be considered as a success in carrying out the duties of a parent. The success in question is to see the child grow up and be able to live independently, meaning that the parent succeeds in educating and convincing his child until he can live independently away from his parents. In this phase, lower intermediate-adult men experience empty nest syndrome because they have high emotional intelligence than intermediate-adult women who have low emotional intelligence (Singh *et al.* 2017).

Empty nest syndrome is more commonly experienced by women than men, research conducted by Helen (in Newman 2008) found that there are differences between men and women in responding when in the phase of empty nest syndrome. Women have a greater chance of experiencing empty nest syndrome because they have more of a role in raising children, such as spending a lot of time at home with children and taking care of children more, while men work more outside the home. Yulia (2021) confirmed this, in her research proving the difference in empty nest syndrome in intermediate adult men and women. When viewed from the average value, he confirmed that empty nest syndrome in males has a value of 61.20 and empty nest syndrome in females has a value of 74.45. This can happen because Moms are more open to their feelings and are more likely to state that they have empty nest syndrome than fathers. From this explanation, it can be seen that it is possible that men can also experience the empty nest syndrome phase.

The results of the categorization of the empty nest syndrome scale show that as many as 60% of 100 parents experience low empty nest syndrome, which means that parents can face this transitional phase as well as be able to overcome their psychosocial problems such as loneliness, anxiety, feeling loss, sadness, and feeling the presence of feelings of emptiness when left by the child to wander. The low empty nest syndrome experienced by parents indicates that they have good resilience in the face of this transitional phase. Judging from the results of the categorization of the resilience scale shows that as many as 90% of 100 parents have high resilience, which means that parents in facing this transitional phase. They are able to rise up and adapt to difficult and misery situations so as to avoid the effects of stress that can damage the individual (Wagnild & Young 1993).

Parents who have a job will be inferior to experiencing empty nest syndrome compared to parents who do not have a job. Ghafur & Hidayah (2014) explained that women who only do work at home and have no other activities outside the home will experience a more severe empty nest syndrome because they "feel no longer needed" which is so severe. In this study, the average respondent had a job and other activities or activities or activities outside of their job, so it can be seen in the data categorization table that no parent has low resilience and there are only 10% of parents who have moderate resilience. The study also showed there were only 2% of parents who had high empty nest syndrome.

The results of the description of the subject in the long-standing qualification left by the wandering child there are as many as 96% of parents who are in the range of 2–10 years, meaning that it is possible that most of the parents feel the empty nest syndrome than earlier, for instance, when parents release their children to go wandering. According to research conducted by Romadhoni (2018), it is known that the length of time left by children wandering is in the range of 2–4 years of 96.4.

In this study, it was found that the mean of empty nest syndrome based on the number of children who went wandering in the range of 1–3 children was 1.27 and the mean in the range of 4–5 children was 4.25. The results showed that parents with a number of children wandering 4–5 children had a higher rate of empty nest syndrome compared to parents who had a number of wandering children as many as 1–3 children. This allows that parents who have a number of children of 4–5 people and all their children go wandering can make parents feel more emptiness because the absence of one child settling at home makes parents have a sense of emptiness toward their daily activities. Unlike the parents who have

3 children but only 1 child goes to wander parents can still get support and satisfaction from other children who are still at home.

The findings of this study also show that the coefficient of resilience variables has a contribution of 0.108 meaning that the effective contribution of resilience in empty nest syndrome is 10.8%, of which 89.2% is influenced by other factors. Other factors that also influence empty nest syndrome are intellectual maturity, development, personal happiness, productivity, confidence, and self-confidence (Ghafur & Hidayah 2014).

Researchers realize that this study has limitations, namely a limited number of research subjects. This condition occurs because it is difficult to find tech-savvy respondents, while data collection is mostly carried out online. In the current era, in which almost everybody uses technology, most parents already use Android-based cellphones or smartphones, but it turns out that not all parents understand to properly handle the cell phone.

5 CONCLUSIONS AND IMPLICATIONS

Based on the results of the studies that have been carried out, it can be concluded that there is a negative relationship between the resilience variable and the empty nest syndrome variable. This suggests that the higher the resilience, the lower the perceived empty nest syndrome. Other results of this study showed that the effective contribution of resilience to the possible appearance of empty nest syndrome was 10.8% whereas the other 89.2% were influenced by other factors.

This study implies that parents who are left by wandering children are expected to be more prepared in facing the transition phase of the empty nest syndrome because basically as parents, they must have known for sure that there will be a time when the child grows up and will live his own life. By being able to pass the empty nest syndrome phase, it is hoped that parents will find it easier to live it such as doing resilient actions, for example, maintaining good relations with children and families.

For subsequent researchers, it is advisable to consider other variables that affect empty nest syndromes such as marital status, spirituality, family resilience, and other. Moreover, more research samples were used to further explore the dynamics of the empty nest syndrome and get results that can be generalized widely.

REFERENCES

Abraham, S. (2012). Ageing Successfully for Managing Empty Nest Syndrome. *India Research & Development Journal, 18*(2) May. Kalady: University of Sanukrit.

Akmalah, N. (2014). Psychological Well-being pada Ibu Usia Dewasa Madya yang Berada pada Fase Sangkar Kosong. *Jurnal Psikologi Industri Dan Organisasi, 3*(2), 1–10. journal.unair.ac.id/filerPDF/jpkk23a00850cafull.pdf

Becker, G., & Newsom, E. (2005). Resilience in the Face of Serious Illness Among Chronically ill African Americans in Later Life. *Journals of Gerontology–Series B Psychological Sciences and Social Sciences, 60* (4). https://doi.org/10.1093/geronb/60.4.S214

Bouchard, G., & McNair, J. L. (2016). Dyadic Examination of the Influence of Family Relationships on Life Satisfaction at the Empty-Nest Stage. *Journal of Adult Development, 23*(3), 174–182. https://doi.org/10.1007/s10804-016-9233-x

Chang, Y., Guo, X., Guo, L., Li, Z., Yang, H., Yu, S., Sun, G., & Sun, Y. (2016). Comprehensive Comparison Between Empty Nest and Non-empty Nest Elderly: A Cross-sectional Study Among Rural Populations in Northeast China. *International Journal of Environmental Research and Public Health, 13*(9). https://doi.org/10.3390/ijerph13090857

Cicchetti, D., & Rogosch, F. A. (2002). A Developmental Psychopathology Perspective on Adolescence. *Journal of Consulting and Clinical Psychology, 70*(1), 6–20. https://doi.org/10.1037/0022-006X.70.1.6

Connor, K. M., & Davidson, J. R. T. (2003). Development of a New Resilience Scale: The Connor-Davidson Resilience scale (CD-RISC). *Depression and Anxiety*, *18*(2), 76–82. https://doi.org/10.1002/da.10113

Darmayanthi, P. K. N., & Lestari, D. (2019). Proses Penyesuaian Diri pada Perempuan Usia Dewasa Madya Yang Berada Pada Fase Sarang Kosong. *Jurnal Psikologi Perkembangan*, *6*(1), 838–848.

Dharmawati, M. A. (2016). Upaya-Upaya Mencegah Sindrom Sarang Kosong Pada Lanjut Usia Perempuan Di Banguntapan Bantul. *Pembagian Harta Waris Dalam Adat Tionghoa Di Kecamatan Ilir Timur I Kota Palembang*, *1*(14 June 2007).

Fikri, L. A. (2021). Hubungan Antara Resiliensi Dan Kualitas Hidup Lansia Selama Pandemi Covid-19. *Skripsi*. Fakultas Psikologi, Universitas Islam Indonesia. https://dspace.uii.ac.id/handle/123456789/33803

Folke, C., Carpenter, S. R., Walker, B., Scheffer, M., Chapin, T., & Rockström, J. (2010). Resilience Thinking: Integrating Resilience, Adaptability and Transformability. *Ecology and Society*, *15*(4). https://doi.org/10.5751/ES-03610-150420

Ghafur, J. & Hidayah, F. S. (2014). Seri Pengabdian Masyarakat 2014. *Jurnal Inovasi dan Kewirausahaan*, *3* (2), 120–125.

Handono, O. T., Psikologi, F., Ahmad, U., & Yogyakarta, D. (2013). Hubungan Penguatan Terhadap Disiplin Anak Usia Dini Di Paud Pembina 1 Kota Bengkulu (Studi Deskriptif Kuantitatif Di PAUD Pembina 1 Kota Bengkulu). *EMPATHY Jurnal Fakultas Psikologi*, *1*(2), 79–89.

Hendriani, W. (2018). *Resiliensi Psikologis* (Kencana (ed.); pertama). Prenadamedia Group.

Hurlock, E. B. (2011). *Psikologi Perkembangan: Suatu Pendekatan Sepanjang Rentang Kehidupan Edisi Kelima*. Jakarta: Erlangga, Edisi 5.

Iacoviello, B. M., & Charney, D. S. (2019). Cognitive and Behavioral Components of Resilience to Stress. In *Stress Resilience: Molecular and Behavioral Aspects*. Academic Press. 23–31. https://doi.org/10.1016/B978-0-12-813983-7.00002-1

Kumalasari, F., Pengajar, S., & Psikologi, F. (2012). Hubungan Antara Dukungan Sosial Dengan Penyesuaian Diri Remaja Di Panti Asuhan Latifah Nur Ahyani. *Jurnal Psikologi Pitutur*, *1*(1), 21–31.

Lazarus, R. S. (1993). Coping Theory and Research: Past, Present, and Future. *Psychosomatic Medicine*, *55* (3), 234–247. https://doi.org/10.1097/00006842-199305000-00002

Mabruroh, M. I. (2018). Shalat Berjama'ah Dalam Mengurangi Empty Nest Syndrome Pada Lansia Di Yayasan Al-Jenderami Selangor Malaysia. *Skripsi*. Fakultas Dakwah dan Komunikasi, Universitas Islam Negeri Sunan Kalijaga.

Mansoor, A., & Salma Hasan, S. (2019). Empty Nest Syndrome and Psychological Wellbeing among Middle Aged Adults. *Pakistan Journal of Social and Clinical Psychology*, *17*(1), 55–60.

Martínez-Martí, M. L., & Ruch, W. (2017). Character Strengths Predict Resilience Over and Above Positive Affect, Self-efficacy, Optimism, Social support, Self-esteem, and Life Satisfaction. *Journal of Positive Psychology*, *12*(2). https://doi.org/10.1080/17439760.2016.1163403

Nabila, S. (2021). Efikasi Diri Skripsi Dan Resiliensi Pada Mahasiswa Yang Sedang Mengerjakan Skripsi Di Universitas X. *Skripsi*. Fakultas Psikologi dan Ilmu Sosial Budaya, Universitas Islam Indonesia. https://dspace.uii.ac.id/handle/123456789/31370

Newman, S. (2008). Empty Nest: Who Is Needier, Parent or Child? *Psychology Today*. https://www.psychologytoday.com/us/blog/singletons/200809/empty-nest-who-is-needier-parent-or-child

Oktaria, F. (2019). Hubungan Konsep Diri Dengan Empty-Nest Pada Ibu Yang Memiliki Anak Merantau Di Desa Waekerta Kabupaten Buru Maluku. *Skripsi*. Fakultas Psikologi, Universitas Negeri Malang.

Reivich, K & Shatte, A. (2002). *The Resilience Factor: 7 Essential Skills for Overcoming Life's Inevitable Obstacles*. Broadway Books.

Romadhoni, D. K. (2018). Hubungan Antara Kepuasan Pernikahan Dengan Empty-Nest Syndrome. *Skripsi*. Fakultas Psikologi, Universitas Muhammadiyah Malang.

Santrock, J. W. (2012). Life-Span Development, Perkembangan Masa Hidup (Edisi Ketigabelas) Jilid 1. *Erlangga*.

Shakya, D. (2014). Empty Nest Mirrored in a growing mind: A Case Report. *Journal of Psychiatrists' Association of Nepal*, *2*(2), 43–45. https://doi.org/10.3126/jpan.v2i2.9726

Silaban, S. M. (2019). Hubungan Antara Kepuasan Pernikahan Dengan Kecenderungan Empty Nest Syndrom Pada Dewasa madya Di Kelurahan Pekan Tanjung Beringin Dusun VI. *Skripsi*. Fakultas Psikologi, Universitas Medan Area. http://repository.uma.ac.id/bitstream/123456789/10757/1/148600143-SantiMariniSilaban-Fulltext.pdf

Simon, V. P. (2009). Studi Deskriptif Sindrom Sarang Hampa. *Skripsi*. Fakultas Psikologi, Universitas Sanata Dharma.

Singh, V., Dubey, B., & Patankar, W. W. (2017). A study on empty nest syndrome among middle-aged men and women. *International Journal of Current Research*, 9(04), 48838–48842.

Sugiyono, D. (2016). *Metode Penelitian Pendidikan Kuantitatif, Kualitatif Dan R&D / Sugiyono*. In Bandung: Alfabeta.

Tugade, M. M., & Fredrickson, B. L. (2004). Resilient Individuals Use Positive Emotions to Bounce Back from Negative Emotional Experiences. *Journal of Personality and Social Psychology*, *86*(2), 320–333. https://doi.org/10.1037/0022-3514.86.2.320

Wagnild, G. M., & Young, H. M. (1993). Development and Psychometric Evaluation of the Resilience Scale. *Journal of Nursing Measurement*, 1(2), 165–178.

Wang, G., Hu, M., Xiao, S. Y., & Zhou, L. (2017). Loneliness and Depression Among Rural Empty-nest Elderly Adults in Liuyang, China: A cross-sectional study. *BMJ Open*, *7*(10), 1–8. https://doi.org/10.1136/bmjopen-2017-016091

Wardani, R. (2012). Kesejahteraan Psikologis Dan Dukungan Emosional Pasangan Pada Ibu "Empty-Nester" Di Kota Bandung. *Prosiding SNaPP: Sosial, Ekonomi Dan Humaniora*, *3*(1).

Yulia, Y. (2021). Perbedaan Empty Nest Syndrome Pada Pria dan Wanita Dewasa Madya Di Desa Serai Wangi Kecamatan Talang Muandau. *Skripsi*. Fakultas Psikologi, Universitas Islam Riau.

Families Mental Health and Challenges in the 21st Century – Arifin et al. (eds)

The role of organizational commitment to job flow experience

Djudiyah*, Wildan Zarief & Ratih Eka Pertiwi
Fakultas Psikologi Universitas Muhammadiyah Malang, Jawa Timur, Indonesia

ABSTRACT: Working in a job is a necessary phase where individuals may cater to their basic needs. Working in a job can instill ease and enjoyment when individuals develop their intrinsic motivation for the intended job. Furthermore, the motivation to continue within a job is perceived to be beneficial for the magnitude of happiness and enjoyment at work. This study aims to look into the effect organizational commitment related to the workers' job flow experience. This research employed 150 participants sampled with the convenience sampling technique with the criterion as follows: service sector workers all across Indonesia, 20–40 years old of age, and have been working for 1 year or more within the same institution. The instruments used in the research are the work-related flow inventory (WOLF) scale and the organizational commitment scale by Allen and Meyers. This research employed a multivariate linear regression method to look at the effect each dimension's organizational commitment has on the employees' job flow experience. This research found that organizational commitment has a significant effect as high as 25.6% on job flow experience.

Keywords: Organizational commitment, job flow experience, workers

1 INTRODUCTION

A balance between the ability workers have and the work challenge they receive is required to work optimally. The optimal work experience is frequently described as a sensation of being engrossed in their job or "being in the zone". This experience leads to a psychological condition where the working individual can focus on their job, followed by the ability to work unhindered, as if the action is flowing. Generally, this sensation is followed by the individual losing their sense of time, so they feel that the working hours pass quickly. The concept of optimal work experience is defined by Csikszentmihalyi (1990) as "flow experience".

It was Csikszentmihalyi who is referred to as the first person who coined the term psychological condition of flow. This theory is utilized to describe a situation where an individual manages to deeply concentrate when they are doing activities. The ability to fully engross in an activity offers a feeling of satisfaction with their achievement and a sense of happiness. The obtained elation from the flow process is intrinsic or from their own self and directly related to what their activity is (Larson & Csikszentmihalyi 2014). These experiences have been studied in several subjects such as sports, arts, and the professional world. The flow experience is known as a booster of the mental quality of work an athlete or an employee in various areas (Hancock *et al.* 2019). In the professional world, this idea of flow is known as a job flow experience.

*Corresponding Author: djudiyah@umm.ac.id

DOI: 10.1201/9781003402381-4

Job flow experience is a phenomenon of mental flow experience that occurred when a person does their job in a professional setting. Maeran & Cangiano (2013) explained the various work-related characteristics an individual has which enhance the flow situation. The characteristics are divided into nine main dimensions; (1) the balance between skills and work challenge, (2) an immersion between work and consciousness, (3) the clarity of work goals, (4) real and clear feedback, (5) concentrating on their task, (6) assuming the control of their work, (7) loss of sense of ego, (8) the transformation of time perception, and (9) autotelic experience (feeling satisfied for their upcoming activities). If some of these characteristics emerge during job experience, it is determined that the individual is experiencing a job flow experience. Job flow experience will encourage the person to gratefully do their job, feel motivated to do their job again, and avoid employer turnover or an employee resigning from an institution (Kim *et al.* 2019).

Csikszentmihalyi (1990) stated that when an individual often enters a flow condition, the working experience becomes autotelic (the motivation to work because work is a reward itself). Maeran & Cangiano (2013) stated that job flow experience is one of the factors that could drive a person to delightfully complete their job and encourage them to feel satisfied with their job. This result summarised that a person who often enters a flow situation during their working hours could develop their motivation to work and produce an internal feeling of satisfaction.

One of the stimulations for an individual entering a flow state during their job is a high concentration level to a merging action and their consciousness (Maeran & Cangiano 2013). Goleman (2013) explained that a focus condition and strong concentration level could promote discipline and their activity commitment. In the professional world, it is called "organizational commitment".

The definition of organizational commitment is a person's willingness and ability to participate, assist and become a part of an organization. Macey & Schneider (2008) defined organizational commitment as a bond between an individual to their organization. Organizational commitment is a binding force between employees and an institution. Organizational commitment is also an emotional tie between the employees and organizational elements include the responsibility inside it (McShane & Glinow 2008). In conclusion, organizational commitment has an important role in maintaining the employees' mental condition and assisting in managing the employees' quality of focus and work performance.

Organizational commitment plays a substantial role in developing the mental quality of organization's members. A strong commitment quality in a working area ideally could assist internal satisfaction. This satisfaction offers unambiguous and focused feedback about the employee's performance which could trigger a flow situation. Indirectly, there is a probability that when an employee determines an intrinsic job commitment, it could trigger the job flow experience at work.

A study by Bakker & van-Woerkom (2017) discovered that when an individual receives unambiguous feedback and clear goals during a working condition, they will be more motivated to complete a task that has a high difficulty level. The study revealed that when a person was experiencing a flow condition at work, they will feel satisfied and proactively increase their work performance. Macey & Schneider (2008) explained that someone's willingness to proactively do a task at work could establish the flow situation and increase work performance. Furthermore, if a person is open to learning new things and has a high level of desire to complete a challenging assignment, they will more easily experience flow. The study concluded that if an employee is willing to be more involved in their job, it will encourage them to feel satisfied with their job.

Farina *et al.* (2018) found that a person with a high initiative level will be more responsible for their own growth and more committed to their job quality. This research revealed that in general, an organization hopes that its members would have their own initiative and be responsible for their job. Moreover, a member of an organization would seek self-growth, satisfaction, and enjoyment while doing their job.

Jackson (2018) revealed that there is an engagement between members of an organization and their attachment to job commitment. The study disclosed that there was a positive correlation between organizational commitment, job satisfaction, and job engagement of organization members. The research explained that an individual's organizational commitment would cause the employee to be more engaged with their job. This engagement elicits a satisfying feeling in an individual in their job.

Tsai (2018) explained that an ambiguous psychological contract will decrease the organizational commitment and the effectiveness of the job. A strong psychological contract with the organization would induce the flow state more easily. The result of the study revealed that achieving a creative performance in a flow state will be easier with a stronger psychological contract with the organization. The study also demonstrated that an excellent job commitment could be a facilitator for the flow state at work. Gota (2017) disclosed that a strong psychological capital would increase the organizational commitment and the flow state at work. The result revealed that there was a correlation between the psychological capital of job commitment and job engagement. Once again, the outcome illustrates that there is a modality in job commitment that could facilitate the individual's job engagement and situation at work. Engagement or engagement at work is one of the factors which formed flow at work.

Besides influencing job satisfaction, organizational commitment also involves engagement in a job such as job flow experience. In a working environment, having a strong commitment is one of the factors that form focus, satisfaction and an immersive condition with their job elements. An individual who has a great job commitment will have intrinsic job goals and a desire to self-develop and undertake more complicated tasks. This desire is manifested in a flow state at work, which contributes to the employee's work productivity and well-being.

Based on the analysis above, it could be concluded that the employees' commitment to the organization could influence the flow state at work. Organizational commitment is identical to someone's ability to dedicate their time, place, and environment to their job. Moreover, flow is a condition where a person manages to fully concentrate on their job and gladly immerse in their job. This research is aiming to prove that the influence of individual commitment facilitates the individual's job flow experience state. The result of this study could be a literature addition for Industrial and Organizational Psychology and Positive Psychology. The practical benefit of this research is a thought for a company to consider utilizing the positive psychology approach (flow) as a development tool for the employees' job quality and well-being.

1.1 *Job flow experience*

Job flow experience or flow at work is a psychological element that could lead a person to work optimally and evokes the intrinsic motivation to complete their task. The condition of flow at work is a circumstance where a working person experiences a merging sensation with their task, hyper-focus with the activities so they feel time pass quicker and the job activity flows from one component to another (Csikszentmihalyi & LeFevre 1989).

Bakker (2008) illustrated that the flow state is an optimum working condition that is characterized by immersion in job activities, feeling of contentment, and intrinsic job motivation at work. The immersion is a merge between an individual's mental state and their job. This state is followed by a feeling of contentment when they do their job, which led to intrinsic achievement. This achievement feeling would create an internalized motivation for someone to continue doing their job.

Chu & Lee (2012) described that flow as a subjective mental sensation and process of an individual that has a critical role so this person would enjoy their activity and repeatedly do it again later. The enjoyment sensation while working would be an intrinsic motivation that assists a person to improve their motivation to complete the task and improve the positive anticipation feeling about their task.

The flow state is defined by Bakker & van-Woerkom (2017) as a natural and automatic mental process that establishes a peak experience while doing a job task. This automatic process triggers effectiveness in doing their job because there is an uninterrupted flow of mental processes at work. When a person experiences the flowing mental process, a disruption to their job is low, hence they maximize their focus.

Wolfigiel & Czerw (2017) explained that the flow state is an autotelic mental state, a deep satisfaction feeling while voluntarily doing an activity because it is considered worthy. The autotelic state is defined as someone considering their activities as a reward itself, and identical to the growth of intrinsic motivation while doing the activity.

In conclusion, flow is a mental state where an individual feel that they are merging with their activity and their job is flowing with time. This condition is described as an optimum working condition because it is automatic, voluntary, rewarding, and develops intrinsic motivation for the job. Moreover, the job flow experience is an illustration of a specific flow state when a person does their job in their profession. With this positive quality, it is expected from individuals in the organization to voluntarily and optimally enhance their work performance.

1.2 *The aspect of flow*

Further research by Bakker (2005) revealed that there were three main aspects when a person enters a flow state. Absorption, enjoyment, and intrinsic motivation are the main aspects of the formation of a flow state. Absorption is a condition where a person's consciousness is absorbed by their activity. Generally, the psychological energy of an individual merges with their current activity without any distraction or disruption. An internal disruption will vanish and an external distraction is automatically ignored. Enjoyment is a condition where a person feels happy and feels relieved while doing their work. This condition is an affective condition that emerges from the positive cognitive judgment of the result of their job. Enjoyment and voluntarism are factors that enhance someone's motivation to do their activity. Intrinsic motivation is a motivation that emerges within someone. This motivation appears without any external validation such as praise or token reward. Intrinsic motivation is formed because of individuals' happiness and voluntarism while performing their activity.

1.3 *The conditions required for flow*

According to Hasty (2019), there were six conditions required for someone experiencing flow. First, having a clear aim. Having a definite goal while performing an activity eases a person to determining structure and direction. Clear goals facilitate the mental evaluation of their position at the job level. Second, knowing how to do the activity. Third, a person who will enter the flow state has depth knowledge of their upcoming task. When the person understands how to perform the task, they could predict the obstacles and advantages they will encounter in their task. Fourth, there is feedback and mini-goals from their job outcome. In general, an individual who experiences a flow state will realize there is some feedback about their job performance. This feedback will not come immediately after they complete the job, but after completing a small but achievable activity. Fifth is the balance between skills and challenges. An equilibrium between skills and challenges is a form factor for a flow state, in which if someone has a proficient skill, followed by an equal challenge, they will attract them to solve the challenges. Concurrently with interest, a person also could intensely do their task without any difficulties. Sixth, an environment with minimum distraction. A conducive space while working is one of the mediators to experiencing a flow state more easily. If a person could concentrate and focus on their job, a flow state will occur automatically.

1.4 *The factors which influence flow*

Csikszentmihalyi (1990) stated that two main factors affect the flow of an individual. The factors are personal skill and activity challenge. Personal skill is described as an ability that someone has to do their task whereas an activity challenge is an illustration of how hard or easy a task is given to an individual in their job.

Figure 1 could assist in explaining the correlation between the factor of personal individual skill and the challenge of their activities.

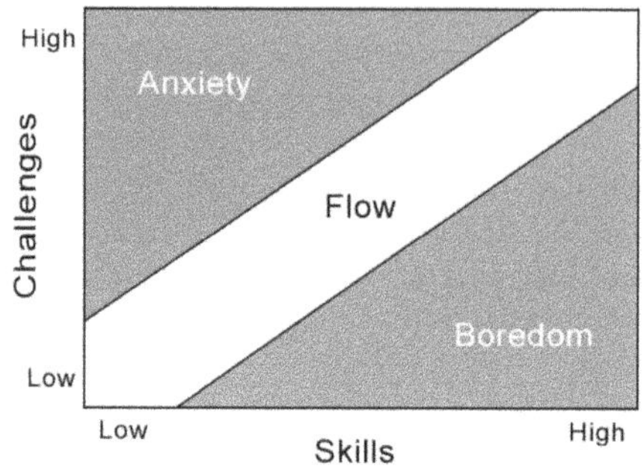

Figure 1. Flow: Challenge vs Skill diagram. Adapted from "Reflective design-in-use: Co-designing an assistive remote communication system with individuals with cognitive disabilities and their families." by M. D. Schmidt, 2007, University of Colorado.

Flow is an optimum working condition that could be achieved by being attentive with balancing the skills and challenges. If their skill is equal to the obtained challenge, the flow state will be achieved. However, if the skill is well below the obtained challenge capacity, someone would feel anxious doing their job because they are unable to self-actualize while doing the task. Conversely, if their challenge is far below their ability, they are unable to grow and feel bored and apathetic to their job.

1.5 *Organizational commitment*

Meyer & Allen (1991) defined organizational commitment as a condition where an individual's affective, cognitive, and continuous attachment to the organization. Meyer and Allen implied that an individual who joins and commits to an organization has a particular attachment according to a symbiotic relationship between employee and company. This attached commitment is based on someone's mindset regarding their decision to join an organization.

Ghosh & Swamy (2014) described that organizational commitment is an individual condition where a person becomes a part of and performs their responsibility in an organization according to their contract. This contract offers a sense of feedback from an individual to the organization. The symbiotic feeling is a commitment of an individual to strengthen the organization. Mesa *et al.* (2019) explained that organizational commitment is an individual's attitude toward the organization, which consists of trust, acceptance of the organization's vision, and desire to maintain the organization's identity. This desire would lead a person to stay and continue working in the organization.

Armenic & Aranya (1983) defined organizational commitment as an identity power and individual engagement with the components of the organization. Individuals who have engagement power and identity in organizations could be referred to as committed individuals. This term occurs because of their role and their strong desire to stay and settle with the responsibility of the institution. Wuryani (2013) explained that organizational commitment is a situation where a person is a strong supporter of organizational value and forms an engagement with that value. It means that the individual who possesses a sense of unity with the organizational vision is more likely to dedicate themselves to the organization.

From the analysis above, it is summarized that an organizational commitment is a drive that someone has elicits a sense of attachment, responsibility, and involvement with the organization. This drive emerges because there is a desire within someone to support and enact the organization's vision aligned with their beliefs.

1.6 *The dimensions of organizational commitment*

Organizational commitment is a situation where an individual offers themselves an opportunity to support, involve and become one of the parts of organizational identity. Meyer & Allen (1991) in their theory of organizational commitment stated that there are three main dimensions components:

Affective Commitment: a job commitment that someone has that emerges from the emotion of fondness and pride of becoming a part of the organization. Affective Commitment occurs when an individual has the same vision as the organization and feels satisfied when doing the activity within the organization's responsibility

Continuance commitment: a commitment that emerges from the relationship of necessity between an individual and their tasks in the organization. The relationship revealed how the company benefits the individual if the individual performs the work and responsibilities of the organization. The more benefit the company obtains from the job, the stronger the individual's intention to stay in the organization.

Normative commitment: a commitment based on the feeling of responsibility to the organization. This commitment appears because there is a normative drive within the self and organizational environment for a person to settle in the organization. This commitment is related to the individual appraisal of its vicinity to assess the positive and negative aspects of staying in the company.

1.7 *The influence of organizational commitment and job flow experience*

An organizational environment is a place where an individual could develop their potential; soft skills and hard skills, together with others to achieve particular goals. A strong organization is an institution with integrity in its surroundings for all the resources. One of the crucial components in organizational development is the individuals inside the organization. The organization could increase its productivity and preserve the attachment of the members having strong goals and dedication to the organization. The individual's commitment to the organization is the main source to strengthen or weaken the structure and performance of the organization. Commitment can be a factor where the individuals could adjust their vision with the organization's goals (Mesa *et al.* 2020). The equilibrium between individuals' vision and organizations affects the individual's ability to more easily actualize themselves. Moreover, if the individual feels satisfied while working in the organization, the individuals will complete their responsibility voluntarily (Wuryani 2013). Job voluntarism incites individuals' sincerity and reinforces intrinsic satisfaction while working (Armenic & Aranya 1983).

An employee with high affective commitment feel proud and elated to be a part of the organization. Their visions and missions are aligned with the organization's vision and mission (Meyer & Allen 1991). They have the urge to stay and accept the vision and maintain the institution's reputation. High-level commitment would assist the individual to develop loyalty to the company and increase organizational productivity (Mesa *et al.* 2020). Armenic & Aranya (1983) explained that the stronger the individuals' commitment to the organization, the higher the individual sincerity to finish their organizational responsibilities with all their skills.

Employees with high continuance commitment perceive that there is a symbiotic relationship between them and the organization where they work. The higher the advantage for both organizations and individuals, the higher the level of continuance commitment. When an employee has high continuance commitment, they will feel responsible for their organization. They will try to stay working in the organization because they feel appreciated by others. They feel responsible for the highs and lows of the organization (Meyer & Allen 1991). The ability and responsibility of the job are a correlation between the individuals' skills and challenges. Job flow experience will occur when the individuals' ability is equal to their responsibility's challenges. If the individuals are sincere, they will utilize their skills maximally to complete their challenges (Bakker 2008).

In conclusion, individual commitment to the organization has an important role to support the job flow experience situation. Commitment is a driving power of an individual to attach to their organizational responsibility. With this attachment, it will drive the individual to decide their goals according to the organization's vision and strengthen the individuals' effort to complete their tasks as well as possible. The balance between the determination to offer their best for the organization followed by the challenge of their task will strengthen the job flow experience situation of individuals.

1.8 *Hypothesis*

Based on the theoretical analysis above, the hypothesis of this research are:

Hypothesis 1: Job Flow Experience is predicted by organizational commitment
Hypothesis 2: Job Flow Experience is predicted by affective commitment
Hypothesis 3: Job Flow Experience is predicted by continuance commitment
Hypothesis 4: Job Flow Experience is predicted by normative commitment

2 METHODS

2.1 *Research design*

In this research, correlational quantitative research was utilized by a psychological scale to observe the current condition objectively and measure with a numerical system (Daniel 2016). Correlational research studied the correlation between variables and the effect between variables (Simon & Goes 2011).

2.2 *Research subject*

The subjects in this research were both male and female employees who currently working in a company. The sampling technique for this study was the convenience sampling technique. The convenience sampling technique was selected because the data sampling method was randomized and everyone who fulfilled the requirement could participate in this study. Etikan *et al.* (2016) explained that convenience sampling is a non-probability sampling method where the sample is selected spontaneously adjusted to the time, condition, and criteria of the participants:

1. Currently working in a service companies
2. Have worked for the company at least for a year
3. Productive age of 20–40 years old

2.3 *The research variable and instrument*

The independent variable for this study was organizational commitment whereas the dependent variable was job flow experience. Organizational commitment is defined as the individual attachment to their company while job flow experience is an individual emotion that merged with their activity. This study aims to examine the attachment of an individual who has the same vision with their company influencing their situation while working especially the emotional immersion with their job.

The data collection method was using the psychological scale. There were two psychological scales utilized in this research: the organizational commitment scale of affective, continuance, and normative and the Work-Related Flow Inventory (WOLF).

The organizational commitment scale was created based on the three aspects of organizational commitment theory by Allen & Meyer (1990). The organizational commitment was assessed by examining three main aspects of affective, continuance, and normative. The scale consisted of 24 items which have three dimensions with 15 favorable items and nine unfavorable items. Subjects were asked to respond to the scale by selecting one from a five-point Likert Scale (1 = Strongly disagree, 5 = strongly agree). The result of Cronbach's Alpha test revealed that the reliability for affective commitment was 0.87, 0.75 for continuance commitment, and 0.79 for normative commitment, which indicates that the scale was reliable. The result of the alpha coefficient was provided by research by Allen & Meyer (1990). This study utilized the adapted scale and modified it by the authors to suit the condition of the job context in Indonesia. Later, the modified scale was tested and try-out, and the try-out result produced 16 items, and the alpha coefficient for each dimension; affective commitment = 0.730, continuance commitment = 0.718, and normative commitment = 0.701, which indicated a reliable instrument.

WOLF scale was utilized to assess the job flow experience phenomenon. The scale was developed by Bakker (2008), using the three flow factors of Csikszentmihalyi (1990): absorption, work enjoyment, and intrinsic motivation and was translated to Bahasa. The scale measured the flow phenomenon, especially in individuals who work in an organization. There were 13 items and all of them were favorable items. The participants were selecting one of a five-point Likert scale (1 = never, 5 = always) to answer the questions. The instrument try-out outcome revealed that the alpha coefficient was 0.83 which displayed the high reliability of the scale.

2.4 *The research procedure and data analysis method*

There were three stages of this study which were preparation, research implementation, and data analysis. The preparation stage started with the authors reviewing the theoretical data related to the study. Next, the implementation stage comprised reviewing the previous studies, determining the research participants, developing the scale according to the theory, and collecting the data through Google Forms. The last stage was analyzing all the obtained data and processing it to become the final validation result of the hypothesis. The data analysis was conducted with IBM SPSS Statistic 25 software with multivariate linear regression.

3 RESULT

The hypothesis was tested with the multivariate linear regression method to examine the effects of the organizational commitment dimensions of flow experience at work. Table 1 revealed the result of the multivariate linear regression test:

Table 1. The outcome of multivariate linear regression of organizational commitment and Job flow experience.

	Variable	B	P (Sign)	R^2
1	Organizational Commitment—Job Flow Experience	22.596	0.000	.256
2	Affective Commitment—Job Flow Experience	.264	.222	
3	Continuance Commitment—Job Flow Experience	.479	.021	
4	Normative Commitment—Job Flow Experience	.540	.004	

From Table 1, the following hypotheses are concluded:

Hypothesis 1: The variables of organizational commitment and job flow experience were submitted to multivariate linear regression. The result revealed that there was a significant effect of job flow experience toward organizational commitment in the creative industry's employees (β = 22.596, p = 0.000 ($p<0.01$)). The statistical outcome suggested that hypothesis one was accepted. Furthermore, the influence of organizational commitment to job flow experience was 25.6%.
Hypothesis 2: The multivariate linear regression method was utilized to assess the effect of job flow experience on affective commitment. There was no significant effect between job flow experience and affective commitment (β = 0.264, p = 0.222 ($p>0.05$)), so hypothesis number two was rejected.
Hypothesis 3: The result of multivariate linear regression (β = 0.479, p =0.021 ($p<0.05$)) revealed that the continuance commitment predicted the job flow experience, thus the third hypothesis was accepted.
Hypothesis 4: The outcome of a multivariate linear regression between normative commitment and job flow experience (β = 0.540, p = 0.004 ($p<0.05$)) suggested that there was a significant influence between the two variables, so the fourth hypothesis was accepted.

Based on the multivariate linear regression results, it could be concluded that organizational commitment had a significant effect on job flow experience. Although affective commitment has no significant influence on job flow experience, continuance commitment and normative commitment did have a significant influence on job flow experience.

4 DISCUSSION

This research aims to prove the hypothesis that job flow experience is predicted by organizational commitment in the workers of service companies. The result of this study revealed that there is a positive impact of organizational commitment to the job flow experience. It means that someone's higher organizational commitment would increase the job flow experience that someone encounters at work. Research about organizational commitment and job flow experience is limited both in Indonesia and abroad. This study is inspired by research by Gota (2017) and Tsai (2018). Both studies found that there was a relationship between commitment and individuals' attachment that could increase focus and psychological performance at work. Organizational commitment and flow at work are factors that boost the psychological modality at work (Gota 2017). Furthermore, the individual's attachment to their organization could drive and facilitate the occurrence of a mental state of flow at work (Tsai 2018). Job flow experience is defined as an individual mental process situation that easily merges and flows with their job activity. The research hypothesis is designed based on the assumption of if the employees experience high emotional (affective) attachment to their organization, they would more easily experience flow at work.

This study is adapting a framework from research by Kim *et al.* (2019) that utilized organizational commitment as a mediator variable of the formation of job flow experience in the staff of hotels and tourism. Our research revealed that organizational commitment influences the formation of job flow experience. Continuance commitment and normative commitment have a significant effect on the job flow experience, however, affective commitment did not. The main cause of this outcome was the employees that have high affective commitment felt joy and proud becoming a part of an institution. They felt that their vision and mission are similar to the organization, so their feeling is combined with the job flow experience. Consequently, they are unable to differentiate between affective commitment and job flow experience at work.

This research is modified to study the effect of specific affective commitment to flow experience at work in the context of Indonesian working culture. The assumptions that the authors have was proven through the quantitative method that organizational commitment has a positive effect on the employees' job flow experience in the service industry. Besides, the organizational commitment and job flow experience were the factors that formed psychological modality, organizational commitment is also a significant influence in reinforcing the job flow experience at service works. In conclusion, this research supports the notion that if an individual at work experiences attachment, they will obtain a positive experience while working and produce optimum performance.

Despite this research that revealed a promising result of organizational commitment and job flow experiences, there are several unexplored gaps that might influence the result. The first gap is this research do not consider external factor that might modify the employees' commitment such as their position, official title, or education background of the participants. The employees' position and official title could provide extra information on whether these external factors would affect their commitment. The higher level of education might enhance the organizational commitment of the staff which led to the job flow experience increasing. The second gap is the environmental working situation of the employee. The focus and conducive subjective condition are mediator factors of experiencing flow at work (Goleman 2013; Maeran & Cangiano 2013). If the working environment of an individual promotes subjective comfort and calmness at work, it would be easier for them to commit to their job and experience the flow state. The third gap is the working culture of the participants' organizations or companies. Achieving the flow state is easier when the employees' received immediate feedback from their job (Bakker & van-Woerkom 2017). The working culture is the main pillar for feedback on the employees' performance. If their working place has a tight culture for the employees, it would facilitate the occurrence of immediate positive feedback which leads to the formation of a flow state at work.

The last gap is the years of experience of the participants. All the participants are required at least have worked for their companies minimum of a year. However, the more experienced an employee has, the easier someone to enter the flow state at work (Bonaiuto *et al.* 2016). The participants' ages were between 20 and 40 years which is categorized as the early adulthood period. This is the age stage where someone began to enter the professional world so their experience balancing the challenge and skill at work is limited, hence assessing their job flow experience correctly is complicated. There are several limitations of this research. The participants of this study are the employees of the service industry, which did not differentiate on what kind of service industry they work at. By choosing one specific type of service industry, it might offer different results compared to the non-specific one. This is related to the fact that job flow experience is related to what their job activity is. The next limitation is partially using each dimension of organizational commitment to examine the correlation with job flow experience. The last limitation is the data is collected by online questionnaire method, so there is a chance for the participants to select only the appropriate answers or Faking Goods. Despite the limitations and gaps, this study offers new knowledge for psychological research, especially about job flow experience. The topic of job flow experience is quite new and unique in Indonesia, so other researchers could use this study as a reference. Future research should consider utilizing job flow experience and organizational commitment to assess the employees' well-being at the workplace. Furthermore, this study could prove that the acceptance and immersive feeling in the workplace would help the individual work in optimum mode. The result of the research would assist the development of organizational commitment and flow in the workplace, so the company could increase the job satisfaction level and tackle the turnover problem in the professional world.

5 CONCLUSION AND IMPLICATION

This research examined the direct effect of organizational commitment on job flow experience. The results revealed that organizational commitment had a significant effect on the job flow experience of the employees in the service industry. The continuance commitment and normative commitment significantly influence the job flow experience, whereas the job flow experience did not significantly influence by affective commitment.

This study indicated that in Indonesia, the staff in the service industry who experienced an immersion in their working environment discovered that it was easier for them to work optimally. The Human Resources Department (HRD) of service companies could consider the outcome of this research to develop the employees' organizational commitment and design the work culture of the organization around the capacity to facilitate the staff's flow state. This research revealed that the percentage of employees' organizational commitment in Indonesia was significantly low. Job flow experience assists in rising the comfortable feeling at work so it helps to decrease the turnover rate that happens quite often in the professional world. There are several suggestions for future research. First, the researcher should consider utilizing the mixed method, where half of the participants' responses are collected online, and the other half of the responses are collected offline to increase the control of the participants. The number of participants also could be increased. Second, other factors that might affect flow at work such as culture, age span, and gender should be considered as variables. Third, future studies could be conducted on the employees of the non-service industry to observe whether they are also experiencing flow or not. Lastly, for the next research, the geographical data can be added to observe the influence of culture on individual perception when a flow state occurred at work.

REFERENCES

Allen, N. J., & Meyer, J. P. (1990). The Measurement and Antecedents of Affective, Continuance and Normative Commitment to the Organization. *Journal of Occupational Psychology*, *63*(1), 1–18. https://doi.org/10.1111/j.2044-8325.1990.tb00506.x

Armenic, J. H., & Aranya, N. (1983). Organizational Commitment: Testing Two Theories. *Industrial Relations*, *38*(2). https://doi.org/10.7202/029355ar

Bakker, A. B. (2005). Flow Among Music Teachers and Their Students: The Crossover of Peak Experience. *Journal of Vocational Behavior*, *66*(1), 26–44. https://doi.org/10.1016/j.jvb.2003.11.001

Bakker, A. B. (2008). The Work-related Flow Inventory: Construction and Initial Validation of the WOLF. *Journal of Vocational Behavior*, *72*, 400–414. https://doi.org/10.1016/j.jvb.2007.11.007

Bakker, A. B., & van-Woerkom, M. (2017). Flow at Work: A Self-determination Perspective. *Journal of Occupational Health Science*, *1*, 47–65. https://doi.org/10.1007/s41542-017-0003-3

Bonaiuto, M., Mao, Y., Roberts, S., Psalti, A., Ariccio, S., Cancellieri, G. U., & Csikszentmihalyi, M. (2016). Optimal Experience and Personal Growth: Flow and the Consolidation of Place Identity. *Frontiers in Psychology*, *7*, 1–12. https://doi.org/10.3389/fpsyg.2016.01654

Chu, L. C., & Lee, C. L. (2012). Exploring the Impact of Flow Experience on job Performance. *The Journal of Global Business Management*, *8*(2), 150–158. https://doi.org/10.3389/fpsyg.2016.01654

Csikszentmihalyi, M. (1990). *Flow: The Psychology of Optimal Experience*. Harper & Row.

Csikszentmihalyi, M., & LeFevre, J. (1989). Optimal Experience in Work and Leisure. *Journal of Personality and Social Psychology*, *56*(5), 815–822. https://doi.org/10.1037//0022-3514.56.5.815

Daniel, E. (2016). The Usefulness of Qualitative and Quantitative Approaches and Methods in Researching Problem-solving Ability in Science Education Curriculum. *Journal of Education and Practice*, *7*(15), 91–100.

Etikan, I., Musa, S. A., & Alkassim, R. S. (2016). Comparison of Convinience Sampling and Purposive Sampling. *American Journal of Theoretical and Applied Statistics*, *5*(1), 1–5. https://doi.org/10.11648/j.ajtas.20160501.11

Farina, L. S. A., Rodrigues, G. R., & Hutz, C. S. (2018). Flow and Engagement at Work: A Literature Review. *Psico-USF*, *23*(4), 633–642. https://doi.org/10.1590/1413-82712018230404

Ghosh, S., & Swamy, D. R. (2014). A Literature Review on Organizational Commitment – A Comprehensive Summary. *Int. Journal of Engineering Research and Applications*, *4*(12), 4–14.

Goleman, D. (2013). *Focus: The Hidden Driver of Excellence*. NY: A&C Black.

Gota, G. (2017). *The Relationship Between Psychological Capital, Work Engagement and Organizational Commitment Amongst Employees at a Selected Food Processing Plant in The Western Cape* [University of Western Cape Master Thesis]. http://hdl.handle.net/11394/5991

Hancock, P. A., Kaplan, A. D., Cruit, J. K., Hancock, G. M., Macarthur, K. R., & Szalma, L. J. (2019). A Meta-analysis of Flow Effects and the Perception of Time. *Acta Psychologica*, *98*, 1–18. https://doi.org/10.1016/j.actpsy.2019.04.007

Hasty, R. (2019). *Flow Theory, Creative Flow States & How to Get in the Zone*. https://hastyreader.com/flow-psychology-mihaly-csikszentmihalyi/

Jackson, C. (2018). *Relationships Between Job Satisfaction, Organizational Commitment, and Teacher Engagement*. Walden University.

Kim, M. J., Bonn, M., Lee, C. K., & Kim, J. S. (2019). Effects of Employees' Personality and Attachment on Job Flow Experience Relevent to Organizational Commitment and Consumer-oriented Behaviour. *Journal of Hospitality and Tourism Management*, 41, 156–170. https://doi.org/10.1016@j.jhtm.2019.09.010

Larson, R., & Csikszentmihalyi, M. (2014). *The experience sampling method. Flow and the foundations of positive psychology*. Jossey-Bass.

Macey, W. H., & Schneider, B. (2008). The Meaning of Employee Engagement. *Society for Industrial and Organizational Psychology*, *1*(1), 3–30. https://doi.org/10.1111/j.1754-9434.2007.0002.x

Maeran, F., & Cangiano, F. (2013). Flow Experience and Job Characteristics: Analyzing the Role of Flow in Job Satisfaction. *TPM*, 20(1), 13–26. https://doi.org/10.4473/TPM20.1.2

McShane, S. L., & Glinow, M. A. V. (2008). *Organizational Behavior: Emerging Realities for the Workplace Revolution* (4th ed.). The McGraw Hill Companies.

Mesa, A. F., Llopos, O., Granero, A. G., & Peñuela, J. O. (2020). Enhancing Organisational Commitment Through Task Significance: The Moderating Role of Openness to Experience. *European Management Journal*, *38*(4), 602–612. https://doi.org/10.1016/j.emj.2019/12/010

Meyer, J. P., & Allen, N. J. (1991). A Three-component Conceptualization of Organizational Commitment. *Human Resource Management Review*, *1*(1), 61–89. https://doi.org/10.1016/1053-4822(91)90011-Z

Simon, M. K., & Goes, J. (2011). *Dissertation & Scholarly Research: Recipes for Success*. Createspace Independent Pub. https://doi.org/10.13140/RG.2.1.5089.0960

Tsai, Y. H. (2018). Modeling Creativity Performance and the Psychological Contract in Technology Industry. *Corporate Management Review*, *38*(2), 95–122. https://doi.org/10.3966/102873102018123802004

Wolfigiel, B., & Czerw, A. (2017). A New Method to Measure Flow in Professional Tasks – A FLOW-W Questionnaire (FLOW at Work). *Polish Psychological Bulletin*, *48*(2), 220–228. https://doi.org/10.1515/ppb-2017-0025

Wuryani, E. (2013). Peranan Komitmen Organisasi Dalam Meningkatkan Kinerja Perusahaan Berdasarkan the Indonesian Institute for Corporate Governance. *Jurnal Ilmu Ekonomi Dan Sosial*, *1*(3), 287–298.

Families Mental Health and Challenges in the 21st Century – Arifin et al. (eds)
© 2024 The Author(s), ISBN 978-1-032-51442-0

Marriage satisfaction in couples with different ethnics

Marchelina Febe*
Psychology, Bunda Mulia University, DKI Jakarta, Indonesia

ABSTRACT: One of the changes in the social structures which is commonly happening nowadays is amalgamation. This process means a combination of two ethnicities in a marriage. It can happen when there's a culture included in a social structure among a group of people. It's the same as this observation which discusses marriage amalgamation between Ambonese/South Moluccans and Javanese, in which there are various myths, prejudice, and polemic. The goal of the observation is to see how the adaptation process of the two ethnics tied in marriage amalgamation. The methods used are phenomenology with a sample of 2 families with children. The main data collection methods are observation, interview, and analysis from the interviews and direct observation found in real life. Based on this result, it's concluded that amalgamation process is not easy; however, with trust, communication, and understanding, there will be a huge satisfaction amalgamation marriage.

Keywords: Amalgamation, adaptation, prejudice

1 INTRODUCTION

Recently, we probably often hear the word 'amalgamation' in some subtopics of culture observation. This topic is undoubtedly interesting, considering Indonesia is the largest archipelagic state in the world which has a strategic location geographically and we must be proud to be a part of Indonesia. With such quality, there are certainly lots of ethnicities with various backgrounds that enable the amalgamation process. Amalgamation is a unity, a combination between two races or ethnicities to be one knot of marriage (Wikipedia, n.d.). In the amalgamation process, there are often some communication problems because there are different cultural traditions or perceptions. These will 'restrict' the marriage into a 'we' group and a 'you' group that can possibly lead to some misunderstanding because of stereotypes. When they remain unresolved, they will transform into a prejudice toward an individual, that will be challenging to change (Pamungkas 2014). These cases can be diverse, depending on our point of view toward what we experienced about certain individuals or ethnicities.

We can see it when there are 'foreigners', such as Chinese crossbreed, they will face some problems in amalgamating, socializing, and adapting, which will affect their behavior in communicating (Abdi *et al.* 2015). In the history, Chinese was considered to have a higher position than native Indonesians, according to Dutch colonial law. Based on this historical fact, there will be a formation of prejudiced stereotypes when these societies interact with one another (Suwarno *et al.* 2019). In the past, Chinese was commonly more advanced and dominant in the economy, persistent and conscientious in working, although their social interaction was not that much. However, the indigenous Indonesians were quite left behind as they were hypocritical and less earnest, though they loved lots of activities with high social interactions. This kind of condition has made a huge social discrepancy. Nowadays, though,

*Corresponding Author: marchelinafebe@gmail.com

DOI: 10.1201/9781003402381-5

these two ethnicities already unite a lot in marriage (marriage amalgamation), so there are lots of self-adaptation from each individual needed to maintain harmony because of the personalities' differences (Juditha 2005). All the prejudice and causes are better to be noticed since the beginning of the acquaintanceship, so hopefully, there won't be long-last conflicts along the marriage life. Concerning that, this observation was written to be a source of additional knowledge for each individual who is going to or has been in an amalgamation marriage.

Amalgamation itself originated from an old American term in 1967 that referred to a combination of two different races (American race—white complexion, African race—dark complexion). However, nowadays, the amalgamation term has more comprehensive meanings. From an industrial point of view, for example, an amalgamation consists of two or more companies (Mulianingsih 2020; Ronit Hacohen Lis 2008) different from a merger, as a merger refers to the entity (Sargeant & Lewis 2018). Another example is from the cultural point of view, where amalgamation is a mix between two people from two different ethnicities. This mix can happen in a marriage, social interaction, assimilation, and immigration (Fauzi & Basid 2019). Marriage amalgamation itself has become a normal thing since some ethnic groups have done it. What to concern more is every person who goes through it. Why? It's because it will be different from the other individual, either in personality, background acts, or ethics history background. The mix of two individuals in a marriage with different perspectives, genders, and different cultural backgrounds and/or ethnicities will surely have a long adjustment process of self-adaptation (Arif 2016).

There are four types of adaptation that exist in an amalgamation marriage which couples can go through (Romano 2008). The first one is the obedience type in which one person is obedient to a significant other who has different ethnicity, accepts him or her without any demand for change. The second type is the compromising type where there is an agreement between the couple by understanding through the compromise process and discussing to find a mutual deal together. Not rarely happen in this type, the couple sacrifices each other's principles which are against their cultural backgrounds, in order to create harmony and maintain the commitment they have made. For this type, there is a possibility of a new culture because of the ethnic blending process. The third type is the elimination type where there is an agreement on decisions or solutions for the problems which does not come from compromise or discussion but from one of the couple yields to follow the culture/tradition of the significant other. When there is no compromise or discussion, then there will be no new culture formed, which results in lacking of culture. Finally, the last type is a consensus type where the couple agrees with each other in applying the culture/tradition of both of them, appreciating each other, and doing the tradition or culture by adjusting it with the problems they might face. Couples with the second and fourth types are suggested to adapt more to minimize future conflicts. Why is that so? When individuals do the amalgamation, they include their own families which have different ethnicities/cultures, and conflicts can appear not only from the marriage internally but also from the family. In addition to the differences in ethnic/cultural backgrounds, small conflicts can grow bigger when we are not able to solve them (Juditha 2005). By doing this observation, the causes of future conflicts can be considered further, as well as the self-adapt solutions needed.

2 METHODS

2.1 *Research design*

The observation topic here is satisfaction in marriage of couples of different ethnicities. This observation uses the phenomenology approach which is a part of qualitative observation. The phenomenology approach focuses on or understands the meaning behind the phenomena which is explained in detail (Fatchan 2009). In phenomenology observation, the observer is not allowed to give meanings or prejudices on their own based on personal perception of phenomenon. The observer also does not use any theory to focus on the phenomena as it is, with full concern (Tyagi *et al.* 2021). The goal of the phenomenology observation is for understanding the responses toward the existing individuals/humans/groups of people/citizens as understandable experiences in interacting.

2.2 *Research subject*

Subjects of this observation are couples of different ethnicities (focusing on couples of Chinese and native Indonesians) who have been married for at least 5 years. In collecting the subjects, we use a systematic purposive sampling technique where we take 2 subjects of couple with and without children, who do amalgamation or cross-cultural marriage, and come from Chinese–Javanese–Ambonese ethnics. We limit the criteria of how long the marriage has been to see how they do self-adaptation, solve conflicts, understand each other, give concerns, couple adaptation, finance adaptation, family adaptation, responsibility, willing to sacrifice, and also relations/communications through the marriage.

2.3 *Instruments*

Phenomenology observation basically relates to interpretations of reality. Phenomenology looks for the answers to the meanings of the phenomenon. Basically, there are two main focuses of this observation, which are as follows: (a) Textural description: subjects' experiences about a phenomenon, which is an objective aspect, factual data, and things happen empirically. (b) Structural description: how subjects experience and interpret their experiences. This description is a subjective aspect. It is about opinions, valuations, feelings, expectations, and other subjective responses from the observation's subjects related to their experiences (Hamid 2013).

In this observation, the observer uses these two things as a measurement focus in doing interviews.

Besides that, the observer also uses NRich Marital Satisfaction Scale Questionnaire from Fowers and Olson (1993). The questionnaire has 15 question items consisting of two big measurement scales, which are Marital Satisfaction and Idealistic Distortion. These scales are used to count the presentations of the satisfaction of the marriage (Enrich Marital Satisfaction; EMS).

2.4 *Data analysis*

We use SPSS v. 26 to test the validity and reliability of EMS, also Microsoft Excel to see the EMS result.

2.5 *Observation procedure*

In this observation, the data collection method is phenomenology. Here are the steps we take:

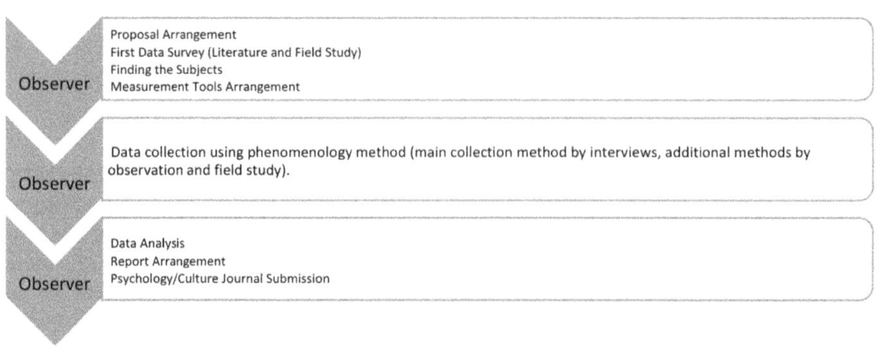

3 RESULT

3.1 *Reliability and validity of EMS*

Based on the test result using SPSS v. 26, it is known that EMS has Cronbach alpha of 0.898 from 15 question items which shows the reliability of this measurement is excellent. Besides that, all 4 subjects (2 couples) who filled out the questionnaires are valid.

In the validity test, the results are from 0,211 to 0,876. Based on this, it is concluded that there are 3 invalid items, which are item no. 2 with a score of 0,88; item no. 3 with a score of 0,124, and item no.6 with a score of 0,124.

Meanwhile, based on the results of the EMS calculation using the following formula: EMS score = PCT – [(.40 × PCT)(ID × .01)] that we calculated by Excel, the data was obtained as follows:

Name	Age	Status	LoM	Ethnic	Total	RS ID	RS MS	P ID	P MS	EMS
S1	35	Husband	6	Javanese	54	18	37	79	75	51.3
S1	34	Wife	6	Chinese	55	19	38	87	80	52.16
S2	29	Wife	6	Chinese	63	19	43	87	95	61.94
S2	33	Husband	5	Ambonese	62	20	43	93	95	59.66

It shows that positive relationship results are significant for the interviews' respondents. It is shown in the table above that the total score obtained is on a high scale of 54–63, whereas the maximum score is 75 (however, there are 6 items of reversed scores.). After counting to find the EMS score, a high score on a scale of 51.3–61.64 is also obtained, which is equivalent to the respondents' interviews results stating that they have been satisfied with the marriage they have been going through.

The calculation is also supported by structured interview results as follows:

On subjects/couple no. 1:
Problems that have occurred in the marriage which is almost 7 years going were mostly based on individuals' backgrounds and their communication methods. Since there is a different culture affects how each couple faces problems daily can result in a dispute. However, it is no longer a problem since each individual understands their significant other well, for example, how they adapt to the voice volume, intonation, and diction when they communicate. It is concluded that the solution/key to the problem is to ameliorate the communication. By improving it during the first years, the respondents have been satisfied with how they have been going through.

On subjects/couple no.2:
This couple has already been married for almost 6 years. During those years, they needed 3 years more or less to do the self-adaptation. Problems they mostly face are related to their self-adaptations toward each other's personality and habits as well as communication to let out their thoughts during this process. In the beginning, it seemed hard and they used to argue with each other, however, after 3 years, this couple is able to understand each other, adapt and understand their significant other's personality and habit, so they have been satisfied with the marriage they have.

4 DISCUSSION, CONCLUSION, AND IMPLICATION

It is relevant to the previous observation that satisfaction in the marriage can be achieved by having excellent communication between each other (Soraiya *et al.* 2016). As what commonly happens when a man and a woman have been married, there will be a new relationship (before having children and building a relationship with them). This husband-wife relationship is the beginning of another new relationship. There have been many broken relationships since the start because of the lack of communication between husbands and wives (Aziz 2017). Communication is an important aspect of a relationship since it will connect many aspects of life, such as finance, religion, parenting and even uttering thoughts and expectations. Miscommunications will certainly create conflicts, just as what the respondents experienced. Due to this, one of the importance of resolving the conflicts is the honesty in communication to discuss the conflicts and look for their way outs (Lestari 2014). Based on the adaptation process of these couples, both of them are included in a consensus where they both agree, appreciate, and accept the tradition/culture of their significant others to reach complete happiness and harmony in their marriages.

REFERENCES

Abdi, F. F. Hafiar, H & Novianti. E. (2015). Perilaku Komunikasi Etnis Tionghoa Peranakan Dalam Bisnis Keluarga (Studi Fenomenologi Mengenai Perilaku Komunikasi Etnis Tionghoa Peranakan Dalam Bisnis Keluarga Di Jakarta). *Jurnal Komunikasi*, *9*(2).

Arif, M. (2016). Analisis Faktor Integrasi Sosio-Kultural-Historis pada Masyarakat Multikultural. *SOSIO DIDAKTIKA: Social Science Education Journal*, *3*(2), 126–134. https://doi.org/10.15408/sd.v3i2.4374. Permalink/DOI

Aziz, A. (2017). Relasi Gender dalam Membentuk Keluarga Harmoni (Upaya Membentuk keluarga Bahagia). *Jurnal harkat: Media Komunikasi Gender*, *13*(1), 27–37.

Fatchan, A. (2009). *Metode Penelitian Kualitatif. Beserta Contoh Proposal Skripsi, Tesis dan Disertasi.* Malang: Jenggala Pustaka.

Fauzi, A., & Basid, A. (2019). *The Potrait of Race Amalgamation in the Movie of Guess Who Directed by Kevin Rodney Sullivan. 50*. https://doi.org/10.4108/eai.27-4-2019.2285313w

Fowers. B.J. & Olson. D.H. (1993). ENRICH Marital Satisfaction Scale: A Brief Research and Clinical Tool. *Journal of Family Psychology*, *7*(2), 176–185.

Hamid, F. (2013). Pendekatan Fenomenalogi (Suatu Ranah Penelitian Kualitatif). *Penelitian Fakultas Saintek UIN Sunan Kalijogo Yogjakarta*, *1*(1), 1–15.

Hazzan, O., & Lis-Hacohen, R. (2016). *The MERge Model for Business Development: The Amalgamation of Management, Education and Research*. Springer.

Juditha, C. (2005). *Stereotip dan Prasangka dalam Konfl ik Etnis Tionghoa dan Bugis Makassar*, *25*, 87–104.

Lestari, Sri. (2014). *Psikologi keluarga: Penanaman Nilai dan Penanganan Konflik dalam Keluarga* (edisi 1). Jakarta: Prenadamedia Group.

Mulianingsih, S. (2020). Pengaruh Strategi Manajemen Perubahan dan Kepemimpinan Transformasional dalam Merubah Budaya Organisasi. *Papatung: Jurnal Ilmu Administrasi Publik, Pemerintahan Dan Politik*, *3*(3), 42–52.

Pamungkas, S. A, Surwandi, S. & Rohmadi. M. (2019). Subordination of Women in Short Story of Kompas Newspaper. *Journal of Budapest International Research and Critics Institute-Journal (BIRCI-Journal)*, *2*(2).

Romano, D. (2008). *Intercultural Marriage: Promises and Pitfalls* (Intercultural Press (ed.); 3rd ed.). Nicholas Brealey.

Sargeant, M., & Lewis, D. (2018). *Employment Law: Business Restructuring*. 280–310. https://doi.org/10.4324/9781315203263-10

Soraiya, P., Khairani, M., Rachmatan, R., Sari, K., & Sulistyani, A. (2016). Kelekatan dan Kepuasan Pernikahan Pada Dewasa Awal di Kota Banda Aceh. *Jurnal Psikologi Undip*, *15*(1), 36–42.

Suwarno, S., Abdulsyani, A., & Pairulsyah, P. (2019). Interaksi dan Asimilasi Masyarakat Keturunan Tionghoa dengan Masyarakat Pribumi (Studi pada Masyarakat Keturunan Tionghoa di Kelurahan Tanjungkarang Pusat Bandar Lampung). *Jurnal Sosiologi FISIP UNILA*, *21*(2), 117–200. http://jurnalsosiologi.fisip.unila.ac.id/index.php/jurnal/article/view/44.

Tyagi, N., Jha, R. S., Chaudhary, A., & Batar, S. (2021). Women in Dual Role: A Sociological Perspective. *Ilkogretim Online*, *20*(1). doi: 10.17051/ilkonline.2021.01.187

Wikipedia. (n.d.). *Amalgamsi.* Retrieved from: https://id.wikipedia.org/wiki/Amalgamasi

Families Mental Health and Challenges in the 21st Century – Arifin et al. (eds)
© 2024 The Author(s), ISBN 978-1-032-51442-0

The validation process of questionnaire emotion regulation on Indonesian adolescents

Rika Hardani*, Diana Setiyawati* & Yuli Fajar Susetyo*
Department Psychology, Gadjah Mada University, Yogyakarta, Indonesia

ABSTRACT: This study intends to validate the Indonesian translation of the original Emotion Regulation Questionnaire (ERQ) and adapt it for use with adolescents. The instrument adaptation procedure will follow the WHO's instrument adaptation recommendations. An investigation using Python Semopy programming to execute reliability and validation tests is then performed. The participants in this study were 582 high school students from Banten, Indonesia. The results demonstrate that the reliability value of the Indonesian ERQ is higher than the original ERQ. Cronbach's alpha was calculated as 0.85 for the reliability test findings, with a 95% confidence range of [CI = 0.832, 0.868]. In conclusion, the Indonesian version of the ERQ instrument demonstrates a fit model suitable for use in adolescents in Eastern cultures, including Indonesia. It is believed that this study will improve the quality of research on adolescent emotion regulation.

Keywords: Emotion, Validation, Phyton, Indonesia, Adolescent

1 INTRODUCTION

The role of the family in the development of children is crucial. According to Defrain and Assay (2007), based on research in the world, it is found that there are similarities in the characteristics of a strong family. These parallels include the existence of appreciation and compassion, dedication, goodwill, fun times spent with one another, spiritual well-being, and the capacity to handle stress and disasters appropriately. Families can influence children in both positive and negative ways.

Families who successfully carry out their functions efficiently will provide good results on child development in many ways, for example, more frequent physical activity in children so that they are healthier and watch TV less often, have better self-emotional regulation (McManus *et al.* 2012; Meyer *et al.* 2014; Morris *et al.* 2007, 2013, 2017; Rutherford *et al.* 2015), have better mental health conditions (Briggs *et al.* 2019), and promote good quality sleep in children (El & Kelly 2017).

Meanwhile, families who fail to carry out their roles efficiently will also affect the child's development in negative things such as nonsuicidal self-injury (NSSI), consuming alcoholic beverages (Hasking *et al.* 2020), bullying tendencies, difficulty in controlling anger and sharing problem-solving skills (Hisar *et al.* 2021), high rates of depression in adolescents (Lin *et al.* 2008), suicide attempts in adolescents (Ram *et al.* 2020), and children's low self-regulation skills and aggressiveness (Ramsden & Hubbard 2002).

*Corresponding Authors: rikaheru97@mail.ugm.ac.id, Diana@ugm.ac.id and fajar_psy@ugm.ac.i

DOI: 10.1201/9781003402381-6

Based on experimental research conducted by Gross and John (2003), the first strategy (Reappraisal) was found to be more effective than the second strategy (Suppression) in terms of impact on memory. On the other hand, emotion suppression strategies will reduce behavioral expression but will not mitigate emotional experience and impair memory. This second strategy will also increase the physiological response.

Previous studies on emotion regulation have been carried out in many countries. However, until now, many problems still arise due to the low ability to manage emotions, especially in adolescents. Thus, research on emotional management in adolescents is still very much needed and continues to be developed. Furthermore, to be able to analyze good quality, an instrument suitable for Eastern cultures, including Indonesia, is required.

Well-known instruments are used for research on emotion regulation, for example, the Emotion Regulation Questionnaire (ERQ), which was first developed by Gross & John (2003). This study aims to adapt the ERQ instrument in Indonesian and validate it in adolescents. In addition, this research will provide an ERQ instrument that will be adjusted using WHO guidelines (2022) and considering cultural sensitivity (Borsa *et al.* 2012). The novelty of the research is that the data analysis process will use the Python Semopy package programming language, which is still relatively new (Igolkina & Meshcheryakov 2020) and is open source (no need to pay).

2 METHODS

Before the research began, the researcher contacted James Gross to ask permission to adapt the ERQ into Indonesian. James Gross gave his permission via email on November 23, 2021. The researcher has also obtained permission from the research ethics committee of the psychology faculty at Gadjah Mada University. The ethics committee approval letter has been issued with number 2676/ UN1/ FPSi.1.3/ SD/ PT. 01.04/2022.

The ERQ adaptation process uses the guidelines issued by WHO (2022) and is equipped with a measuring instrument adaptation protocol carried out by Borsa *et al.* (2012). In general, six steps are carried out. These steps include translating the original ERQ into Indonesian, synthesizing the translation results, evaluating the synthesis results, readability tests, back-translation, and pilot studies. After the completion of ERQ adaptation process, it is continued by trying the Indonesian version of ERQ in the adolescent age group. This stage is done online using a Google Form.

The data was then analyzed using the Python Semopy package (Igolkina & Meshcheryakov 2020). The analysis carried out was the reliability test and the Indonesian version of the ERQ validity test.

3 RESULTS

Adapting the Indonesian version of ERQ involves experts in the field of psychology who have good Indonesian and English skills. Therefore, at each stage, it is carried out carefully and pays attention to all inputs given by these experts.

The participants involved in this study were 582 high school students from schools in the Banten region, Indonesia. The age of the participants ranged from 14 years to 17 years. By filling out the ERQ questionnaire, the researcher obtained informed consent from parents and approval from students to participate in the study. The time for filling out the questionnaire is about 5 minutes.

Statistical analysis using Python Semopy package. The coding script used is as in the following figure.

```
!pip install pingouin

import pandas as pd
!pip install semopy

import semopy
from  semopy import Model

!pip install factor_analyzer

import pingouin as pg

from factor_analyzer import FactorAnalyzer

from factor_analyzer import (ConfirmatoryFactorAnalyzer, ModelSpecificationParser)

Adata = pd.read_excel('/content/drive/MyDrive/Colab Notebooks/Dataset Colab/SectionC.xlsx')
Adata.head()     #Pandas's command that show the data table to confirm the data

ca= pg.cronbach_alpha(data=Adata)
alpha = pd.DataFrame(ca, columns=['Value'])
items = ['Chronbach Alpha', 'CI(0.95)']
alpha['Items'] = items
alpha.head()

fa = FactorAnalyzer(n_factors=2,rotation= "promax",method='ml')

fa.fit(Adata)

ev= fa.get_eigenvalues()
pdev=pd.DataFrame(ev)
pdt= pdev.T
dfT = pdt.set_axis(["Original EV", "Common Factor EV"], axis=1)
dfT
```

Figure 1. Script coding python semopy.

```
efa= fa.loadings_
df = pd.DataFrame(efa, columns = ['Reappraisal','Suppression'])
df


model_dict = {"Reappraisal": [ "P1","P3","P5","P7","P8","P10"],
              "Suppression": ["P2","P4","P6","P9"]}
model_spec = ModelSpecificationParser.parse_model_specification_from_dict(Adata, model_dict)
cfa = ConfirmatoryFactorAnalyzer(model_spec,max_iter = 500,disp=False)
cfa.fit(Adata.values)
cfalo = cfa.loadings_
dcfa = pd.DataFrame(cfalo)
dcfa


Reapp_data=Adata[["P1","P3","P5","P7","P8","P10"]]
ca_reapp= pg.cronbach_alpha(data=Reapp_data)
alpha = pd.DataFrame(ca_reapp, columns=['Value'])
items = ['Chronbach Alpha of Reappraisal', 'CI(0.95)']
alpha['Items'] = items
alpha.head()


Suppres_data=Adata[["P2","P4","P6","P9"]]
ca_suppres= pg.cronbach_alpha(data=Suppres_data)
alpha = pd.DataFrame(ca_suppres, columns=['Value'])
items = ['Chronbach Alpha of Suppression', 'CI(0.95)']
alpha['Items'] = items
alpha.head()


desc = """
        Emotion_Regulation =~ Suppression + Reappraisal
        Suppression =~ P2+P4+P6+P9
        Reappraisal =~ P1+P3+P5+P7+P8+P10
        """
model = Model(desc)
print(desc)


res = model.fit(Adata)
stats = semopy.calc_stats(model)
print(stats.T)


from semopy import Optimizer
opt = Optimizer(model)
obj_function_value = opt.optimize()


from semopy.inspector import inspect
inspect(opt,std_est= True)


g = semopy.semplot(model, 'Emotion-26073.pdf',std_ests=True)
```

Figure 1. (Continued)

The reliability and validity test used Exploratory Factor Analysis (EFA) and Confirmatory Factor Analysis (CFA) methods (Yong & Pearce 2013). The reliability test was carried out with the Cronbach's alpha coefficient of the ERQ that had been adapted to get the ERQ Cronbach's alpha value of 0.85. For the Reappraisal aspect, the Cronbach alpha value = 0.87, and the Cronbach alpha value for the Suppression aspect = 0.76. This value is higher than the original Cronbach alpha ERQ value (Reappraisal = 0.79 and Suppression = 0.73). The Pearson correlation coefficient was used for calculations. This analysis indicates that if Cronbach's Alpha value > 0.70, the questionnaire is declared reliable or consistent (Bland & Altman 1997). The Cronbach alpha value of the adaptation ERQ shows a value above 0.7. According to Bland & Altman (1997), the questionnaire is said to be consistent if the Cronbach alpha value is more significant than 0.7.

Adapted ERQ confidence level, 95% [CI = 0.832, 0.868]. The Reappraisal aspect has a 95% confidence level [CI = 0.855, 0.887], whereas hile the Suppression aspect has a 95% confidence level [CI = 0.727, 0.791]. The reliability test was also carried out using the EFA method. This test was conducted to determine how many factors make up the ERQ of this adaptation. Kline (2015) stated that when conducting the EFA test, many of the factors present in the measuring instrument were assumed to be unknown to the researcher. The results of the EFA test found two factors or constituent aspects. This outcome is identical to the initial ERQ, which has two aspects.

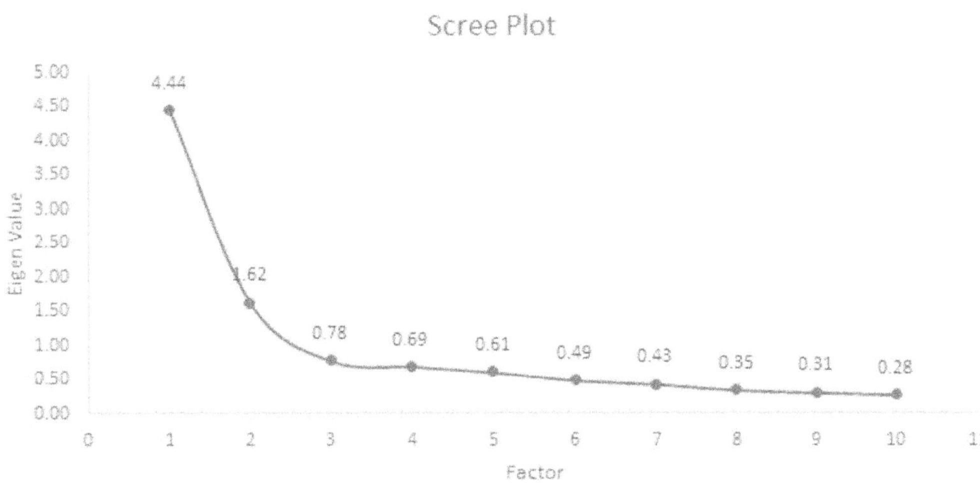

Figure 2. The scree plot.

Furthermore, the adapted ERQ validity test was carried out. To determine the accuracy of an instrument in measurement, it is necessary to perform a validity test. CFA is one technique that may be utilized to evaluate the validity (CFA). According to Marsh *et al.* (2014), the CFA test can be used to find out whether a particular construct has a consistent value or not.

For the Reappraisal aspect, the CFA value is 0.51–0.83, and the Suppression aspect is 0.50–0.81, while the CFA value of the ERQ as a whole is 0.50–0.81. According to Kline (2015), the higher the loading factor value of the CFA, the better.

In addition to using the CFA method, the method of calculating the Root Mean Square Error of Approximation (RMSEA) and Comparative Fit Index (CFI) can also be used to assess the validity of an instrument. The RMSEA value of the adapted ERQ is 0.07. The Comparative Fit Index (CFI) is 0.95.

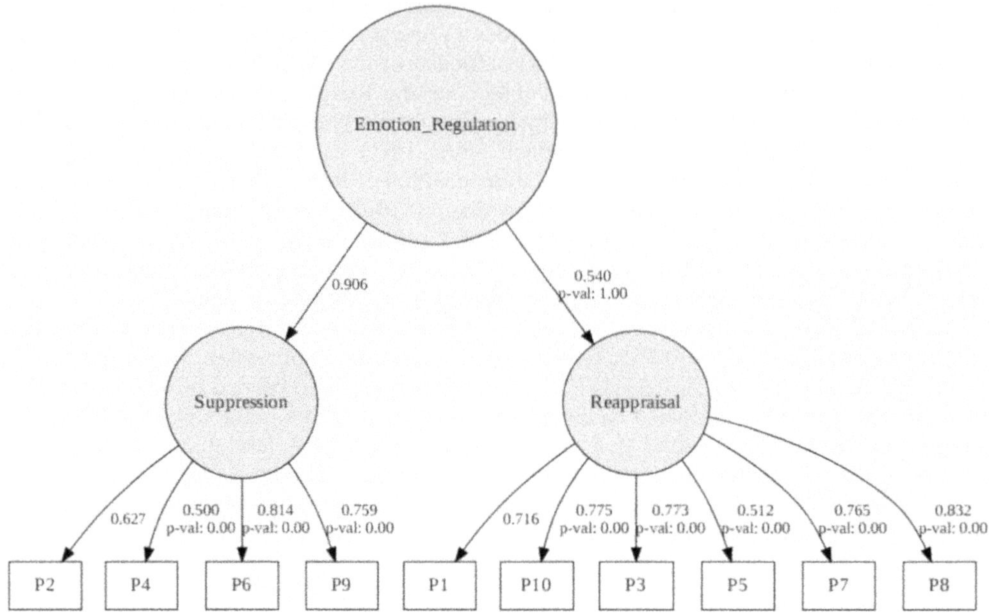

Figure 3. The ERQ Indonesia version correlated factors.

4 DISCUSSION

Thompson (1994) defines emotion regulation as a modulating process occurrence, duration, and intensity of internal states of feeling (both positive and negative) and physiological processes associated with emotions. Emotion regulation refers to the ability to observe, assess, understand, and change emotional reactions to achieve goals and achieve adaptive functioning (Gratz & Roemer 2004). Good emotional self-regulation skills will have an essential role in academic success (Ivcevic & Brackett 2014; Villavicencio & Bernardo 2013), and will improve the social and emotional health of children and adolescents (Morris *et al.* 2017), academic achievement, family and peer interactions, and mental health (Dahl 2004).

On the other hand, the lack of ability to regulate emotions indicates difficulties in responding to emotions, including challenges in being aware of explaining and accepting conditions that require emotional responses (Gratz & Roemer 2004). Several studies have shown that difficulty in managing emotions is correlated with a variety of disorders, including eating disorders (Harrison *et al.* 2010), substance abuse (Tull *et al.* 2012), and alcoholism (Fox *et al.* 2008), problematic behaviors such as adolescent aggressiveness (DeWall *et al.* 2007; Sani *et al.* 2017), tend to self-harm (Hasking *et al.* 2020), maladaptive behavior eating disorders, self-injury, and substance abuse (Buckholdt *et al.* 2015).

It is challenging to manage emotions well to get the most out of them. Individuals may use various strategies to deal with emotional changes. Gross and John (2003) presented two plans for emotion regulation. The first strategy is to reassess the conditions at hand. This strategy tries to manage emotions by changing the perspective on the situation that occurs to reduce its emotional impact. The second strategy is to suppress the feelings that arise.

The ERQ adaptation process is carried out carefully at each stage. First, the translation stage was carried out by three experts. The second stage is the synthesis carried out by two experts. Then the expert judgment stage is carried out by three experts. Finally, seven high school students carried out the readability test, and three people carried out the back translation. Experts at each stage are different people and performed at other times and places. The detailed process carried out at each location can further improve the accuracy of the measuring instrument (Borsa *et al.* 2012).

The use of the Python Semopy package to analyze data is carried out because of advantages such as having a complete package, shorter script code, and convenient for joint work or collaboration (Igolkina & Meshcheryakov 2020). Moreover, it is more interesting because it is open source that means it does not cost money to use this programming language. This is probably because the items asked in the ERQ questionnaire are easy for participants to understand.

The results of the analysis using the Python Semopy package also showed accurate results. Working on the analysis process also does not take too long. Python Semopy package can also use Google Colab to make it more practical and easier for researchers. The many advantages of this programming language show that Python Semopy package has the potential to be used in various research fields, including much social research. Overall, the study results indicate that the ERQ adapted to Indonesian is suitable and can be used for research on emotion regulation in Indonesian adolescents.

5 CONCLUSION AND IMPLICATION

The ERQ instrument is relatively easy to understand by adolescent participants, so the value of the statistical test results shows relevant results. The ERQ instrument, adapted into Indonesian, shows its suitability for use in research on emotions in Indonesian adolescent participants.

The limitation of the research is that it only includes participants from a specific region. Suggestions for further study can take participants from a wider area. It can also be done by expanding the age range of the participants, for example, the age of children or early adulthood.

REFERENCES

Bland, J.M., & Altman, D.G. (1997). Statistics notes: Cronbach's alpha. *BMJ*, 314(7080), 572.

Borsa, J. C., Damásio, B. F., & Bandeira, D. R. (2012). *Cross-cultural Adaptation and Validation of Psychological Instruments: Some Considerations. Paidéia (Ribeirão Preto)*, 22, 423–432.

Briggs, E. C., Fairbank, J. A., Tunno, A. M., Lee, R. C., Corry, N. H., Pflieger, J. C., Stander, V. A., & Murphy, R. A. (2020). Military Life Stressors, Family Communication, and Satisfaction: Associations with Children's Psychosocial Outcomes. *Journal of Child & Adolescent Trauma*, 13(1), 75–87. https://doi-org.ezproxy.ugm.ac.id/10.1007/s40653-019-00259-z

Buckholdt, K. E., Parra, G. R., Anestis, M. D., Lavender, J. M., Jobe-Shields, L. E., Tull, M. T., & Gratz, K. L. (2015). Emotion Regulation Difficulties and Maladaptive Behaviors: Examination of Deliberate Self-harm, Disordered Eating, and Substance Misuse in Two Samples. *Cognitive Therapy and Research*, 39(2), 140–152.

Dahl, R. E. (2004). Adolescent Brain Development: A Period of Vulnerabilities and Opportunities. Keynote address. *Annals of the New York Academy of Sciences*, 1021(1), 1–22.

DeFrain, J., & Asay, S. M. (2007). Strong Families Around the World: An Introduction to the Family Strengths Perspective. *Marriage & Family Review*, 41(1–2), 1–10.

DeWall, C. N., Baumeister, R. F., Stillman, T. F., & Gailliot, M. T. (2007). Violence Restrained: Effects of self-regulation and its Depletion on Aggression. *Journal of Experimental Social Psychology*, 43(1), 62–76.

El, S. M., & Kelly, R. J. (2017). Family Functioning and Children's Sleep. *Child Development Perspectives*, 11 (4), 264–269. https://doi-org.ezproxy.ugm.ac.id/10.1111/cdep.12243

Fox, H. C., Hong, K. A., & Sinha, R. (2008). Difficulties in Emotion Regulation and Impulse Control in Recently Abstinent Alcoholics Compared with Social Drinkers. *Addictive Behaviors*, 33(2), 388–394.

Gratz, K. L., & Roemer, L. (2004). Multidimensional Assessment of Emotion Regulation and Dysregulation: Development, Factor Structure, and Initial Validation of the Difficulties in Emotion Regulation Scale. *Journal of Psychopathology and Behavioral Assessment*, 26(1), 41–54.

Gross, J. J., & John, O. P. (2003). Individual Differences in Two Emotion Regulation Processes: Implications for Affect, Relationships, and Well-being. *Journal of Personality and Social Psychology*, 85(2), 348.

Harrison, A., Sullivan, S., Tchanturia, K., & Treasure, J. (2010). Emotional Functioning in Eating Disorders: Attentional Bias, Emotion Recognition, and Emotion Regulation. *Psychological Medicine*, 40(11), 1887–1897.

Hasking P, Dawkins J, Gray N, Wijeratne P, Boyes M. (2020). Indirect Effects of Family Functioning on Nonsuicidal Self-injury and Risky Drinking: The roles of Emotion Reactivity and Emotion Regulation. *Journal of Child & Family Studies*, 29(7), 2070–2079. doi:10.1007/s10826-020-01722-4

Hisar, T., Kızılkurt, Ö. K., & Dilbaz, N. (2021). The Effect of Anger and Family Functions on Bullying Behavior in Individuals with Substance Use Disorder Between the Ages of 15 and 25. *Alpha Psychiatry*, 22 (1), 61–66. https://doi-org.ezproxy.ugm.ac.id/10.5455/apd.96949

Igolkina, A. A., & Meshcheryakov, G. (2020). Semopy: a Python Package for Structural Equation Modeling. *Structural Equation Modeling: A Multidisciplinary Journal*, 27(6), 952–963.

Ivcevic, Z., & Brackett, M. (2014). Predicting School Success: Comparing Conscientiousness, Grit, and Emotion Regulation Ability. *Journal of Research in Personality*, 52, 29–36.

Kline, R. B. (2015). *Principles and practice of structural equation modeling*. Guilford publications.

Lin, H., Tang, T., Yen, J., Ko, C., Huang, C., Liu, S., & Yen, C. (2008). Depression and its Association with Self-esteem, Family, Peer, and School Factors in a Population of 9586 Adolescents in Southern Taiwan. *Psychiatry & Clinical Neurosciences*, 62(4), 412–420. https://doi-org.ezproxy.ugm.ac.id/10.1111 /j.1440-1819.2008.01820.x

Marsh, H. W., Morin, A. J., Parker, P. D., & Kaur, G. (2014). Exploratory Structural Equation Modeling: An Integration of the Best Features of Exploratory and Confirmatory Factor Analysis. *Annual Review of Clinical Psychology*, 10(1), 85–110.

McManus, B., Mandic, C., Carle, A., & Robert, S. (2012). The Effect of Parent-Child Function on Physical Activity and Television Viewing among Adolescents with and without Special Healthcare Needs. *International Journal of Disability, Development & Education*, 59(3), 305–319. https://doi-org.ezproxy.ugm.ac.id/10.1080/1034912X.2012.697749

Meyer, S., Raikes, H. A., Virmani, E. A., Waters, S., & Thompson, R. A. (2014). Parent Emotion Representations and the Socialization of Emotion Regulation in the Family. *International Journal of Behavioral Development*, 38(2), 164–173.

Morris, A. S., Silk, J. S., Steinberg, L., Myers, S. S., & Robinson, L. R. (2007). The Role of the Family Context in the Development of Emotion Regulation. *Social Development (Oxford, England)*, 16(2), 361–388. https://doi.org/10.1111/j.1467-9507.2007.00389.x

Morris, A. S., Criss, M. M., Silk, J. S., & Houltberg, B. J. (2017). The Impact of Parenting on Emotion Regulation During Childhood and Adolescence. *Child Development Perspectives*, 11(4), 233–238.

Ram, D., Koneru, A., & Gowdappa, B. (2020). Relationship between Life Skills, Repetitive Negative Thinking, Family Function, and Life Satisfaction in Attempted Suicide. *Indian Journal of Psychiatry*, 62 (3), 283.

Ramsden, S. R., & Hubbard, J. A. (2002). Family Expressiveness and Parental Emotion Coaching: Their role in Children's Emotion Regulation and Aggression. *Journal of Abnormal Child Psychology*, 30(6), 657–667.

Rutherford, H. V., Wallace, N. S., Laurent, H. K., & Mayes, L. C. (2015). Emotion Regulation in Parenthood. *Developmental Review*, 36, 1–14. https://doi.org/10.1016/j.dr.2014.12.008

Sani, S. R. H., Tabibi, Z., Fadardi, J. S., & Stavrinos, D. (2017). Aggression, Emotional Self-Regulation, Attentional Bias, and Cognitive Inhibition Predict Risky Driving Behavior. *Accident Analysis & Prevention*, 109, 78–88.

Thompson, R. A. (1994). Emotion regulation: A theme in search of a definition. In N. A. Fox & J. J. Campos (Eds.), *Monographs of the Society for Research in Child Development: Vol. 59. The development of emotion regulation: Biological and behavioral considerations* (2–3, Serial No. 240, pp. 25–52). Chicago, IL: University of Chicago Press.

Tull, M. T., Weiss, N. H., Adams, C. E., & Gratz, K. L. (2012). The Contribution of Emotion Regulation Difficulties to Risky Sexual Behavior Within a Sample of Patients in Residential Substance Abuse Treatment. *Addictive Behaviors*, 37(10), 1084–1092.

Villavicencio, F. T., & Bernardo, A. B. (2013). Positive Academic Emotions Moderate the Relationship between Self-regulation and Academic Achievement. *British Journal of Educational Psychology*, 83(2), 329–340.

WHO (World Health Organization). Diunduh Maret 2022. Guide to Adaptation and Scale Up – WHO | World Health https://cdn.who.int › docs

Yong, A. G., & Pearce, S. (2013). A Beginner's Guide to Factor Analysis: Focusing on Exploratory Factor Analysis. *Tutorials in Quantitative Methods for Psychology*, 9(2), 79–94.

Families Mental Health and Challenges in the 21st Century – Arifin et al. (eds)
© 2024 The Author(s), ISBN 978-1-032-51442-0

The effect of self-disclosure in social media on psychological well-being in adolescents during the Covid-19 pandemic

Azhar Kania Farahghina Kamila*, Hudaniah, M.Si* & Dian Caesaria Widyasari, M.Sc
Faculty of Psychology, University of Muhammadiyah Malang, Malang, Indonesia

ABSTRACT: During the Covid-19 Pandemic, life aspects changed, including adolescents. Their daily activities were limited, which may disturb their psychological well-being reflected through six aspects: autonomy, environmental mastery, personal growth, positive relation with others, purpose in life, and self-acceptance. However, adolescents could disclose themselves through social media to manage psychological well-being in limited life during the Covid-19 Pandemic. This study aims to determine the influence of self-disclosure on social media on psychological well-being in adolescents during the Covid-19 Pandemic. This study is a quantitative correlational design. A total of 316 adolescents who were active users of Instagram between the age of 13 and 21 years were selected through accidental sampling. Two instruments used were the Revised Self-Disclosure Scale and Ryff's Psychological Well-Being Scale. The results indicated that self-disclosure on social media significantly affects psychological well-being among adolescents during the Covid-19 Pandemic and based on multiple linear regression, indicated that self-disclosure on social media significantly affects all aspects of psychological well-being among adolescents, except for autonomy. This study implies that adolescent social media users need to control self-disclosure activities so they do not develop into oversharing, which can harm psychological well-being or self-security.

Keywords: Psychological well-being, self-disclosure, social media, adolescents, Covid-19

1 INTRODUCTION

Indonesia is one of the countries that has been exposed to Covid-19. Lasted for two years, the Covid-19 pandemic has also limited the movement of teenagers. Social restrictions due to the Covid-19 pandemic have seized many opportunities for teenagers to undergo normal development. In this adolescent phase, they are still looking for identity and want to form a self-image that can be shown to the general public. However, this pandemic reduces their chances of getting it. Restricted interactions during a pandemic can have a detrimental effect on the psychological well-being of adolescents because they experience social isolation, depression, anxiety, and an increase in maladaptive behavior (O'Sullivan *et al.* 2021). During the Covid-19 pandemic, 82.67% of adolescents had moderate levels of psychological well-being, 17.32% of adolescents were in a low category, and no results were found indicating that adolescents had a high level of psychological well-being (Soputan & Mulawarman 2021). Quite different from before the Covid-19 pandemic, it is known that adolescents' psychological well-being level is 16.7% in the high category, 67.6% in the medium category, and 15.7% in the low category (Sugiyo *et al.* 2019). The difference in the results from these data lies in the pre-pandemic period. There are groups of adolescents who have a high level

*Corresponding Authors: azharkania@gmail.com and hudaniah@umm.ac.id

DOI: 10.1201/9781003402381-7

of *psychological well-being*. Meanwhile, during the pandemic, no results indicated that adolescents' psychological well-being was in the high category.

An excellent psychological well-being condition is essential for adolescent growth and development. This is because individuals with good psychological well-being will function fully and tend to be able to respect themselves (Choi & Lim 2016). *Psychological well-being* itself is defined as the realization and full achievement of the potential of individuals who can accept their past along with all the shortcomings and advantages (self-acceptance), demonstrate an independent attitude (autonomy) and can establish positive relationships with others (positive relations with others), able to master the environment (environmental mastery), have a purpose in life (purpose in life), and able to develop themselves (personal growth) (Ryff 1989). Psychological well-being factors include age, social and economic status, culture, gender, social support, religiosity, and personality (Dewi & Mulyo 2017).

The personality factor that influences psychological well-being is self-disclosure. Self-disclosure is a form of communication in which a person discloses information about himself that is usually hidden. Through self-disclosure, individuals will gain an understanding of themselves, have the capability to overcome difficulties, improve communication skills and strengthen relationships with others (DeVito 2012). Through self-disclosure, individuals can more easily communicate, expand relationships, reduce guilt, and increase social acceptance (Gainau 2019). Research by Qoniah (2021) found that self-disclosure significantly positively influences the psychological well-being of orphanage teenagers during the Covid-19 pandemic.

Meanwhile, another study conducted by Sulistyorini (2007) revealed a correlation between self-disclosure and each aspect of psychological well-being. Her research found that the level of self-disclosure has a significant and positive relationship with five of the six aspects of psychological well-being. These aspects are self-acceptance, autonomy, positive relations with others, purpose in life, and personal growth. On the other hand, the study also found that self-disclosure had no significant correlation with aspects of environmental mastery in the psychological well-being variable.

Research by Joinson & Levine (in DeVito 2012) found that reciprocity in self-disclosure activities has a higher speed and rate in online media than in face-to-face. On social media, self-disclosure is implemented by uploading personal information in text, photos, and videos (Ampong *et al.* 2018). There are various social media to facilitate individuals' self-disclosure. Instagram is one of the most popular and frequently visited social media and a means for self-disclosure (Kusyanti & Safitri 2016). When Instagram users upload images, there is a column to write captions. Users can use captions to pour out their hearts according to the moments shared through the image. In addition, an Instagram story feature allows users to be creative in sharing their stories. The purpose of using Instagram stories is to express emotions. Through self-disclosure, the user can release something that is stuck in him so that it brings peace (Johana *et al.* 2020). Thus, the social media referred to in this study refers to Instagram.

The latest research by Luo & Hancock (2020) shows that self-disclosure on social media significantly influences psychological well-being. In line with this research, a study by Dogan (2016) shows that networking on social media is an essential predictor of adolescents' psychological well-being, happiness, and life satisfaction. Supporting this research, Soputan & Mulawarman (2021) found that self-disclosure on social media affects psychological well-being in high school students. However, Matthes *et al.* (2021) found that online self-disclosure could not predict psychological well-being during a pandemic, although it revealed a positive reciprocal relationship between happiness and psychological well-being. This may indicate that online self-disclosure can impact psychological well-being if happiness is a mediating variable.

The existence of a gap in the results of previous research on the effect of self-disclosure in social media on psychological well-being in adolescents during the Covid-19 pandemic became the background for researchers to examine "The Effect of Self-disclosure in Social Media on Psychological well-being in Adolescents during the Covid-19 pandemic." The problem in this study is whether self-disclosure in social media can affect adolescents' psychological well-being during the Covid-19 pandemic. So this study aims to determine whether there is an effect of self-disclosure on social media on psychological well-being in adolescents during the Covid-19 pandemic. This study hypothesizes a significant influence between self-disclosure on social media and psychological well-being in adolescents during the Covid-19 pandemic.

2 METHODS

This study is a quantitative correlational design. A total of 316 adolescents who were active users of Instagram between the age of 13 and 21 years were selected through accidental sampling. Among them, 93% were girls, and 76.6% were in ages 17–21 years old. Self-disclosure on social media was measured using the Revised Self-Disclosure Scale (RSDS) derived from Leung's (2002) research which is composed of 5 aspects of self-disclosure introduced by Wheeless (1986). The aspects of self-disclosure are control of depth, honesty-accuracy, amount of disclosure, valence/positiveness-negativeness, and intent of self-disclosure. This scale consists of 15 valid items with a reliability level of 0.754. Psychological well-being was measured by Ryff's Psychological Well-Being Scale (1989). This scale is based on six aspects; autonomy, environmental mastery, personal growth, positive relations with others, purpose in life, and self-acceptance. There are 38 valid items and have a reliability of 0.845. Data were analyzed using SPSS 25.0, and multiple linear regression was used to predict the relationship between the two variables' dimensions.

3 RESULT

This study involved 316 participants aged 13–21 who are active users of social media, especially Instagram, accessing Instagram at least once a day during the Covid-19 pandemic.

Based on Table 1 in model 1, with variable X being self-disclosure and variable Y being psychological well-being, having a significance of p $0.000 < 0.05$, it means that self-disclosure on social media has a significant effect on psychological well-being. Thus, the hypothesis in this study is accepted. Multiple linear regression testing of self-disclosure on each dimension of psychological well-being found that in model 2, with the variable X being self-disclosure and the variable Y being autonomy, having a significance of p $0.100 > 0.05$, so there was no effect of self-disclosure on autonomy. In model 3, with variable X being self-disclosure and variable Y being environmental mastery, having a significance of p $0.000 < 0.05$, it means that self-disclosure has a significant effect on environmental mastery. In model 4 with variable X being self-disclosure and variable Y being personal growth and having a significance of p $0.000 < 0.05$ it means that self-disclosure has a significant effect on personal growth. In model 5 with variable X being self-disclosure and variable Y being positive

Table 1. Multiple linear regression test on each model.

Model	X	Y	R	Sig.
1	Self-Disclosure	Psychological Well-Being	0.374	0.000
2	Self-Disclosure	Autonomy	0.171	0.100
3	Self-Disclosure	Environmental Mastery	0.346	0.000
4	Self-Disclosure	Personal Growth	0.324	0.000
5	Self-Disclosure	Positive Relations with Others	0.404	0.000
6	Self-Disclosure	Purpose in Life	0.322	0.000
7	Self-Disclosure	Self-Acceptance	0.331	0.000

relation with others has a significance of p 0.000 < 0.05, which means that self-disclosure has a significant effect on positive relations with others. In model 6 with variable X being self-disclosure and variable Y being purpose in life and having a significance of p 0.000 < 0.05, then self-disclosure has a significant effect on purpose in life. Lastly, in model 7 with variable X being self-disclosure and variable Y being self-acceptance, having a significance of p 0.000 < 0.05, which means that self-disclosure has a significant effect on self-acceptance.

4 DISCUSSION

Based on the results of the data analysis, it can be known that self-disclosure on social media has a significant influence on psychological well-being in adolescents during the Covid-19 pandemic. The results of this study follow the latest research conducted by Nguyen & Ho (2022), which showed that there was a relationship between online self-disclosure and well-being in adolescents in Vietnam, moderated by online social support. Supporting these findings, Adzhani, Baihaqi, & Kosasih (2020) found that self-disclosure influences subjective well-being in adolescent Instagram users. The relationship between self-disclosure and psychological well-being is further strengthened by Kelana's research (2017), which shows that the relationship between the two variables is significant. It is known that during the Covid-19 pandemic crisis, people are limited to doing activities outside the home, which causes boredom and even stress and reduced face-to-face interactions with other people. So that in order to maintain psychological well-being as well as relationships with other people, the thing to do is interact through social media. As found in a study in China, online self-disclosure mediates the relationship between interactions on WeChat (a chat platform from China) and the well-being of students during the lockdown due to the Covid-19 pandemic (Amosun *et al.* 2021). It can be explained that when adolescents interact through social media during the lockdown, they want to be open, known about their health conditions to others, and hear or give motivational words to help relieve their worries and depression.

Furthermore, based on testing regarding the effect of self-disclosure on each dimension of psychological well-being, it was found that self-disclosure has a significant effect on all dimensions except autonomy. The dimensions of psychological well-being that are significantly influenced by self-disclosure are environmental mastery, personal growth, positive relations with others, purpose in life, and self-acceptance.

The absence of the influence of self-disclosure on social media on the autonomy dimension, which is described by Ryff (1989) as an independent attitude in regulating behavior and making life choices, being confident in their abilities, and being able to overcome the problems they face is because they are still in the process of becoming independent both emotionally, behaviorally, and value. This is one of the fundamental developmental tasks in adolescence, as described by Steinberg (2014). It is not easy for adolescents to gain independence. This is because of the difficulty of breaking the infantile bonds that have been attached and comfortable during childhood. Therefore, other factors play a role in developing adolescent autonomy, such as the quality of communication with family. This is based on research by Liu Q *et al.* (2019). It was found that more open communication and rarely having communication problems in the family will lead to better autonomy. Open communication with the family creates a positive environment; this allows adolescents to develop autonomy. They can freely express their feelings and views, make decisions independently, engage in autonomous behavior, and are not influenced by external impulses (Bi X & Wang 2021). So based on previous research, the family has an essential role in the formation or development of autonomy in adolescents, both based on their parenting style or communication patterns. Therefore, the opportunity for families to influence autonomy in adolescents can be more excellent than doing self-disclosure on social media alone.

According to the perspective of positive psychology, when a person is willing to share experiences that make him feel depressed with others, such as trauma and depression, they can get better social support, be able to face challenges, and increase integrity in their environment (Ko & Chen 2009). Following this, this study proves that self-disclosure on social media influences environmental mastery in adolescents during the Covid-19 pandemic. These dimensions describe how individuals can master the environment, manage it, take advantage of opportunities around them effectively and determine something reasonable and appropriate according to personal needs (Ryff 1989). In line with this research, Hasanah & Minerty (2018) found a positive relationship between self-disclosure and social interaction in adolescents.

This study also found that self-disclosure on social media influences the personal growth dimension. It is about the individual's ability to develop self-potential, be open to new experiences, and continue to develop knowledge and behavior from time to time. Self-disclosure during adolescence has been associated with positive developments such as well-being (Ko & Chen 2009). In line with this, the increase in self-disclosure involvement during adolescence is supported by the development of social cognitive processes in adolescents, such as perspective-taking and advanced mentalities that support the ability to understand and interpret the feelings, intentions, and desires of others (Frith & Frith 2012). Vijayakumar & Pfeifer (2020) also revealed that self-disclosure is a fundamental interpersonal process for adolescents to achieve developmental goals.

Gainau (2019) states that when someone wants to build relationships with other people, self-disclosure is a crucial ability to have. This statement can be strengthened by the results of this study, which found that self-disclosure on social media influences positive relationships with others. This dimension describes the ability to establish positive relationships with others, indicated by warm relationships with others, having empathy, giving affection and intimacy, and being able to give and receive in human relationships. This is supported by research conducted by Lumsden (1996), who found that self-disclosure helps a person to be able to communicate with others, increase self-confidence and build close relationships with others. Moreover, during the pandemic, face-to-face interactions with other people are limited. So the thing that can be done to maintain positive relationships with others is to interact online through social media.

Furthermore, self-disclosure on social media influences the dimension of purpose in life. This dimension describes the extent to which individuals believe that living in the present and the past is meaningful so it must have meaning or purpose in life. A study by Satrio and Budiani (2018) found a correlation between the meaning of life and self-disclosure on Instagram. Self-disclosure helps a person to recognize himself and can guide him toward the goal of life, which is whom I want to be (Handayani *et al.* 1998). When someone self-discloses on social media, they have communicated something, be it thoughts, feelings, life experiences, or anything related to personal life, which can positively impact the meaning of life and well-being in everyday life.

Another benefit that can be obtained when someone self-discloses on social media is that it can increase self-acceptance. According to DeVito (2012), when individuals self-disclose, they will know other people's opinions of themselves, and it turns out that most people can accept themselves based on other people's opinions. When other people reject an individual, the individual tends to reject himself. Through self-disclosure and support from people, individuals may be able to feel positive responses so that individuals can build a positive self-concept. This follows Hanafiah's research (2012) which found a relationship between the level of self-acceptance and self-disclosure in adolescents. Thus, this study supports previous research because there is an effect of self-disclosure on social media on the self-acceptance dimension in which individuals have a positive attitude toward themselves, and are willing to accept whatever has happened in their past and accept good or bad things in their lives.

The results of this study need to be developed in future studies because the current findings can provide an overview of how self-disclosure on social media affects psychological well-being in adolescents. Because psychological well-being has an essential role in the sustainability of the development, values, direction, and life goals chosen by adolescents for their future.

Meanwhile, there are limitations in this study where the number of female participants is far more than males, and the scale used in this study has not been adjusted to the conditions of the Covid-19 pandemic, so it still uses general statements. In addition, this research design is correlational, which is limited in testing the causal relationship between research variables.

5 CONCLUSION AND IMPLICATION

This study's findings are that self-disclosure on social media significantly affects the dimensions of psychological well-being, namely environmental mastery, personal growth, positive relations with others, purpose in life, and self-acceptance. These results also identify that the higher the self-disclosure on social media, the higher the environmental mastery, personal growth, positive relations with others, purpose in life, and self-acceptance among adolescents during the Covid-19 pandemic. In addition, it is also known that there is no influence of self-disclosure on social media on the autonomy dimension of adolescents during the Covid-19 pandemic.

This study implies that adolescent social media users need to control self-disclosure activities so they do not develop into oversharing, which can harm psychological well-being or self-security. Parents need to monitor teenagers' activities on social media to remind them about things that are appropriate or not to be shared so that the psychological well-being of teenagers can be fulfilled. Social media developers should be able to monitor more about the content that users will share more so that they can provide warnings or withdraw sensitive content that can interfere with the welfare or safety of users. Practitioners such as counselors, psychologists, counseling guidance teachers, or others can use social media to conduct counseling (cybercounseling) on adolescents because teenagers may be able to more easily express themselves so that practitioners can help improve psychological well-being. In addition, it is hoped that this research can be used as a reference for future research related to self-disclosure and psychological well-being. Furthermore, researchers are expected to expand the discussion on each dimension of self-disclosure and psychological well-being, find the influence or difference of the positive and negative impacts of self-disclosure on social media on psychological well-being, and strive for the number of male and female participants who are more balanced and use a scale that is by the conditions when collecting data.

REFERENCES

Adzhani, S. N., Baihaqi, M. I. F., & Kosasih, E. (2020). Persepsi Dukungan Sosial Sebagai Mediator Pengungkapan Diri Dan Kesejahteraan Subjektif Pada Pengguna Instagram. *MEDIAPSI, 6*(1), 60–70.

Amosun, T. S., Jianxun, C., Rufai, O. H., Muhideen, S., Shahani, R., Shah, Z., & Koroma, J. (2021). WeChat Usage During COVID-19 Pandemic Lockdown: The Mediating Role of Online Self-disclosure on Quality of Friendship and Well-being. *Global Knowledge, Memory and Communication.*

Ampong, G. O. A., Mensah, A., Adu, A. S. Y., Addae, J. A., Omoregie, O. K., & Ofori, K. S. (2018). Examining Self-disclosure on Social Networking Sites: A Flow Theory and Privacy Perspective. *Behavioral Sciences, 8*(6), 1–17.

Bi, X., & Wang, S. (2021). Parent-Adolescent Communication Quality and Life Satisfaction: The Mediating Roles of Autonomy and Future Orientation. *Psychology Research and Behavior Management, 14*, 1091.

Choi, S. B., & Lim, M. S. (2016). Effects of Social and Technology Overload on *Psychological Well-being* in Young South Korean Adults: The Mediatory Role of Social Network Service Addiction. *Computers in Human Behavior, 61*, 245–254.

DeVito, J. A. (2012). *The Interpersonal Communication Book* 13th Edition. New York: Pearson.

Dewi, D. S., & Mulyo, M. (2017). Psychological Well Being Pada Siswa Tunanetra. *Jurnal Psikologi Pendidikan dan Perkembangan*, 11–23.

Dogan, U. (2016). Effects of Social Network Use on Happiness, *Psychological Well-being*, and Life Satisfaction of High School Students: Case of Facebook and Twitter. *Egitim ve Bilim, 41*(183), 217– 231

Frith, C. D., & Frith, U. (2012). Mechanisms of Social Cognition. *Annual Review of Psychology*, *63*, 287–313.
Gainau, M. B. (2019). *Self-disclosure* Effect on Cultural Context of Papuan Teenagers. *International Journal of Social Sciences and Humanities*, *3*(2), 62–70.
Hanafiah, N. A. (2012). *Hubungan Antara Self Esteem, Self Confidence, Dan Self Acceptance Dengan Self Disclosure Pada Remaja* (Skripsi, Universitas Negeri Malang).
Handayani, M. M., Ratnawati, S., & Helmi, A. F. (1998). Efektifitas Pelatihan Pengenalan Diri Terhadap Peningkatan Penerimaan Diri Dan Harga Diri. *Jurnal Psikologi*, *25*(2), 47–55.
Hasanah, U., & Minerty, P. B. (2018). Hubungan Antara Self Disclosure Dengan Interaksi Sosial pada Remaja di Kota Banda Aceh. *Journal of Healthcare Technology and Medicine*, *4*(2).
Johana, K., Lestari, F. D., & Fauziah, D. N. (2020). Penggunaan Fitur Instagram Story Sebagai Media Self Disclosure dan Perilaku Keseharian Mahasiswi Public Relations Universitas Mercu Buana. *Jurnal Ilmu Manajemen Terapan*, *1*(3), 280–289.
Joinson, A.N. (2001). *Self-disclosure* In Computer-Mediated Communication: The Role Of Self-Awareness and Visual Anonymity. *European Journal of Social Psychology*, *31*, 177–192.
Kelana, P. J. (2017). *Hubungan Antara Kesejahteraan Psikologis Dengan Keterbukaan Diri Individu Yang Menjalani Hubungan Jarak Jauh*. Fakultas Psikologi. Universitas Surabaya.
Ko, H. C., & Chen, T. K. (2009). Understanding the Continuous Self-disclosure of Bloggers from the Cost-benefit Perspective. In *2009 2nd Conference on Human System Interactions* (pp. 520–527). IEEE.
Kusyanti, A., & Safitri, O. (2016). How Do I Look: *Self-disclosure* of Instagram Users in Indonesia. *Journal of Education and Social Sciences*, *5*(2), 242–247.
Leung, L. (2002). Loneliness, Self-disclosure, and ICQ ("I seek you") use. *CyberPsychology & Behavior*, *5*(3), 241–251.
Liu, Q., Lin, Y., Zhou, Z., & Zhang, W. (2019). Perceived Parent–adolescent Communication and Pathological Internet use Among Chinese Adolescents: A Moderated Mediation Model. *Journal of Child and Family Studies*, *28*(6), 1571–1580.
Lumsden, G., & Lumsden, D. (1996), *Communicating with Credibility of Confidence*. Boston: Wadsworth Publishing Company.
Luo, M., & Hancock, J. T. (2020). Self-disclosure and Social Media: Motivations, Mechanisms and Psychological Well-being. *Current Opinion in Psychology*, *31*, 110–115.
Matthes, J., Koban, K., Neureiter, A., & Stevic, A. (2021). Longitudinal Relationships Among Fear of COVID-19, Smartphone Online Self-disclosure, Happiness, and Psychological well-being: Survey Study. *Journal of medical Internet research*, 23(9), e28700.
Nguyen, H. T., & Ho, T. T. Q. (2022). Online self-disclosure and well-being Among Vietnamese Adolescents: online social support as a mediator. *Mental Health and Social Inclusion* *26*(4), 339–346 https://doi.org/10.1108/MHSI-01-2022-0003
O'Sullivan, *et al.* (2021). A Qualitative study of Child and Adolescent Mental Health During the COVID-19 Pandemic in Ireland. *International Journal of Environmental Research and Public Health*, *18*(3), 1062.
Qoniah, H. (2021). *Religiusitas dan Self Disclosure Terhadap Psychological Well-being Pada Remaja Panti Asuhan Selama Pandemi Covid-19. Skripsi*. Universitas Muhammadiyah Malang. Malang
Ryff, C. D. (1989). Happiness is Everything, or is it? Explorations on the Meaning of Psychological Well-being. *Journal of Personality and Social Psychology*, *57*(6), 1069.
Satrio, H. P., & Budiani, M. S. (2018). Hubungan Pengungkapan Diri Melalui Media Sosial Instagram Dengan Makna Hidup Pada Mahasiswa Fakultas Bahasa dan Seni Universitas Negeri Surabaya. *Character: Jurnal Penelitian Psikologi*, *5*(2).
Soputan, S. D. M., & Mulawarman, M. (2021). Keterbukaan Diri Sebagai Prediktor Kesejahteraan Psikologis Siswa. *Jurnal Bikotetik (Bimbingan dan Konseling: Teori dan Praktik)*, *5*(1), 33–42.
Steinberg, L. (2014). *Adolescents*, 10th Edition. New York: McGraw-Hill
Sugiyo, S., Muslikah, M., Antika, E. R., & Sutikno, U. G. (2019, December). Profil *Psychological well-being* pada Siswa Kelas X SMA Negeri 12 Semarang Tahun Pelajaran 2018/2019. In *Proceeding of International Conference on Islamic Education (ICIED)*. *4*(1), pp. 116–120.
Sulistyorini, Anjar. (2007). *Hubungan Antara Tingkat Self Disclosure Dengan Psychological Well Being Pada Wanita Penderita Kanker Payudara Stadium Lanjut*. Skripsi Thesis, Universitas Airlangga.
Vijayakumar, N., & Pfeifer, J. H. (2020). Self-disclosure During Adolescence: Exploring the Means, Targets, and Types of Personal Exchanges. *Current Opinion in Psychology*, *31*, 135–140.
Wang, G., Zhang, Y., Zhao, J., Zhang, J., & Jiang, F. (2020). Mitigate the Effects of Home Confinement on Children During the COVID-19 Outbreak. *The Lancet*, *395*(10228), 945–947.
Wheeless, L. R., Nesser, K., & Mccroskey, J. C. (1986). The Relationships of *self-disclosure* and Disclosiveness to High and Low Communication Apprehension. *Communication Research Reports*, *3*(1).

Families Mental Health and Challenges in the 21st Century – Arifin et al. (eds)
© 2024 The Author(s), ISBN 978-1-032-51442-0

The effect of hardiness on Indonesian teachers' subjective career success

Tri Muji Ingarianti
Faculty of Psychology, University of Muhammadiyah Malang, Malang, Indonesia
Doctoral Program of Psychology, University of Airlangga, Surabaya, Indonesia

Yogha Setiawan* & Devina Andriany
Faculty of Psychology, University of Muhammadiyah Malang, Malang, Indonesia

ABSTRACT: Teachers need career goals, understanding, and opportunities to achieve their subjective career success. However, as problems may hinder their career path, they need to have a strong, resilient personality to address and overcome stressors in their career and eventually evaluate their career and achieve their subjective career success. This paper investigates the effect of hardiness on Indonesian teachers' subjective career success. This study applied a quantitative method. Three hundred twenty Indonesian teachers were recruited as participants using accidental sampling. The data were analyzed using a simple regression test. The result shows that hardiness affects Indonesian teachers' subjective career well-being ($p<0.05$). To be more specific, hardiness was found to significantly affects dimensions of subjective career success, including recognition, meaningful work, influence, authenticity, personal life, growth and development, and satisfaction, but not the quality of work.

Keywords: teacher, hardiness, subjective career success

1 INTRODUCTION

Individuals are basically free to choose their careers. Career is important because it is associated with one's psychological and financial aspects to achieve career success (Heslin 2005). Most people still view career success using measurable factors like salary, position, and promotion, a view known as objective career success (Spurk *et al.* 2019). People nowadays increasingly view their careers using subjective factors like work satisfaction, work–life balance, purpose, sense of meaning, and quality of work (Heslin 2005; Hupkens *et al.* 2021; Ingarianti *et al.* 2020).

Teaching is one of the careers that is considered ideal and a teacher is a role model for his/her students (Mundiri & Bariroh 2019). Being a teacher is often seen as a noble profession as it strives for the country's high-quality education and establishes a foundation for the future generation. UNESCO Institute of Statistics (UIS) report a 50% increase between 2000 and 2019. There are 96 million teachers worldwide, and in Indonesia alone, there are 2.9 teachers from elementary to high school levels, including those with Institutional and honorary teacher statuses. It is important for teachers to know their career level, and their career achievement and to have a career goal in order to achieve their career success and satisfaction.

*Corresponding Author: yoghas84@webmail.umm.ac.id

DOI: 10.1201/9781003402381-8

Data published by UNESCO in Global Education Monitoring 2016 shows that Indonesian teachers are in the lowest position among 14 developing countries. One of the possible causes is the teachers' poor perspectives on their careers. Supporting the data, Utami (2019) found that although teachers are large in number, most of them lack personal and professional qualities. The recruitment strategy also becomes a hindrance in this profession. Further, teachers lack the motivation to improve their career achievement and mismatch in their duties (Dewi & Rusdinal 2020). Teachers with a proper understanding of their profession tend to be self-regulated, enhance positive emotions, and optimize students' learning process. Teachers with higher self-regulation tend to exhibit lower burnout levels, allowing them to optimize their careers and eventually achieve career success (Heydarnejad et al. 2021; Karamooz & Narafshan 2017).

A survey by Ikatan Guru Honorer shows that, overall, teachers' performance is still low, implying no difference in performance between certified and non-certified teachers. Honorary teachers' performance is also lower than those with civil servant status, and no significant difference between those with bachelor's and master's degrees (Salam & Heriyanto 2018). Institutions concerning teachers' careers need to pay attention to employees' value adjustment to their work environment, one of the factors associated with career development and satisfaction (Vargas & Teixeira 2018).

Teachers' career is the key factor determining an education's success, as it is reported to affect the educational quality at school (Adabi *et al.* 2017). One of the problems faced by Indonesian teachers is abundant, time- and energy-consuming administrative tasks. We interviewed two teachers, a senior high school and a junior high school teacher, with more than ten years of work experience through a phone call on 17 and 18 November 2021. Our interview results showed that in addition to their teaching duties, they were busy with administrative tasks. Subject teachers' administrative tasks include designing lesson plans, writing teaching journals for learning purposes, employee administrative tasks, administrative tasks of employee performance target (SKP), and making monthly, semestral, and annual learning evaluations and reports. Meanwhile, school counselors face demands for designing service plans and information services, consisting of individual and group counseling, compiling students' data, designing service evaluation, i.e., analysis and follow-ups, and making monthly, semestrial, and annual reports.

These administrative data are supposed to be teachers' guidelines to ensure a more directed learning process and sustainable teachers' careers. Nevertheless, failure in completing these tasks may result in suboptimal learning process achievement, causing poor performance that adversely affects teachers' career success. Suboptimal achievement of the organizational goal may result in teachers' poor careers as educators (Hendri 2010).

Teachers' poor competence, particularly pedagogical and professional ones, will likely hinder their career development. This is in line with data published on the npd.kemdikbud website (regional educational scale), showing that teachers' competence test results in the last five years are still below the minimum standard (i.e., 55) (Hendri 2010; Leonard 2015). Several factors may be responsible for teachers' poor competence. Teachers' poor technological skills force the students to engage in a traditional learning process and put themselves in overwhelming administrative tasks.

Internal factors contributing to teachers' poor competence include low motivation to develop their careers, low career satisfaction, job stress, and poor commitment (Jabar 2012; Kusniyati 2019; Sennen 2017).

Limited access to performance and capacity improvement, e.g., training and scholarship, important opportunities for teachers' development, may also affect teachers' competence. Technology development emerges as a challenge, which a teacher should properly address. Proper digital mastery is also important for a teacher's career, as teachers need to have this skill to perform their tasks during their learning process (Afriansyah 2020). Teachers' careers may also be affected by their surrounding environment, as it will likely affect their mindset, activity, and career orientation (Widodo *et al.* 2020). Furthermore, enacted policies may also cause social dynamics among contract teachers, as they expect a profession that prioritizes length of service above other professional responsibilities—A condition that results in unclear work division between civil service teachers and contract teachers (Huang & Revina 2020).

Teachers' poor competence and professionalism are also accounted for by their low salaries. Some teachers view their salaries as insufficient, causing poor well-being and motivation to achieve career success. Regarding this salary issue, some teachers work in other fields outside their working hours as teachers, which eventually affects their teaching activities (Mulyasa 2008; Saripudin 2014). This condition affects their work-life balance and puts more stress on such individuals, hindering their career development (Riffay 2019). Moreover, public recognition of teachers is still low. Some teachers even do not value their own careers, implying their low self-esteem. This condition reflects one's low career success (Saripudin 2014).

Viewing this phenomenon, we focus on subjective career success in this study. Career success is one's work outcome stemming from psychological aspects, skills, and work experience (Ng *et al.* 2005). Subjective career success (SCS) judgment is one's judgment based on their own feelings (Judge & Mueller 2007). In General, SCS is a reaction to and evaluation of one's career (Pan & Zhou 2015). In order to achieve their SCS, an individual needs to have a career goal, proper career understanding, and perceived opportunities (Parker & Arthur 2002). Teachers perceiving SCS will likely have higher career opportunities, better career plans and strategies, and better characteristics (Supa 2013).

A personality trait is also reported to contribute to SCS. It is unique and different for every individual and deemed important for teachers' performance during learning and interactions with their students. In the context of teachers' tasks, administrative jobs, pedagogical, professional, and social competence may influence their traits and achieve balanced career success, a balanced achievement of subjective and objective career success (Ramdhani *et al.* 2020). Previous studies report that the big five personality traits serve as predictors of SCS, including neuroticism, conscientiousness, extroversion, agreeableness, and openness. Proactive traits and locus of control are reported to affect SCS (Ng *et al.* 2005; Smidt *et al.* 2018). One of the strongest yet understudied predictors of SCS hardiness (Cunningham 2018).

This variable is proposed by Kobasa, who states that individuals facing stress may exhibit frustration and self-protection as reactions to stressors. In this regard, one's hardiness may turn problems into a solution to their problems. These stressors could be internal and external. The former includes burnout, stress, and lack of motivation, while the latter includes overwork, coworkers' conflict, and organizational social support. (Ingarianti *et al.* 2020; Ivancevich *et al.* 2006; Prasetyo *et al.* 2018). Hardiness is a relatively new topic in the field of psychology, and even fewer studies link this variable to subjective career success, especially in the Indonesian context. One of the studies was conducted by Cunningham (2018), who reported a relationship between hardiness and SCS among World War II veterans, athletes, and commercial deep-sea divers. It is interesting to examine this variable in teachers, as this profession is similar to WW2 veterans, athletes, and commercial deep-sea divers. Further, stressors in teachers' jobs and their responsibility to implement subjective aspects in their careers make me assume that hardiness affects teachers' subjective career success.

Based on the description above, hardiness is expected to be associated with SCS. The present study examines these two variables among Indonesian teachers by putting hardiness as the independent variable and subjective career success as the dependent variable. From a theoretical perspective, the present study will extend the literature on psychology, especially the development of subjective career success. This study may also serve as a reference to develop the knowledge of subjective career success. From a practical perspective, this study provides teachers with a reference on how to achieve their subjective career success and explains teachers' subjective career success from hardiness personality traits. In the organizational context, teachers' SCS may, in turn, improve their quality.

2 METHODS

2.1 *Research design*

This study applied a quantitative method., a method used to analyze the relationship between variables, presented in numerical or percentage (Creswell 2014). In this study, a correlational design was applied to observe the relationship between the two variables without making any intervention, addition, or manipulation.

2.2 *Participants*

Participants in this study were recruited using the accidental sampling technique (Sugiyono 2018). Participants' inclusion criteria were set following Law no. 14 of 2005 on Teacher and Lecturer, particularly related to competence and teaching certificate in Article 8. This article stipulates that teachers should have a qualification, competence, and teaching certificate and be psychologically and physically healthy. This study's inclusion criteria include teachers with qualifications, competence, teaching certificate, and at least five years of working experience. In addition, participants in this study were also teachers with Appointment Decrees enacted by their Institution according to Regulation of Minister of National Education no. 35 of 2005. This study recruited 320 respondents in total.

2.3 *Research variables and instruments*

Hardiness in this study was examined as the independent variable. It refers to a personality trait indicated by resistance against stressors or stress-triggering stressors. This variable was measured using commitment, control, and challenge (Kobasa *et al.* 1982). Hardiness was measured using the Dispositional Resilience Scale (DSR-15) developed by Kobasa (1979), which was adapted by Barthone (2007). This scale employed a 4-point Likert scale, ranging from 0 (strongly disagree) to 3 (strongly agree). The indicator of commitment consists of two unfavorable and three favorable items (e.g., "I find my life meaningless." Meanwhile, the indicator of control (CO) consists of one unfavorable and four favorable items (e.g., I can't do much to change the future.) Lastly, the indicator of challenge consists of three unfavorable and two favorable items (e.g., "I like a daily schedule with only a few changes") The DRS-15 scale exhibited a reliability test result of 0.83 with commitment, control and challenge indicators showing a score of 0.77, 0.71, and 0.70, respectively.

The subjective career success (SCS), this study's dependent variable, refers to one's reactions to achieving career satisfaction and one's overall perception of career achievement (Shockley *et al.* 2015). Subjective career success comprises eight aspects, authenticity, growth and development, influence, meaningful work, quality of work, recognition, personal life, and satisfaction. The subjective career success was measured using Shockley *et al.*'s (2015) Subjective Career Success Inventory (SCSI). This scale employed a 5-point Likert scale, ranging from 1 (strongly disagree) to 5 (strongly agree). The SCIS consists of twenty-four items, with three items measuring three aspects. Some items read, "I have met the highest standard quality in my job, I feel my job meaningful."

2.4 *Procedure and data analysis*

The present study was conducted in three stages. In the first stage, by identifying problems and collecting supporting sources, we prepared a relevant theoretical review. After that, research instruments were selected to measure research variables. As previously mentioned, this study employed Kobasa's (1979) Dispositional Resilience Scale (DSR-15) which was adapted from Barthone and Shockley *et al.*'s (2015) Subjective Career Success Inventory (2015). In the second stage, instruments were distributed to 320 respondents between 30 November and 16 December 2021. After finishing the data collection process, the next step was the data analysis stage. The data were analyzed using simple linear regression with IBM SPSS 25. After finishing the data analysis process, conclusions were drawn based on the research result.

3 RESULT

This study involved 320 Indonesian teachers as respondents. Table 1 presents the analysis results.

Table 1. Respondents description.

Category	Frequency	Percentage (%)
Sex		
Male	106	33.1
Female	214	66.9
Length of Service		
<10 years	101	31.6
10–15 years old	77	24.1
>15 years old	142	44.4
Domicile		
Java	280	87.5
Kalimantan	40	12.5
Educational Background		
D4	6	1.9
S1	295	92.2
S2	19	2.9
Employment Status		
Civil Servant	150	46.9
Institutional Teacher	89	27.8
Honorary Teacher	81	25.3
Income		
<Rp.5 Million	249	77.8
>Rp. 5 Million	71	22.2

As presented in Table 1, most respondents were female (n = 214, 66.9%), most respondents have worked for more than 15 years (44.4%). Most of them came from Java Island (n = 280, 87.5%), hold a bachelor's degree (n = 295, 92.9%), were in civil service (n = 150, 469%), had an income below 5.000.000 (n = 249, 77.8%).

The demographic data were then tested using ANOVA to see the effect of demographic data on Subjective career success. The test result is presented in Table 2.

Table 3 shows the coefficient of determination, indicated by R-square analysis result of 3.8% (recognition), 3.5% (Quality of work), 3.4 (meaningful work), 5.1% (influence), 5.5% (authenticity), 3.7% (personal life), 4.3% (growth and development), and 3.5% (Satisfaction).

Table 2. ANOVA test result.

No.	ANOVA Test	F	p-value	Conclusion
1	Gender	1.252	0.264	No significant difference between male and female teachers' SCS.
2	Length of Service	3803	0023	A difference in subjective career success was noticed among the length of service groups.
3	Domicile (Java, Kalimantan)	0617	0433	No difference in subjective career success was found between groups residing in Java and Kalimantan.
4	Education (D4,S1, S2)	2087	0126	No difference in subjective career success was noticed in terms of educational background.
5	Employment status (Civil Service, Institutional Permanent Teacher, Honorary Teacher)	3095	0047	There is a difference in SCS among teachers with civil service, institutional permanent teachers, and honorary teachers.
6	Income (<5.000.000, >5.000.000)	1033	0310	No difference in SCS was noticed in terms of income.

Table 3. Multiple linear regression analysis result.

	B	*SE B*	β	*t*	*p*	R^2	*F*
Recognition							
Constant	9825	0607	0194	16238	<.001	0038	12,429***
Hardiness	0078	0022		3526	<.001		
Quality Work							
Constant	9256	0632	0188	14648	<.001	0035	11,649***
Hardiness	0079	0023		3413	0001		
Meaningful Work							
Constant	10740	0553	0168	19409	<.001	0034	11,343***
Hardiness	0068	0020		3368	0001		
Influence							
Constant	8523	0650	0225	13106	<.001	0051	16,994***
Hardiness	0098	0024		4122	<.001		
Authenticity							
Constant	10301	0548	0235	18811	<.001	0055	18,551***
Hardiness	0086	0020		4307	<.001		
Personal Life							
Constant	9820	0650	0191	15114	<.001	0037	12,100***
Hardiness	0082	0024		3479	0001		
Growth and Development							
Constant	10420	0573	0207	18195	<.001	0043	14,205***
Hardiness	0079	0021		3769	<.001		
Satisfaction							
Constant	9967	0587	0230	16985	<.001	0053	17,688***
Hardiness	0090	0021		4206	<.001		

Regression method: *stepwise*
$^*p < .05$; $^{**}p < .01$; $^{***}p < .001$.

4 DISCUSSION

This study aims to see the effect of hardiness on subjective career success among Indonesian teachers. The discussion is presented in two sections: the effect of hardiness on overall SCS and the effect of hardiness on SCS's eight dimensions. The effect of Hardiness on overall SCS. Respondents in this study have different demographic backgrounds. The analysis result shows a significance value of 0.000 ($p < 0.005$), indicating that hardiness affects Indonesian teachers' subjective career success. In other words, hardiness serves as one of the personality traits that serve as a source of resistance against stressors, which may eventually improve teachers' subjective career success. Kobasa (1979), who proposes the concept of hardiness, defines this variable as a personality trait that acts as a source of resistance against stressful events.

Hardiness was found to contribute to Indonesian teachers' SCS, as these two variables are interrelated. Individuals with high hardiness may address their career problems more effectively. Stressors commonly faced by teachers include burnout, low career self-efficacy, and lack of social support from family and coworkers. Individuals with an adequate level of hardiness may minimize stressful events or career problems. As Cunningham (2018) asserts, hardiness helps individuals address their career problems and stressors. In the same vein, Wallnas and Jendle (2017) report that Swedish teachers with higher hardiness levels tend to be able to solve their problems and subjectively evaluate their careers related to the problems. Being a teacher is a noble yet difficult profession. They are responsible for educating their students and preparing them for the future. Teachers face various challenges throughout their careers. Teachers' problems become more complex, considering that today's situation demands teachers to stay updated with the development. Both personal and professional problems may affect teachers' careers, and failure to address these problems will likely affect their subjective career success. Therefore, it is necessary to have hardiness as one of the sources of resistance to overcome their life problems.

Hardy teachers will likely evaluate their careers to find solutions and set their subjective success standards. They will be able to evaluate their career positively and not see it as a stressor that negatively affects their life experience. Purwati and Mahfud (2019) found that individuals with higher hardiness levels will likely be able to solve their problems and turn their job stress into an opportunity to achieve subjective success. In other words, hardy teachers will likely put themselves differently when setting their subjective career success.

In the following section, we discuss the effect of hardiness on eight dimensions of subjective career success. First, we found a significant effect of hardiness on recognition. This dimension refers to others' recognition of one's career performance, respect, and social status. Hardy individuals tend to be motivated to solve their career problems by setting their career goals, which may lead to career success (Heckhausen *et al.* 2019).

Second, hardiness was found to significantly affects meaningful work. This dimension reflects one's engagement in their career, either individually or in a group, to create meaningful value for their career. Individuals engaged in their careers may see their careers as a calling (Shockley *et al.* 2016). In this regard, hardier individuals will likely exhibit more meaningful contributions to achieve their subjective career success and not compare their current job to others.

Third, hardiness was found to affect influence significantly. This dimension refers to other people's and organizations' influence on one's career. One's career contributes to the organization and affects their career performance. Hardiness is associated with influence and leads to career adaptation as individuals have a deeper understanding of their careers (Hupkens *et al.* 2021; Van der Heijden *et al.* 2021).

Fourth, hardiness significantly affects authenticity. This dimension represents one's career direction suitability with personal needs. This career direction is determined based on one's personal needs and preference that supports career development. Individuals with higher hardiness levels tend to view their career path as a growth, challenges, and opportunities to achieve their subjective career success. Individuals with a modern career mindset will likely interpret their career success based on their values without comparing them to other individuals (Hupkens *et al.* 2021; Shockley *et al.* 2016).

Fifth, hardiness was found to have a significant role in personal life. Personal life represents one's career activity that affects their personal activities. Personal life is closely associated with one's career. Hardy individuals will likely have a good relationship with their family regardless of their career, which allows them to maintain their good relationships with their family and coworkers. Hardy individuals tend to be able to overcome stressors in their careers, which is helpful to achieve their subjective career success.

Sixth, hardiness plays a significant role in growth and development. This dimension represents individuals' career growth following their skills. This indicates that new experience is needed to attain career progress by developing a perception of future career success. In this regard, hardy individuals will see their career experience as a process to develop their skills (Cunningham 2018; Hupkens *et al.* 2021). These variables imply that hardy individuals will be able to evaluate their careers in order to achieve their subjective career success.

Seventh, a significant result was found between hardiness and satisfaction. This dimension represents one's satisfaction with an individual's positive feelings about their career. However, individuals who do not actively contribute to their organization may see their careers as a stressor (Mayrhofer *et al.* 2016). One's hardiness could help identify stressors in career and improve their social status throughout their career. Satisfaction could be helpful in achieving subjective career success by seeing their career satisfaction without comparing it to others. Lastly, an insignificant result was found regarding the effect of hardiness on teachers' quality of work. This dimension represents one's quality of work. Teachers' stressors include abundant administrative tasks and the absence of a standard of quality, as mentioned in government regulation no. 30 of 2019 on civil services. The regulation state that employees' targeted performance is merely performing tasks relevant to their functions, exercising professional development, and performing complementary tasks.

Cunningham (2018) has conducted a study on the role of hardiness in SCS. This study's novelty lies in teachers' involvement as the study respondents. Teachers have unique char acteristics as they have skills to educate their students, adjust to different classroom situations, and act as role models for their students, which is different from Cunningham's (2018) study that focuses on commercial deep-sea divers. A previous study shows a positive correlation between hardiness and all SCS dimensions. Meanwhile, this study found that hardiness has a positive correlation with dimensions 1–7. This difference enriches the existing literature on the role of hardiness in SCS in different careers and serves as the novelty of the role of hardiness in SCS among Indonesian teachers.

5 CONCLUSION AND IMPLICATION

Based on the study result and discussion, this study found the role of hardiness in Indonesian teachers' subjective career success. More specifically, hardiness affects seven dimensions of SCS (i.e., recognition, meaningful work, influence, authenticity, personal life, growth and development, and satisfaction). Hardiness was found not to affect quality work. Hardy teachers will likely achieve their subjective career success.

Based on the study result, teachers are expected to improve their hardiness in developing their careers, properly overcome stressors, and eventually achieve their subjective career success. They may improve their hardiness by having a high enthusiasm to change and by seeing failure as an opportunity to learn and develop their skills. Teachers also need to make career decisions, be responsible for their choice and believe they can control their surroundings despite various unexpected factors. Future studies are recommended to involve other professions, as they may extend the report on the role of hardiness in subjective career success in Indonesia.

REFERENCES

Abele, A. E., & Spurk, D. (2009). The Longitudinal Impact of Self-efficacy and Career Goals on Objective and Subjective Career Success. *Journal of Vocational Behavior*, *74*(1), 53–62. https://doi.org/10.1016/j.jvb.2008.10.005

Adabi, K., Manusia, A., & Indonesia, R. (2017). Effect of Self-efficacy and Organizational Culture on Subjective Career Success in High School Teachers. *Journal of Psychology*, *22*(1), 53–64. http://repository.uinjkt.ac.id/dspace/handle/123456789/38110

Afriansyah, A. (2020). Tantangan Guru Masa Kini. Retrieved from Media Indonesia website: https://mediaindonesia.com/opini/365020/tantangan-guru-masa-kini

Azeem, S. M. (2010). Personality Hardiness, Job Involvement and Job Burnout Among Teachers. *International Journal of Vocational and Technical Education*, *2*(3), 36–40. http://www.academicjournals.org/IJVTE

Bartone, P. (2007). Test-retest Reliability of the Dispositional Resilience scale-15, A Brief Hardiness Scale. *American Psycholgical Society*, *101*(3), 943–944. https://doi.org/10.2466%2Fpr0.101.3.943-944

Bartone, P. T., & Bowles, S. V. (2020). Coping with Recruiter Stress: Hardiness, Performance and Well-being in US Army Recruiters. *Military Psychology*, *32*(5), 390–397. https://doi.org/10.1080/08995605. 2020. 1780061

Boudreau, J. W., Boswell, W. R., & Judge, T. A. (2001). Effects of Personality on Executive Career Success in the United States and Europe. *Journal of Vocational Behavior*, *58*(1), 53–81. https://doi.org/10.1006/jvbe.2000.1755

Coetzee, M., & Schreuder, D. (2011). The Relation Between Career Anchors, Emotional Intelligence and Employability Satisfaction Among Workers in the Service Industry. *Southern African Business Review*, *15* (3), 76–97. https://hdl.handle.net/10520/EJC92935

Creswell, J. W. (2014). *Research Design: Qualitative, Quantitative and Mixed Methods Approaches*, 4th ed. Singapore: Sage Publication.

Cunningham, R. N. (2018). The Role of Personality, Hardiness, Resilience and Grit in Mediating Subjective Career Success in Commercial Deep Sea. *March*, 1–150. http://scholar.sun.ac.za/handle/10019.1/103769

Dahling, J. J., & Lauricella, T. K. (2017). Linking Job Design to Subjective Career Success: A Test of Self-determination theory. *Journal of Career Assessment*, *25*(3), 371–388. https://doi.org/10.1177/10690727166 39689

Dai, L., & Song, F. (2016). Subjective Career Success: A Literature Review and Prospect. *Journal of Human Resource and Sustainability Studies*, *4*(3), 238–242. https://doi.org/10.4236/jhrss.2016.43026

Dewi, A. P., & Rusdinal, R. (2020). Perkembangan Karir Guru. *Jurnal Manajemen, Kepemimpinan, dan Supervisi Pendidikan*, *5*(1), 8–13. https://doi.org/10.31851/jmksp.v5i1.3489

Field, A. P. (2009). *Discovering Statistics using SPSS: (and sex, drugs and rock "n" roll)*, 3rd ed. Los Angeles: SAGE Publications.

Heckhausen, J., Wrosch, C., & Schulz, R. (2019). Agency and Motivation in Adulthood and Old Age. *Annual Review of Psychology*, *70*(August), 191–217. https://doi.org/10.1146/annurev-psych-010418-103043

Hendri, E. (2010). Guru Berkualitas: Profesional Dan Cerdas Emosi Abstrak. *Jurnal Saung Guru*, *1*(2), 1–11.

Heslin, P. A. (2005). Conceptualizing and Evaluating Career Success. *Journal of Organizational Behavior*, *26* (2), 113–136. https://doi.org/10.1002/job.270

Heydarnejad, T., Fatemi, A. H., & Ghonsooly, B. (2021). The Relationship Between Critical Thinking, Self-regulation, and Teaching Style Preferences Among EFL Teachers: A Path Analysis Approach, *7*(1), 96–108. https://doi.org/10.17323/jle.2021.11103

Hidayat, D. R., Wening, C., & Alfan, R. (2019). *Teori dan Aplikasi Dalam Bimbingan dan Konseling Komprehensif.* Jawa Barat: Jejak Publisher.

Huang, A. R., & Revina, S. (2020). The Struggle To Recruit Good Teachers in Indonesia: Institutional and Social Dysfunctions. *RISE: Research on Improving Systems of Education.* 20/041. https://doi.org/10.35489/BSG-RISE-WP2020/041

Hupkens, L., Akkermans, J., & Solinger, O. (2021). The Dynamics Of Subjective Career Success: A Qualitative Inquiry. *Sustainability*, *13*(14). https://doi.org/10.3390/su13147638

Ingarianti, T. M., Fajrianthi, F., & Chusairi, A. (2020). Kesuksesan Karier Subjektif Sebagai Identitas Karier Karyawan. *Buletin Psikologi*, *28*(2), 201. https://doi.org/10.22146/buletinpsikologi.43401

Ingarianti, TM., Suhariadi, F. & Fajrianthi. (2022). Adaptasi Alat Ukur Kesuksesan Karier Subjektif. *Jurnal Ilmiah Psikologi Terapan*, *10*(1), 79–88. https://dx.doi.org/10.22219/jipt.v10i1.19002

Ivancevich, J. M., Robert, K., & Matterson, M. T. (2006). *Perilaku dan manajemen organisasi.* Jakarta: Erlangga.

Jabar, C. S. A. (2012). Membangun Guru Masa Kini. *Teacher and Principalship Development*, 1–11.

Judge, T. A., & Mueller, J. D. K. (2007). Personality and Career Success. *Handbook of career studies*, *9*, 59.

Karamooz, M., & Narafshan, M. H. (2017). The Relationship Between Self-regulated Strategies and Burnout: A Teacher Analysis in the EFL Context of Iran. *Studies in Linguistics and Literature*, *1*(2), 186–196. https://doi.org/10.22158/sll.v1n2p186

Kariyawasam, J. D., & Welmilla, I. (2020). Impact of Big Five Personality Factors on Career Success: A Study Based on Staff-level Employees in a Leading Apparel Firm in Sri Lanka. *Kelaniya Journal of Human Resource Management*, *15*(2), 1–20. https://doi.org/10.4038/kjhrm.v15i2.76

Kobasa, S. C. (1979). Stressful Life Events, Personality, and Health: An Inquiry into Hardiness. *Journal of Personality and Social Pscyhology*, *37*(1), 1–11. https://doi.org/10.1037/0022-3514.37.1.1

Kobasa, S. C., Maddi, S. R., & Kahn, S. (1982). Hardiness and Health: A Prospective Study. *In Journal of Personality and Social Psychology*, *42*(1), 168–177. https://doi.org/10.1037/0022-3514.42.1.168

Kreitner, R., & Kinicki, A. (2005). *Perilaku organisasi* (*5th ed.*). Jakarta: Salemba Empat.

Kusniyati, T. (2019). Faktor-Faktor Yang Mempengaruhi Kepuasan Kerja Guru TK Di Kecamatan Panjatan Kulon Progo. *In Prosiding Seminar Nasional Magister Psikologi Universitas Ahmad Dahlan*, 538–545.

Leonard, L. (2015). Kompetensi Tenaga Pendidik Di Indonesia: Analisis Dampak Rendahnya Kualitas SDM Guru Dan Solusi Perbaikannya. *Formatif: Jurnal Ilmiah Pendidikan MIPA*, *5*(3), 192–201. https://doi.org/10.30998/formatif.v5i3.643

Likoff, L. (2009). *Career Discovery Encyclopedia (j. chambers (ed.)*, 7th edn. New York: Ferguson Publishin.

Maddi, S. R. (2002). The Story of Hardiness: Twenty Years of Theorizing, Research, and Practice. *Consulting Psychology Journal: Practice and Research*, *54*(3), 173–185. https://doi.org/10.1037/1061-4087.54.3.175

Mahardika, M. D., Kistyanto, A., Ekonomi, F., & Negeri, U. (2020). Pengaruh Kepribadian Proaktif Terhadap Kesuksesan Karir Melalui Adaptabilitas Karir. *In Forum Ekonomi*, *22*(2), 185–195. http://journal.feb.unmul.ac.id/index.php/FORUMEKONOMI

Mayrhofer, W., Briscoe, J. P., Hall, D. T. T., Dickmann, M., Dries, N., Dysvik, A., Kaše, R., Parry, E., & Unite, J. (2016). Career Success Across the Globe: Insights from the 5C Project. *Organizational Dynamics*, 45(3), 197–205. https://doi.org/10.1016/j.orgdyn.2016.07.005

Mulyasa. (2008). *Kurikulum Berbasis Kompetensi Konsep, Karakteristik, dan Impelementasi*. Bandung: Remaja rosdakarya.

Mund, P. (2016). Kobasa Concept of Hardiness (a Study with Reference to the 3cs). *International Research Journal of Engineering*, *2*(1), 34–40. https://sloap.org/journals/index.php/irjeis/article/view/243

Mundiri, A., & Bariroh, A. (2019). Amplifikasi Profesi Guru Dalam Proses Pendidikan Transformatif Perspektif Al-ghazali. *Islam Futura*, *18*(1), 159–184. https://doi.org/10.22373/jiif.v18i1.3288

Ndlovu, V., & Ferreira, N. (2019). Students' Psychological Hardiness in Relation to Career Adaptability. *Journal of Psychology in Africa*, *29*(6), 598–604. https://doi.org/10.1080/14330237.2019.1689468

Ng, T. W. H., Eby, L. T., Sorensen, K. L., & Feldman, D. C. (2005). Predictors of Objective and Subjective Career Success: A Meta-analysis. *Personnel Psychology*, *58*(2), 367–408. https://doi.org/10.1111/j.1744-6570.2005.00515.x

Olsson, M. (2011). *Handbook of Success: How to Make Your Life What You Want it to be* (1st ed.). Siforia. https://books.google.co.in/books?id=0XvOJSexGLMC&printsec=copyright#v=onepage&q&f=false

Omondi, A. A. (2020). Effect of Organizational Sponsorship, Career Management Behaviour and Proactive Personality in Predicting Managers Subjective Career Success. *International Journal of Human Resource Studies*, *10*(4), 208219–208219. https://doi.org/10.5296/ijhrs.v10i4.17915

Ouellette Kobasa, S. C., Maddi, S. R., Puccetti, M. C., & Zola, M. A. (1985). Effectiveness of Hardiness, Exercise and Social Support as Resources Against Illness. *Journal of Psychosomatic Research*, *29*(5), 525–533. https://doi.org/10.1016/0022-3999(85)90086-8

Pan, J., & Zhou, W. (2015). How do Employees Construe their Career Success: An Improved Measure of Subjective Career Success. *International Journal of Selection and Assessment*, *23*(1), 45–58. https://doi.org/10.1111/ijsa.12094

Parker, P., & Arthur, M. B. (2002). Bringing "New Science" Into Careers Research. *Management*, *5*(1), 105–125. https://doi.org/10.3917/mana.051.0105

Pordelan, N., & Hosseinian, S. (2021). Online Career Counseling Success: The Role of Hardiness and Psychological Capital. *International Journal for Educational and Vocational Guidance*, *21*(3), 531–549. https://doi.org/10.1007/s10775-020-09452-1

Purwati & Mahfud, Y. (2019). Pengaruh Dukungan Sosial, Kepribadian Hardiness, Dan Efikasi Diri Terhadap Burnout (Studi Pada Guru SMP Negeri 1 Sigaluh Banjarnegara). *Journal of Economic*, *1*(1), 70–77. https://doi.org/10.32500/jebe.v1i1.877

Prasetyo, Y. A., Fathoni, A., & Malik, D. (2018). Analisis Pengaruh Konflik Keluarga-Stress Kerja Dengan Dukungan Sosial Sebagai Variabel Moderasinya (Studi Pada Guru Demak). *Journal of Management*, *4*(4).

Rakhmi, R. F., Solehuddin, & A, A. H. (2020). The importance of hardiness for students in 21st century. *2nd International Conference on Elementary Education*, *2*, 969–976. http://proceedings.upi.edu/index.php/icee/article/view/708%0A

Ramdhani, N. K., Abdullah, T., Retnowati, R., Abidin, Z., & Riyad, M. (2020). Teacher Career Success with Strengthening Innovativity, Training Effectiveness And Personality (Correlation Analysis and Sitorem on Public Elementary School Civil Servant Teachers in Bogor Regency). *PalArch's Journal of Archaeology of Egypt/Egyptology*, *17*(6), 7743–7766.

Rasdi, R. M., Ismail, M., & Garavan, T. N. (2011). Predicting Malaysian Managers' Objective and Subjective Career Success. *International Journal of Human Resource Management*, *22*(17), 3528–3549. https://doi.org/10.1080/09585192.2011.560878

Riffay, A. (2019). Pengaruh Keseimbangan Kehidupan Kerja (Work Life Balance) Dan Kepuasan Kerja Terhadap Komitmen Organisasi Guru SD Negeri Di Kecamatan Kota Masohi. *Wahana Pendidikan*, *5*(3), 40–48. https://doi.org/10.5281/zenodo.3360415

Salam, A. F., & Heriyanto, S. (2018). Pengaruh Sertifikasi Guru Dan Pengembangan Karir Terhadapkepuasan Kerja Guru Akuntansi SMA Di Kota Cilegon. *Jurnal Pendidikan*, *12*(1), 23–32. https://doi.org/10.30957/cendekia.v12i.434.Hasil

Saripudin. (2014). Faktor-faktor Yang Mempengaruhi Kompetensi Profesional Guru Bidang Kompetensi Keahlian Teknik Instalasi Tenaga Listrik Di SMK. *Invotec*, *10*(1), 67–88.

Schultz, D. & Schultz, S. (2010). *Psychology and Work Today: An Introduction to Industrial and Organizational Psychology*. New York: Routledge.

Semeijn, J. H., van der Heijden, B. I. J. M., & De Beuckelaer, A. (2020). Personality Traits and Types in Relation to Career Success: An Empirical Comparison Using the Big Five. *Applied Psychology*, *69*(2), 538–556. https://doi.org/10.1111/apps.12174

Sennen, E. (2017). Problematika Kompetensi Dan Profesionalisme Guru. *Makalah disajikan dalam Prosiding Seminar Nasional HD PGSD Wilayah IV*, *201*, 16–21.

Shockley, K. M., Ureksoy, H., Rodopman, O. B., Poteat, L. F., & Dullaghan, T. Tyan. (2015). Development of a New Scale to Measure Subjective Career Success: A Mixed-methods Study. *Journal of Organizational Behavior*, *37*(1), 128–153. https://doi.org/10.1002/job.2046

Smidt, W., Kammermeyer, G., Roux, S., Theisen, C., & Weber, C. (2018). Career Success of Preschool Teachers in Germany– The Significance of the Big Five Personality Traits, Locus of Control, and Occupational Self-efficacy. *Early Child Development and Care*, *188*(10), 1340–1353. https://doi.org/10.1080/03004430.2017. 1314275

Spurk, D., & Dries, N. (2019). Antecedents and Outcomes of Objective Versus Subjective Career Success: Competing Perspectives And Future Directions. *Journal of Management*, *45*(1), 35–69. https://doi.org/10.1177/0149206318786563.

Sudjana. (2003). *Teknik Analisis Regresi dan Korelasi Bagi Peneliti*. Bandung: Tarsito.

Sugiyono, S. (2018). *Metode Penelitian Pendidikan Kuantitatif, Kualitatif dan R&D*. Bandung: Alfabeta.

Supa, M. (2013). Factors Contributing to the Subjective Career Success Among Islamic Educators in Primary Schools. *Journal of Education and Literacy Studies*, *1*(1), 55–60. https://doi.org/10.7575/aiac.ijels.v.1n.1p.55

Tambunan, T. B. (2017). Pengembangan Karir Guru Menuju Indonesia Emas. *Seminar Nasional Pendidikan Dasar Universitas Negeri Medan*, 227–235.

Utami, S. (2019). Meningkatkan Mutu Pendidikan Indonesia Melalui Peningkatan Kualitas Personal, Profesional, Dan Strategi Rekrutmen Guru. *Prosiding Seminar Nasional Pendidikan FKIP*, *2*(1), 518–527.

Van der Heijden, B. I. J. M., Veld, M., & Heres, L. (2021). Does Age Matter? Examining Career Commitment as a Moderator in the Relationship Between Age-related HR/D Practices and Subjective Career Success for Younger Versus Older Academic Staff. *Human Resource Development Quarterly*, 1–21. https://doi.org/10.1002/hrdq.21463.

Van Dyk, G. (2015). Hardiness as Predictor of Work Readiness: A Preliminary Exploratory Study. *Journal of Psychology in Africa*, *25*(1), 80–82. https://doi.org/10.1080/14330237.2014.997031.

Vargas, R. C., & Teixeira, A. (2018). Satisfaction at Ifes: An Analysis Through Career. *Revista de Administração Mackenzie*, *19*(3). https://doi.org/10.1590/1678-6971/eRAMG180101.

Wallnas, A., & Jendle, H. (2017). Effects of Exercise, Social Support and Hardiness on Occupational Stress in Swedish Teachers. *Psykologi III*, 40. http://oru.diva-portal.org/smash/record.jsf?dswid=-4791&pid=diva2%3A1120081.

Waysman, M., Schwarzwald, J., & Solomon, Z. (2001). Hardiness: An Examination of Its Relationship with Positive and Negative Long Term Changes Following Trauma. *Journal of Traumatic Stress*, *14*(3), 531–548. https://doi.org/10.1023/A:1011112723704.

Widodo, S., Meiliani, M., & Muchiri, M. K. (2020). Investigating Factor Structures Underlying the Construct of Career Anchor Inventory in the Indonesian Context. *Proceeding of the 4th Internasional Conference on Business*, 193–206.

Families Mental Health and Challenges in the 21st Century – Arifin et al. (eds)
© 2024 The Author(s), ISBN 978-1-032-51442-0

Parent–adolescent sexual communication in Indonesia: Study of Z generation

Siti Maimunah*
University of Muhammadiyah Malang, Indonesia
University of Gadjah Mada, Indonesia

Tina Afiatin* & Arum Febriani*
University of Gadjah Mada, Indonesia

ABSTRACT: Sexual communication is still considered taboo. Therefore, this research aims to investigate parent–adolescent sexual communication, barriers to communication, and communication methods from the perspectives of the Z generation. A qualitative survey has been administered to 325 adolescents aging from 12–22 years old. Data analysis was performed by means of MAXQDA software. The result indicated that 'never' was the highest response the participants gave in almost all topics of sexual communication with 82.5% on contraception topic, 48.99% on unexpected pregnancy, 47.7% on HIV/AIDS disease, 34.2% on the prohibition of unmarried relationship, and 54.5% on the sexual relationship. The main reasons adolescents never expected to talk about topics related to premarital sex were due to shyness (64.4%) and fear (26.0%). The fear shown by the respondents existed because of personal anxiety that their parents would get mad at them and blame them for premarital sex. In this research, it was also found that there were two models or communication methods favored by Gen-Z for sex education, namely counseling (44%) and preaching (30%).

Keywords: Gen-Z, premarital sex behavior, parent-adolescent sexual communication

1 INTRODUCTION

Gen-Z, the term proposed by UPCEA (i.e., University Professional and Continuing Education Association), refers to a generation born between 1995 and 2010 (Francis & Hoefel 2018). Gen-Z is raised and nurtured when technological development begins to massively exist in life, in which the web, Internet, smartphone, laptop, and digital media are accessibly free. This kind of situation accentuates the characteristics of this generation to show up differently from the previous generation, even from the millennial one. As a digital native, Gen-Z is not only accustomed to information access but also information inventiveness. On the other hand, Gen-Z is still in need of personal communication. Hughes (2018) advocates that being digitalized is not necessarily impersonal. In some cases, personalization is the key that interests Gen-Z so interpersonal communication is considered the most effective way to get in touch with them. Seemiller *et al.* (2019) also claimed that in some communicational states, Gen-Z prefers to have personal touch instead, including interaction with parents.

*Corresponding Authors: sitimaimunah20@gmail.com, maimunah@umm.ac.id, afiatin04@ugm.ac.id and arum_febriani@ugm.ac.id

DOI: 10.1201/9781003402381-9

Based on the year of birth, Gen-Z is now considered to be around 12–27 years old. In this study, the Gen-Z involved were aged 12–22 years and not married. They are teenagers. Teenagers are identical with many problems, and one of which is premarital sex (Chaves *et al.* 2005; Kotchick *et al.* 2001; Lansford *et al.* 2010; Layton 2011; Price & Hyde 2009; Zimmer-Gembeck & Helfand 2008). The term 'premarital sex' is referred to sexual activity before marriage. UNICEF makes use of 'fornication' term that is closest in meaning to the former. Meanwhile, some psychological researchers refer to the term 'adolescent sexual behavior'. This is because there will be a dichotomy when referring to adolescent sexual behavior, i.e., those who have and who have not committed sexual intercourse, as adolescents are identical with unmarried state or status (Whitaker *et al.* 2000).

Several previous studies have shown the negative effects of premarital sexual behavior. Henric & Bruk (2016) put some of the risks attributable to premarital sex into a range of events: 15.3% in women with an unintended pregnancy; 82% had induced abortion; 20.9% had sexually transmitted diseases; 14.3% reported vaginal discharge; 5.5% had genital ulcers; and 1.1% had inguinal swelling. Meanwhile, Abdullah & Umar (2013) postulate that there are some serious consequences caused by premarital sex on the lives of adolescents, such as HIV/AIDS, regret, unwanted pregnancy, loss of self-esteem, loss of family support, and depression. Although premarital sexual behavior is considered to violate norms and has the potential to cause negative impacts, from year to year the trend of premarital sexual behavior is increasing.

In Indonesia, in spite of the absence of detailed records, there have been some institutions reporting their researches. The Ministry of Health, in 2009, released researches in four big cities in Indonesia, consisting of Central Jakarta, Surabaya, Bandung, and Medan. The results indicated that as many as 35.9% of adolescents who had made a friend with other people had been active in sexual intercourse; with 6.9% of whoever committing premarital sex (Policy of Indonesian Health). Meanwhile, according to Badan Koordinasi Keluarga Berencana Nasional (BKKBN 2004), adolescents aging from 15–19 years old had been through sexual intercourse. Another indicator that explained the increase in adolescent premarital sex behavior is pregnancy, both in cities and in villages. The Ministry of Health provides the data of the proportion on adolescent pregnancy in Indonesia in 2013 as follows: pregnancy happened to a group of adolescents aged between 10–14 years old and 15–19 years old, both in the context of villages and cities. Pregnancy happening to the former group (10–14 years old) showed insignificant rates, constituting 0.02% in cities and 0.03% in villages. Meanwhile, pregnancy happening to the latter group (15–19 years old) indicated higher rates with 1.28% in cities and 2.71% in villages.

There are several factors associated with premarital sexual behavior in adolescents. Individual factors are the most central factors in adolescents who are actively involved in premarital sex such as pubertal development, attitudes, and behavior (Kotchick *et al.* 2001). Other individual factors are personality (Costa *et al.* 1995; Kirby 2002), willingness to take risks (White *et al.* 1994), sensation seeking (Noar *et al.* 2006), impulsivity (Teese & Bradley 2008), and the ability to socialize (Zimmer-Gembeck *et al.* 2004). In addition, there are also external factors that can be correlated with the phenomenon of adolescent premarital sex, namely peer groups (Whitaker & Miller 2000). (Kirby 2001) states that not only peer groups but also assumptions about certain behaviors by peers contribute to adolescent premarital sexual activity. In addition, family factors are also considered quite significant in causing adolescent premarital sexual behavior. Family factors that influence teenagers to engage in premarital sexual behavior include single-parent or step-parent status indicates a higher likelihood of adolescents having premarital sex than those with complete or biological parents (Boislard & Poulin 2011; Kotchick *et al.* 2001; Meschke *et al.* 2000), number of children in the family (French & Dishion 2003), quality of interaction and interaction (Jaccard *et al.* 1996), and parental support (Kotchick *et al.* 2001).

The communication factor is an important part of the relationship between parents and adolescents. In the process of development, adolescents need information related to their sexual maturity. Several studies explain that parental communication about sex plays an important role in delaying early adolescent sexual intercourse, increasing awareness of the use of contraception and condoms, and reducing premarital sex (Diiorio *et al.* 2003; Harris *et al.* 2013; Hutchinson *et al.* 2003; Hyde *et al.* 2013). Unfortunately, few parents are open to discuss sexual topics with their children. Many of them are regarding the topic as taboo and unpleasant. As a result, teenagers, especially Gen-Z, will be disadvantaged by this unfortunate situation.

The characteristics of Gen-Z's parents are different from those of other generations, with significant diversities and inconsistencies. In terms of sex, Gen-Z needs immediate access to information for accurate and rapid self-retrieval of information. Indeed, it is completely easy for them to acquire information by means of search engines and similar stuff. On the other way around, most parents considered sexual topics as taboos so it is unpleasing for them to discuss the topics with their adolescent children.

Information searching through search engines will be highly risky to Gen-Z if it is without any guidance. There are limitations left unconsidered by the generation, such as religious and social norms apart from their essence for living in a society. This causes Gen-Z frequently involved in the violation of the norms as a consequence of the use of media. According to the Association of Internet Providers in Indonesia (APJII) survey in 2017, the Internet users in Indonesia had increased to reach out 143.26 million out of 262 million population throughout Indonesia. Meanwhile, referring to the age group, those aged 19–34 years are frequently accessing the Internet with the number of 49.52%. For the most-accessed service, the chatting feature is dominant with a total of 89.53% (128 million), social media 87.12% (124.82 million), and search engine 74.84% (107.2 million). Through the Internet, adolescents can access any kind of information. Information accessibility has elevated the rates of risk caused by frequent sexual exposure by which adolescents might show up negative transformation, specifically about sexual intercourse.

Agdebe *et al.* (2016) in their research on 399 students showed that around 57.38% of them had sexual intercourse due to internet exposure. While the reasons why they often access the internet are due to the transfer of information from their peers, lack of information from parents and teachers, easy internet access, and satisfaction after obtaining sexual information online. Conditions like this indicate that adolescents really need adequate sources of information in fulfilling sexual information based on their main age period. They must have accurate and reliable information that best conforms to accepted norms in society. A series of systematic reviews by Rogers (2016) of 38 studies that existed between 1986 and 2005 showed that sexual reference communication involving parents and adolescents significantly influences safe sexual behavior. A meta-analysis conducted by Widman *et al.*, (2015) on 52 articles also showed less different results than Rogers, explaining that parent–adolescent sexual communication results in safe adolescent sexual activity. Some of the studies above focused on the role of parents in sexual communication with adolescents. The novelty in this research is looking at sexual communication from the side of Gen-Z as teenagers, the obstacles experienced, and the expected methods of sexual communication.

However, in the Indonesian context, premarital sex behavior is not permissive. Premarital sex is something perceived as violating the existing norms, especially those related to religion and society. Thus, adolescents are supposed to avoid premarital sex behavior and delay committing the behavior until they come to marriage life. Unfortunately, many parents are not completely into giving adequate information related to sexual behavior needed by adolescents. Furthermore, parents also consider sexual topics taboo for adolescents. For that reason, the researchers would like to explore Gen-Z's premarital sex behavior and parent-adolescent sexual communication for accurate referral intervention to adolescents as well as parents. Thus, this research aims to investigate parent–adolescent sexual communication, barriers to communication, and communication methods from the Gen-Z perspectives.

2 METHODS

2.1 *Respondent*

The approach used in this current research is cross-sectional. The subjects involved in the research were those aged 12–22 years old and unmarried. In addition, 325 adolescents are native to some areas of Indonesia, consisting of East Java, Central Java, East Java, Kalimantan, Papua, and West Nusa Tenggara. Moreover, the involved respondents of the research were still officially enrolled in Junior and Senior High Schools and Higher Education.

To set the respondents, the researchers made use of 2 methods, direct and online selections. In direct selection, the researcher cooperated with a particular school with more than 2,000 students. All the students were the tenth graders because the eleventh and twelfth graders were joining an internship program away from the schools.

The total of the tenth graders signified 847 students grouped based on 14 competencies of expertise, 8 programs of expertise, and 28 classes. Out of 28 classes, as many as 10 classes were selected randomly based on the presence list. Considering the presence lists taken from the 10 classes, 191 respondents were recruited. Meanwhile, to recruit the respondents online, the researchers used 'Google form' feature to be distributed through WhatsApp and Facebook.

2.2 *Instrument*

In data collection, the researchers used a qualitative survey to describe the phenomena of adolescent premarital sex and parent-adolescent sexual communication. For directly-selected respondents, they were requested to fill out an open-ended scale. In its counterpart, the online-selected, they were requested to complete the scale provided in 'Google form' and directly submit the scale only if they met the requirements and agreed to be part of the research. Further, they were also allowed to skip, not to fill out, and forward the form to their colleagues who were willing to.

2.3 *Data analysis*

In data analysis, the researcher occupied MAXQDA 2018 software. MAXQDA is one of the software programs to give assistance for qualitative data analysis so that researchers can easily set up, add on, and code any text or visual data forms. By using MAXQDA, the researchers were allowed to encode through manual coding techniques on the data of the survey. The result of data analysis would be presented in the form of percentages and table based on variables used in the research.

3 RESULTS

3.1 *Descriptive statistics*

This research aimed at perceiving the portrayal of premarital sex behavior and parent-adolescent sexual communication. The following is the representative description of the 325 respondents participating in this research:

Alluding to Table 1, it is obvious that the distribution of gender was almost even between the two groups. In age domain, the middle-adolescent group, with the range of 16–18, was shown more dominant with a total of 204 respondents, while in terms of the puberty period, the 12–13 years old group outnumbered its other counterpart groups with the total of 195 respondents. Considering the educational background, the results varied, but it is evident that the majority of them had attended vocational and senior secondary schools. At the time, most of the respondents were living with their two biological parents with a total of 230 persons. These respondents were taken from some areas in Indonesia, such as East Java, Central Java, West Java, Kalimantan, Papua, and West Nusa Tenggara. In the East Java group, many of the respondents were natively from Malang, Batu, and Surabaya. Furthermore, the majority of the respondents were at that time still officially enrolled in Senior or Vocational Secondary School and Higher Education majoring in Indonesian Language Education and Letters, Civil Engineering, Economics and Business, Law, Electrical Engineering, Accounting, Nursing, Management, and so forth.

Table 1. Description of the respondents.

Variable		Total	Percentage
Gender	Female	166	51.1
	Male	159	48.9
Ages	13–15	56	17.2
	16–18	204	62.8
	19–22	65	20
Period of Puberty	10–11	46	14.1
	12–13	195	60
	14–15	78	24
	16–18	4	1.2
	Unidentified	2	0.7
Education	Junior Secondary School	19	5.8
	Senior Secondary School	43	13.2
	Vocational Secondary School	195	60
	Undergraduate	66	20.3
	Fresh Graduate	2	0.7
Living status	Living with both biological parents	230	70.8
	Living with both parents (with step-father/step-mother)	5	1.5
	Living with a single parent	35	10.8
	Living with grandmother and/or grandfather	5	1.5
	Living in the boarding house	32	9.8
	Others	18	5.6
Total of Respondents		325	100

3.2 *Gen-Z's understanding of premarital sex behavior*

The results of research related to the understanding of Gen-Z regarding premarital sexual behavior are obtained in Table 2.

Table 2. Gen-Z's understanding of premarital sex behavior.

Premarital Sex Definition		Reason for Having Premarital Sex		Forms of Premarital Sex		Effect of Premarital Sex	
Premarital relationship	61.9%	Carnality	46.8%	Kissing	70%	Pregnancy	54.0%
Sexual intercourse	59.1%	Promiscuity	24.6%	Touch the genital	68.6%	HIV / AIDS	26.3%
Unmarried Sex	38.8%	Curiosity	23.6%	Sexual intercourse	56.3%	Ostracized	22.8%
Violating religious norm	25.4%	Friend Influence	13.4%	Holding hands	21.2%	Sexual Disease	20.4%

(continued)

Table 2. Continued

Premarital Sex Definition		Reason for Having Premarital Sex		Forms of Premarital Sex		Effect of Premarital Sex	
Violating state law	16.8%	Lack of guidance	13.0%	Masturbation	16.7%	Feeling of sin	19.7%
Unusual behavior	5.8%	Lack of knowledge	13.0%	Making out	6.9%	Embarrassment	12.5%

Not many respondents in this study were actively engaged in premarital sex; only 14.7% showed premarital sexual behavior. Of these, several participants wanted to stop the behavior because they knew they had violated the norm. However, many of the respondents stated their reluctance to stop this behavior because they were afraid of their lover and wanted to please them, and some of them were even addicted to doing this.

3.3 *Parent-adolescent sexual communication among Gen-Z*

As mentioned previously, this study also intended to know about parent-adolescent sexual communication, especially among Gen-Z. Therefore, the researchers divided sexual-referenced communication into some distinct themes, i.e., prohibitions in dating close mates, HIV/AIDS, an unexpected pregnancy, and contraception. The result indicated that 'never' was the most dominant response to each of the questions about sexual topics, as Table 3 shows.

Table 3. Parent-adolescent sexual communication among Gen-Z.

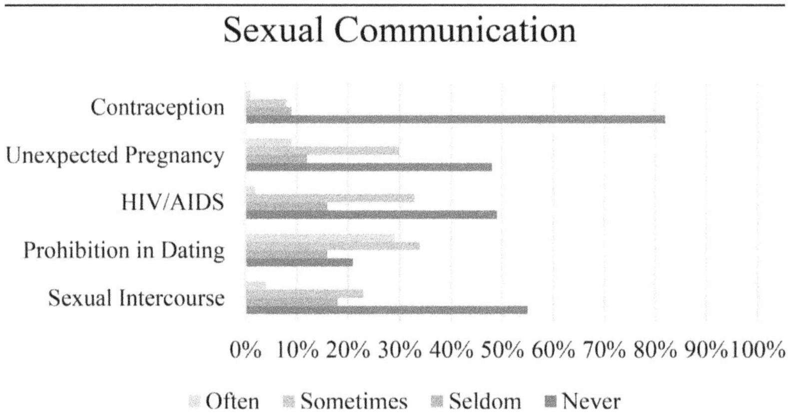

There were some reasons why Gen-Z was reluctant to discuss and talk about sexual topics with their parents. Shyness was the most dominant factor perceived by Gen-Z with a total percentage of 65.2. In addition to being shy, Gen-Z was also fearful with a percentage of 17.1, feeling too early to talk about it at 14%, considering the topics as taboos at 12.8%, considering taboos and feeling fearful getting raging at 5.5%, feeling anxious with 4.3% and considering the topics were too private with 3%.

Emotional states attaching Gen-Z during sexual-referenced communication with parents varied: shyness at 66.4%, fear at 26%, anxiety at 17.4%, nervousness at 10.6%, bad feeling at

5.5%, and heart-beating, disguise, and confusion with 1.7% in each. Among them, nonetheless, some of whom felt fine to have such kind of communication with their parents in spite of low frequency.

Furthermore, with the reference to sharing partners, most of them would rather choose their mothers as their communication partners with 88.8%. On the other hand, the rest 11.2% preferred their fathers.

Generally speaking, adolescents need premarital sex topics in order to save them away from premarital sex behavior. Related to this, material about the consequences of premarital sex was supposed to be given to them. There were 45.1% of adolescents preferred sexual education to be taught to them. Following, the religious norm on premarital sex and the dangerous consequences of premarital sex were in the second and third ranks. In the next ranks, there were followed by ideal ways for intercommunication, sexual diseases, prohibitions in dating, and sexually transmitted diseases, and HIV/AIDS materials. Table 4 described it in detail.

Table 4. Generation Z expectations about themes, methods, and educators of sex education.

Themes of sexual education		Methods of sexual education		Educators of sexual education	
Effect of premarital sex	45.1%	Counseling	44.0%	Teacher	84.2%
Religious norms	30.4%	Preaching	30.0%	Parent	82.0%
Dangers of premarital sex	29.6%	Discussion	24.0%	Family	9.2%
Appropriate relationship	14.8%	Sharing	10.0%	Friend	4.9%
Diseases due to sex	14.4%	Seminar	7%	Doctor	3.2%
Prohibitions in dating	13.2%			Psychologist	1.4%
Reproduction health	10.9%			Religious preacher	1.1%

4 DISCUSSION

This study tries to describe the understanding of premarital sexual behavior among Gen-Z and how much they are involved in premarital sexual behavior. The results showed that respondents in this research know that premarital sex is a behavior that violates religious and social norms. There are 14.7% of the total respondents engaged in premarital sex, which means that not many of them were involved in premarital sex. However, despite the small number of respondents who had premarital sex, some teenagers have shown unusual behavior in dating such as making out and touching vital organs, touching breasts, touching buttocks, fingering, and touching sensitive body parts, as well as oral sex. This enabling condition needs serious attention. One of the ways to prevent teenagers from engaging in premarital sex is parent–adolescent sexual communication. It is known that adolescents who have open and receptive sexual communication with their parents are less likely to engage in risky sexual behavior (Guilamo-Ramos *et al.* 2006; Kotchick *et al.* 1999; Rogers *et al.* 2015). Parent–adolescent sexual communication also contributes significantly to shaping attitudes, and beliefs and predicting decisions to engage in or avoid sexual activity in adolescents (Boyas *et al.* 2012; Clawson & Weber 2003).

In addition, in this study, it was found that carnality is the main reason respondents engage in premarital sex. The other two highest reasons are promiscuity and curiosity. There are other reasons such as peer influence, lack of direction, and lack of knowledge. Based on these reasons, it can be concluded that the reasons that accompany respondents to engage in premarital sexual behavior can be grouped into two, namely internal and external reasons. Internal reasons consist of carnality, curiosity, and lack of sexual knowledge. While external reasons consist of promiscuity, the influence of friends, and lack of guidance or direction from parents. This is supported by research conducted by Adhikari *et al.* (2014) that the influence of parents and peers is interrelated in influencing adolescents to determine their point of view on sexual behavior.

In this research, data about the effects caused by premarital sex is based on Gen-Z's viewpoint. It was shown that the majority of the respondents knew very well about pregnancy consequences after doing it, but not the dangers that might raise in adolescent pregnancy. Their assumption was only focused on being shy after they were pregnant. This denoted that their knowledge on reproduction health was still limited. The Health Demographical Survey of Indonesia in 2007 and 2012 had shown some aspects related to adolescent puberty and reproduction health. The result showed that there was found a trend of proportional increase in both the groups about HIV/AIDS and transmissible infection, along with a decrease in the variable of knowledge about prevention from pregnancy in the group of female adolescents (Fadila *et al.* 2018). In addition, adolescent attitude and behavior were indeed inseparable from cognitive competence and environmental situation. The limitation on understanding and problem-solving skills, with social support from society, would result in an ineffective solution. In common, what they solely believed was that pregnancy occurring after premarital sex could be recovered by early marriage. As a consequence, they would stay with their faith when they found such a thing happened to them; there was still a way to solve it by marriage. Therefore, it was necessary to early prevent adolescents from premarital sex behavior.

Another finding indicated that the rates of puberty age of the respondents ranged on average between 12 and 13 years old. This was in line with Boynton-Jarret & Harville (2012), claiming that the average puberty ages declined during the 20th century and in the early 21st century (with a range of age constituting (10–13 years old), especially within the population along with social and economic development. Puberty age was also proven to correlate with adolescent sexual behavior. Those showing earlier physical maturity tended to be sexually active in advance and had more sexual couples within the adolescent period (Kotchick *et al.* 2001; Lansford *et al.* 2008).

Related to parent–adolescent sexual communication, the researchers grouped the sexual-referenced communication into five themes: contraception, an unexpected pregnancy, HIV/AIDS, prohibitions in dating, and sexual intercourse. The theme-based division was aimed at constructing an obvious portrayal of which themes had been and never been discussed or communicated between parents and adolescents. The result had shown that almost all the themes remained rarely discussed by the two, with the following percentages: contraception (82.5%), sexual intercourse (54.5%), HIV/AIDS (48.9%), unexpected pregnancy (47.7%), and prohibitions on dating (20.9%). Only a few respondents reported frequent discussion with their parents about prohibitions in dating (29.2%).

There were some underlying factors why adolescents never discussed or communicated with parents about sexual topics. The two highest factors were shame (66.4%) and fear (26%), other factors were feeling awkward or taboo, parents do not have time because they are busy at work, communication is not smooth, and the presence of relatives or other relatives. This occurrence should be tolerated as Indonesian culture, perceiving those kinds of topics as taboos so that they should be left undiscussed in public, let alone with parents. This was what made adolescents feel ashamed whenever they would like to discuss sexual topics with their parents. The fear was more likely due to parental anger as their parents would probably suspect them of having committed premarital sex. So, they preferred to remain silent and discuss with their peers.

This condition should bring forth shared attention. When adolescents are ashamed or fearful of communicating with their parents, while the parents never want to initiate the conversation, thus, it is worried that adolescents try to find out other sources of information with a low degree of reliability. Therefore, parental guidance plays significant role to support their children's development, especially in the context of sexual development. Some studies had shown that there was a relationship between parent-adolescent sexual communication and adolescent sexual behavior. Bersamin *et al.* (2006) claimed that parent-adolescent sexual communication is correlated with early-aged sexual behavior. Boislard & Poulin (2011) also found similar results in which adolescents did not spontaneously uncover who they really were before their parents and hide that they were more likely to do early-aged sexual intercourse. Kajula *et al.*, (2014) stated that parent-adolescent sexual communication constituted the most protective medium for adolescent sexual health.

In fact, the respondents had been well aware of their need for adequate information about sex. With the reference to sexual topics expected by the respondents, 45.1% of whom responded by answering the effects of premarital sexual behavior. This was consistent with the previous explanation that the majority of respondents had been aware of premarital sex consequences, with the most-worried effect lying on unexpected pregnancy without further investigation into the danger of early-aged pregnancy. In addition, the religious norm was also what they expected to have (30.4%). The respondents knew that all those things were prohibited based on religious custom, but unfortunately, they did not find the reasons why they were prohibited. Other else materials adolescents would like to receive consisted of premarital sex consequences (29.6%), correct socialization (14.8%), and sexually transmitted diseases (14.4%)—about which they only knew HIV and AIDS—prohibitions in dating peers (with 13.2%), reproduction and health-production (10.9%) dan HIV/AIDS (5.8%).

Furthermore, there was the most-favorite method adolescents would like to have for receiving information for sexual education, with counseling (44%), preaching (30%), discussion (20%), some other information sharing (10%), and seminar (7%). Meanwhile, with the reference to who delivered materials of sexual education, the teachers placed at the first rank (with 84.2%), followed by parents (82%), families (9.2%), friends (4.9%), doctors (3.2%), psychologists (1.4%) and other *ustadz* and *ustadzah* (religious preacher). This finding could be input for further researchers as an advice to set up an intervention for sex education among Gen-Z.

Despite the presence of conventions about sexual education in school, teachers have a lot of chances to elaborate on sexual education in school considering their teaching competence and skill they had. Contrarily, not all parents had sufficient knowledge and good skill at sharing information, especially related to sex. In essence, in education, teacher and parental involvement are of great expectation to achieve the outcomes of sexual education. There were three set parameters to see the three levels of achieving sexual education outcomes (Walker & Milton 2006), namely, micro (individual), meso (organization), and macro (policy). So, it is urgent to sharpen and strengthen the micro-level prior to two others, with family integral to this level. This had become a suggestion for upcoming researchers that intervention in sexual education for adolescents is initiated at micro-level, by approaching to family context. In addition, preaching and counseling methods were still relevant for Gen-Z.

5 CONCLUSION AND IMPLICATION

The result showed that 14.7% of the total respondents were involved in premarital sexual behavior. However, some of the respondents could show portrayals of risky activities that could lead to premarital sex, such as making out and touch on sex organs, breast-touching, butt-touching, fingering, touching, holding sensitive body parts, and oral sex. Related to parent-adolescent sexual communication, the result indicated that almost all sexual topics set were never included in parent-adolescent sexual communication. Only a few of them had ever discussed the topics. The main reasons why they were reluctant to speak about the topics were shyness and fear factors. The fear referred to an anxious state that their parents were going to get mad at them and perceive that they had committed premarital sex. In addition, this research also found that counseling and preaching were two specific methods adolescents (Gen-Z) preferred to have for sexual education with parents being key councelors, especially mothers.

Moreover, teacher and parental roles are necessary for adolescent development, specifically when they have been sexually mature. Evoking teacher and parental roles can be done through a training session with a proper delivery style based on adolescent preference.

REFERENCES

Abdullahi, M., & Umar, A. (2013). Consequences of Pre-marital Sex Among the Youth a Study of University of Maiduguri. *Journal of Humanities and Social Science*. Vol. 10(1), 10–17.

Adhikari, A.B., Cederbaum, J., Sathoff, C & Torro, R. (2014). Direct and Indirect Effects of Maternal and Peer Infleunces on Sexual Intention Among Urban African American and Hispanic females. *Child Adolesc Soc Work J.* Vol. 31, 559–575.

Agdebe, O. C., Kio, J. O., & Ajetunmobi, A.A. (2016). Analysis of the Influence of the Internet on Sexuality Education and Sexual Behaviour Among Adolescents in Nigeria Universities: The Case of University of Lagos. *Journal of Biology, Agriculture and Healthcare*. Vol. 6(16), 74–83.

Bersamin, M.M., Walker, S., Fisher, D.A., & Grube, J.W. (2006). Correlates of Oral Sex and Vaginal Intercourse in Early and Middle Adolescence. *Journal of Research on Adolescence*. Vol. 16, 59–68.

Boyas, J. F., Stauss, K. A., & Murphy-Erby, Y. (2012). Predictors of Frequency of Sexual Health Communication: Perceptions from Early Adolescent Youth in Rural Arkansas. *Child Adolesc Soc Work J.* Vol. 29, 267–284. https://doi.org/10.1007/s10560-012-0264-2

Bronfenbrenner, U. (1986). Ecology of the Family as a Context for Human Development Research Perspectives. *Developmental Psychology*. Vol. 22(6), 723–742.

Chaves, S.L.E., Tortolero, S.R., Markham, C.M., Low, B.J., Eitel, P., & Thickstun, P. (2005). Impact of the Media on Adolescent Sexual Attitude and Behaviors. *Journal of The American Academy of Pediatrics*. Vol. 116(1).

Cilliers, E.J. (2017). The Challenge of Teaching Generation Z. *PEOPLE: International Journal of Social Science*. Vol. 3(1), 188–198.

Clawson, C. L., & Reese-Weber, M. (2003). The Amount and Timing of Parent-adolescent Sexual Communication as Predictors of Late Adolescent Sexual Risk-taking Behaviors. *Journal of Sex Research*. Vol. 40(3), 256–265. https://doi.org/10.1080/00224490309552190

Costa, F. M., Jessor, R., Donovan, J.E., & Fortenberry, J. D. (1995). Early Initiation of Sexual Intercourse: The Influences of Psychosocial Unconventionality. *Journal of Research on Adolescence*. Vol. 5, 93–121.

Diiorio, C., Pluhar, E., & Belcher, L. (2003). Parent-child Communication About Sexuality. *Journal of HIV/ AIDS Prevention & Education for Adolescents & Children*. Vol. 5(3–4), 7–32.

Fadila, W., Darojad, N., Nugroho, A. (2018). Masa Remaja Dan Pengetahuan Kesehatan Reproduksi: Analisis Survei Demografi Kesehatan Indonesia 2007 dan 2012. *Jurnal Kesehatan Reproduksi*. Vol. 9(1), 15–25.

Francis, T., & Hoefel, F. (2018). *True Gen: Generation Z and its Implication for Companies*. McKinsey & Company.

French, D.C., & Dishion, T.J. (2003). Predictors of Early Initiation of Sexual Intercourse Among High-risk Adolescents. *Journal of Early Adolescence*. Vol. 23(3), 295–315.

Guilamo-Ramos, V., Jaccard, J., Dittus, P., & Bouris, A. M. (2006). Parental Expertise, Trustworthiness, and Accessibility. *Journal of Marriage and Family*. Vol. 68(December), 1229–1246.

Harris, A. L., Sutherland, M. A., & Hutchinson, M. K. (2013). Parental Influences of Sexual Risk Among Urban African American Adolescent Males. *Journal of Nursing Scholarship*. Vol. 45(2), 141–150.

Henric, B.V., & Bruk. E.M. (2016). Assessment of Associated Risk Factors and Consequences of Premarital Sex Among High School Female Students. *Journal of Community Health and Epidemiology*. Vol. 2(1), 012–018.

Hughes, J. (2018). Communicating with Generation Z: Everything You Need to Know. *Article* https://www.keystoneacademic.com/news/communicating-with-generation-z-everything-you-need-to-know. Di akses pada tanggal 4 Juli 2019, Pukul 22.15

Hutchinson, M. K., Jemmott J. B., Jemmott, L. S., Braverman, P., & Fong, G. T. (2003). The Role of Mother–daughter Sexual Risk Communication in Reducing Sexual Risk Behaviors Among Urban Adolescent Females: A Prospective Study. *Journal of Adolescent Health*. Vol. 33(2), 98–107.

Hyde, A., Drennan, J., Butler, M., Howlett, E., Carney, M., & Lohan, M. (2013). Parents' Constructions of Communication with Their Children about Safer Sex. *Journal of Clinical Nursing*. Vol. 22(23–24)

Ismail, K. & Hamid, S.R.A. (2016). Communication about Sex-reproduction Health Issues with Adolescents: A Taboo among Malaysian Parents. *European Journal of Social Science Education and Research*. Vol. 6(1), 26–40.

Jaccard, J., Dittus, P.J., & Gordon, V.V. (1996). Maternal Correlates of Adolescent Sexual and Contraceptive Behaviour. *Family Planning Perspective*. Vol. 28(4), 159–185.

Kajula, L.J., Sheon, N., Vries, H.D., Kaaya, F.S., & Aarø, L.E. (2014). Dynamic of Parent-Adolescent Communication on Sexual Health and HIV/AIDS in Tanzania. *AIDS Behav*. Vol. 18, S69–S74.

Kirby, D. (2001). Understanding What Works and What doesn't in Reducing Adolescent Sexual Risk Taking. *Family Planning Perspectives*. Vol. 33(6), 276–281.

Kotchick, B.A., Shaffer, A., Forehand, R., & Miller, K.S. (2001). Adolescent Sexual Risk Behaviour: A Multi-system Perspective. *Clinical Psychology Review*. Vol. 21, 493–519.

Lansford, J.E., Yu, T., Erath, S.A., Pettit, G.S., Bates, J.E., & Dodge, K.A. (2010). Developmental Precursors of Number of Sexual Partners From Ages 16 to 22, *Journal of Research on Adolescence*. Vol. 1;20(3): 651–677.

Layton, E., Dollahite, D.C., & Hardy, S.A. (2011). Anchors of Religious Commitment in Adolescents. *Journal of Adolescent Research*. Vol. 26(3), 381–413.

Meschke, L. L., Zweig, J. M., Barber, B. L. & Eccles, J. S. (2000). Demographic, Biological, Psychological, and Social Predictors of the Timing of First Intercourse. *Journal of Research on Adolescence*. Vol. 10, 315–338.

Noar, S.M., Zimmerman, R.S., Palmgreen, P., Lustria, M., & Horosewski, M.L. (2006). Integrating Personality and Psychosocial Theoretical Approaches to Understanding Safer Sexual Behavior: Implications for Message Design. *Health Communication*. Vol. 19, 165–174.

Price, M. N., & Hyde, J.S. (2009). When Two isn't Better Than One: Predictors of Early Sexual Activity in Adolescence using a Cumulative Risk Model. *Journal of Youth and Adolescence*. Vol. 38, 1059–1071.

Rogers, A. (2016). Parent–adolescent Sexual Communication and Adolescents' Sexual Behaviors: A Conceptual Model and Systematic Review. *Adolescent Res Rev*. DOI 10.1007/s40894-016-0049-5

Rogers, A., Rogers, A. A., S, M., Ha, T., Ph, D., Stormshak, E. A., ... Ph, D. (2015). Quality of Parent – Adolescent Conversations About Sex and Adolescent Sexual Behavior: An Observational Study Quality of Parent e Adolescent Conversations About Sex and Adolescent Sexual Behavior: An Observational Study. *Journal of Adolescent Health*. Vol. 57(2), 174–178. https://doi.org/10.1016/j.jadohealth.2015.04.010

Seemiller, C., & Meghan, G. (2019). *The Gen Z Hub, Leading the Conversation on Generation Z*. Thegenzhub.com/prees.

Teese, R., & Bradley, G. (2008). Predicting Recklessness in Emerging Adults: A Test of a Psychosocial Model. *The Journal of Social Psychology*. Vol. 148(1), 105–126.

Whitaker, D.J., Miller, K.S., & Clark, L.F. (2000), Reconceptualizing Adolescent Sexual Behavior: Beyond did they or didn't they? *Journal of Family Planning Perspectives*. Vol. 32(3), 111–117.

White, K. M., Terry, D.J., & Hogg, M.A. (1994). Safer Sex Behavior: The Role of Attitudes, Norms, and Control Factors. *Journal of Applied Social Psychology*. Vol. 24, 2164–2192.

Widman, L., Choukas-Bradley, S., Seth M. Noar, S.M., Nesi, J., & Garrett, K. (2015). Parent-adolescent Sexual Communication and Adolescent Safer Sex Behavior a Meta-analysis. *JAMA Pediatrics*. Vol. 170 (1), E1–E10.

Zimmer-Gembeck, M. J., & Helfand, M. (2008). Ten Years of Longitudinal Research on U.S. Adolescent Sexual Behavior: Developmental Correlates of Sexual Intercourse, and the Importance of Age, Gender and Ethnic Background. *Developmental Review*. Vol. 28, 153–224.

Zimmer-Gembeck, M.J., Siebenbruner, J., & Collins, W.A. (2004). A Prospective Study of Intraindividual and Peer Influences on Adolescents' Heterosexual Romantic and Sexual Behaviour. *Archives of Sexual Behavior*. Vol. 33(4), 381–394.

Families Mental Health and Challenges in the 21st Century – Arifin et al. (eds)
© 2024 The Author(s), ISBN 978-1-032-51442-0

Becomes a flourishing adult with your general self-efficacy

Alifah Nabilah Masturah* & Hudaniah*
Psychology Faculty, University of Muhammadiyah Malang, Malang, Indonesia

ABSTRACT: A flourishing person believes in his own ability to have broad and satisfying to do their jobs. They believe in their ability to carry out various activities that support their work. The aim of the study was to determine the effect of general self-efficacy (GSE) on flourishing in adults. The method used in this research was a cross-sectional quantitative method with accidental sampling. There were 84 women and 35 men as workers in government and private institutions. The range of participants was 18–53 years old with the majority being 25 and 29 years old as many as 13 participants. Participants in this research were 63 Javanese employees, 26 were multicultural, and other ethnic were 30 employees. Other ethnic participants include employees coming from many cultures such as Madura, Bugis, Banjar, Minahasa, Bima, Melayu, Melanesia, Batak, and Manado. Measurements were taken using general self-efficacy scale (GSES) and flourishing scale (FS). Data analysis used simple linear regression. This study showed that GSE was strongly able to predict flourishing ($B=0.601$; $p<0.01$) with 36.2%. Other results showed that flourishing had a correlation with age ($r = 0.210$; $p<0.05$) and GSE ($r = 0.595$; $p<0.01$), however ethnic ($r = -0.173$; $p>0.05$) and gender ($r = -0.126$; $p>0,05$) had no correlation with flourishing. It means that only age and GSE had a relationship with flourishing.

Keywords: Employees, Flourish, Self-efficacy

1 INTRODUCTION

Employees who grow are termed employees with high flourishing. That is, they are individuals who are able to interpret their life experiences with positive feelings (Seligman 2012). For employees, work is a means of preaching. Thus, the values of everyone who works in its work as worship must be carried out with full responsibility and accepted sincerely (Rosa *et al.* 2020).

The purpose of positive psychology is flourishing, namely, identifying and nurturing talents, potentials, and developing strengths, as well as helping people to live more productive and meaningful lives. Flourishing is an individual's ability to have involvement, meaning, relationship, and achievement as subjective and objective components in him. This explanation can illustrate that flourishing is not just in the mind. Employees who have flourishing will work effectively because individuals and teams develop and make positive contributions (Effendy & Subandriyo 2017). In line with that, it is also known that the internal conditions of individuals, such as emotions and self-concept affect their mental health (Masturah 2018).

Previous research on 258 secondary school teachers in Gauteng, South Africa, showed that high flourishing in the workplace will increase the fit between individuals and the

*Corresponding Authors: alifah_nabilah@umm.ac.id and hudaniah@umm.ac.id

DOI: 10.1201/9781003402381-10

environment, improve employee performance, and increase organizational citizenship behavior, but at the same time, it will reduce their intention to quit (Rothmann *et al.* 2019). This is supported by Osaki's study of 50,000 people in Japan who have ikigai (the meaning of life) are more likely to have jobs and better levels of education than those who do not (Mogi 2017).

Meanwhile, research conducted on 4,493 adult men showed flourishing is a form of mental health that is influenced by self-efficacy, where low self-efficacy turned out to be a good predictor of low mental health and high self-efficacy was a good predictor for mental health is also high (Bavojdan *et al.* 2011).

The ability of humans to change their circumstances in a positive direction and master the challenges of life makes a major contribution to the achievement of mental health (Bandura 1997). This ability is termed general self-efficacy (GSE). That is, the individual's belief in being able to handle a variety of challenging and new tasks, and competently manage stressful situations. Being able to cope with unexpected situations and achieve desired results in the face of challenges creates positivity, self-confidence, resilience, satisfaction, and a sense of accomplishment in a person (Luszczynska *et al.* 2005). Self-efficacy plays an important role in psychological adjustment, psychological problems, physical health, and behavior change strategies that are guided professionally and independently (Maddux 2000).

Another study related to the need for self-efficacy that is generally owned by individuals in growing is a study of 366 elementary and junior high school teachers in Italy. This study resulted in a partial mediating effect of self-efficacy for each group of teachers. As distance learning difficulties during COVID-19 can affect stress perception, improving transverse skills, such as self-efficacy, could better equip teachers to deal with stressful situations in their work (Rabaglietti *et al.* 2021).

The aim of the study was to determine the effect of GSE on flourishing in adults. So, this study is expected to give an overview of that. When adults flourish, they need to know that inner strength is what makes it happen.

2 METHOD

This research was a quantitative approach with a cross-sectional study design. It means the participants are measured for outcome and exposure at the same time. This research used accidental sampling. Participants in this research were 35 men employees and 84 women employees. The range of participant age was 18–53 years old with the majority being 25 and 29 years old as many as 13 participants. Participants in this research were 63 Javanese employees, 26 were multicultural, and 30 employees were of other ethnicities. Other ethnic participants include employees coming from many cultures like Madura, Bugis, Banjar, Minahasa, Bima, Melayu, Melanesia, Batak, and Manado. GSE was measured by the general self-efficacy scale (GSES) Indonesian version adapted from (Novrianto *et al.* 2019). The scale was 10 items on unidimensional with a reliability was 0.832. There were four responses for it with a range from *not at all true* to *exactly true*. Flourishing was measured by *Flourishing Scale* (FS) with 8 items adapted from (Diener *et al.* 2010) with a reliability was 0.85. There were seven responses for it with a range from *strongly disagree* to *strongly agree*. Data analysis used simple linear regression.

3 RESULT

The participants of this research were 119 employees. This study showed that the GSE was strongly able to predict flourishing (B = 0.601; $p<0.01$) with 36.2% effectiveness. Then, Table 1 showed the detail.

Table 1. Regression.

Variable	B	R^2	F	p
GSE	0.601	0.362	66.313	<0.001

Dependent: Flourishing.

The additional result showed in Table 2 that flourishing had a correlation with age ($r = 0.210$; $p<0.05$) and GSE ($r = 0.595$; $p<0.01$); however, ethnicity ($r = -0.173$; $p>0.05$) and gender ($r = -0,126$; $p>0,05$) had no correlation with flourishing.

Table 2. Correlation matrix.

	Variable	1	2	3	4
1	Gender	1.000			
2	Ethnic	0.097			
3	Age	−0.103	0.002		
4	GSE	−0.180	0.120	0.071	
5	Flourishing	−0.126	−0.173	0.210*	0.595**

*p<0.05; **p<0.01.

4 DISCUSSION

The findings in this study are shown in Table 1 that there is a strong relationship between GSE and flourishing with 36.2% effectiveness. A significant positive relationship between GSE, psychological, and functional aspects of well-being was found in a study of 279 students (Kvarme *et al.* 2009). Another similar study conducted on 4,493 adult men showed that low self-efficacy was a good predictor of low mental health and high self-efficacy was a good predictor of high mental health (Bavojdan *et al.* 2011). In addition, a study conducted on 300 employees in a manufacturing organization in Turkey found that self-efficacy has a more positive effect on job satisfaction, task performance, and citizenship behavior when trust in the organization is high (Ozyilmaz *et al.* 2018).

Self-efficacy in general can be grown and learned. This consists of the dimensions of level, generality, and strength. The level dimension refers to an individual's belief in their ability to complete tasks at different levels of difficulty. Each individual has a different perception of the level of difficulty of a task. The generality dimension relates to the wide range of areas of behavior in which individuals feel confident in their abilities. Individuals can rate themselves as efficacy in a series of activities or only in certain areas. The strength dimension relates to the strength of an individual's belief in his abilities. A high strength value describes his determination in the face of various obstacles to complete the given task (Bandura 1997; Ivtzan *et al.* 2018; Monteiro *et al.* 2020).

Results showed that flourishing had a correlation with age and GSE. In line with others that mentioned the effect of age on flourishing (Cruz *et al.* 2019; Keyes & Westerhof 2012; Momtaz *et al.* 2016; Schotanus-Dijkstra *et al.* 2016). This study is also in line with a meta-analysis of 40 studies consisting of 33 population cohorts that identified older adults who received health care services and were found to be at greater risk of having lower GSE than those who did not. In addition, older adults who received acute hospitalization were more likely to have lower GSE than those who received outpatient or periodic controls. This impacts on poorer health-promoting behaviors (Whitehall *et al.* 2021). This means that GSE in adults does not only have an impact on their mental health. However, it also affects their physical health.

Meanwhile, previous research mentioned the effect of gender on flourishing (De La Fuente *et al.* 2020; Ertac & Tanova 2020; Seyranian *et al.* 2018), and effect of ethnicity on flourishing (Jankowiak 2013; Keyes 2007; Said 2020). This is different from the results shown in Table 2 that there is no correlation between flourishing with gender and ethnicity.

5 CONCLUSION AND IMPLICATION

GSE can predict flourishing. Then, age and GSE have a relationship with flourishing, but ethnicity and gender have no relationship with flourishing. It means that GSE could be determined individual flourish. On the other side, flourishing can relate to individual age. Therefore, people need to increase their GSE for being flourish. They can do that with try hard for the best result, straight for the goals, having a coping ability, thinking of a solution, and handling anything in a good way. They will also flourish with maturity. Future research can be exploring others' personalities and/or external situations to increase flourishing.

REFERENCES

Bandura, A. (1997). *Efficacy: The exercise of control.* Freeman.

Bavojdan, M. R., Towhidi, A., & Rahmati, A. (2011). The Relationship Between Mental Health and General Self-efficacy Beliefs, Coping Strategies and Locus of Control in Male Drug Abusers. *Addiction & Health, 3* (3–4), 111.

Cruz, C., Pocinho, R., & Carrana, P. (2019). Flourishing in Old Age. *International Congress of Health and Well-Being Intervention.*

De La Fuente, R., Parra, Á., Sánchez-Queija, I., & Lizaso, I. (2020). Flourishing During Emerging Adulthood From a Gender Perspective. *Journal of Happiness Studies, 21*(8), 2889–2908. https://doi.org/https://doi.org/10.1007/s10902-019-00204-9

Diener, E., Wirtz, D., Tov, W., Kim-Prieto, C., Choi, D., Oishi, S., & Biswas-Diener, R. (2010). New Well-being Measures: Short Scales to Assess Flourishing and Positive and Negative Feelings. *Social Indicators Research, 97*(2), 143–156. https://doi.org/10.1007/s11205-009-9493-y

Effendy, N., & Subandriyo, H. (2017). Tingkat Flourishing Individu Dalam Organisasi PT X dan PT Y. *EXPERIENTIA: Jurnal Psikologi Indonesia, 5*(1), 1–17. https://doi.org/https://doi.org/10.33508/exp.v5i1.1548

Ertac, M., & Tanova, C. (2020). Flourishing Women Through Sustainable Tourism Entrepreneurship. *Sustainability, 12*(14), 5643. https://doi.org/https://doi.org/10.3390/su12145643

Ivtzan, I., Young, T., Lee, H. C., Lomas, T., Daukantait⊠, D., & Kjell, O. N. E. (2018). Mindfulness Based Flourishing Program: A Cross-Cultural Study of Hong Kong Chinese and British Participants. *Journal of Happiness Studies, 19*(8), 2205–2223. https://doi.org/10.1007/s10902-017-9919-1

Jankowiak, W. (2013). Urban Mongols: The Flourishing of Multiple Identities in a Vibrant Ethnic Community. *Chinese Sociological Review, 45*(3), 53–73. https://doi.org/https://doi.org/10.2753/CSA2162-0555450303

Keyes, C. L. M. (2007). Promoting and Protecting Mental Health as Flourishing: a Complementary Strategy for Improving National Mental Health. *American Psychologist, 62*(2), 95. https://doi.org/https://doi.org/10.1037/0003-066X.62.2.95

Keyes, C. L. M., & Westerhof, G. J. (2012). Chronological and Subjective Age Differences in Flourishing Mental Health and Major Depressive Episode. *Aging & Mental Health, 16*(1), 67–74. https://doi.org/https://doi.org/10.1080/13607863.2011.596811

Kvarme, L. G., Haraldstad, K., Helseth, S., Sørum, R., & Natvig, G. K. (2009). Associations Between General Self-efficacy and Health-related Quality of Life Among 12–13-year-old school children: a cross-sectional survey. *Health and Quality of Life Outcomes, 7*(1), 1–8. https://doi.org/https://doi.org/10.1186/1477-7525-7-85

Luszczynska, A., Gutiérrez-Doña, B., & Schwarzer, R. (2005). General Self-efficacy in Various Domains of Human Functioning: Evidence From Five Countries. *International Journal of Psychology*, *40*(2), 80–89. https://doi.org/https://doi.org/10.1080/00207590444000041
Maddux, J. E. (2000). *Self-efficacy: The power of believing that you can.* Teoksessa
Masturah, A. N. (2018). Personality and Positive Emotion of Student. *Proceedings of International Conference Asian Association of Indigenous and Cultural Psychology (AAICP)*, 767–772.
Mogi, K. (2017). *The book of Ikigai Untuk Hidup Seimbang, Lebih Bahagia, Dan Panjang Umur (Terjemahan)*. Penerbit Noura.
Momtaz, Y. A., Hamid, T. A., Haron, S. A., & Bagat, M. F. (2016). Flourishing in Later Life. *Archives of Gerontology and Geriatrics*, *63*, 85–91. https://doi.org/https://doi.org/10.1016/j.archger.2015.11.001
Monteiro, F., Pereira, M., Canavarro, M. C., & Fonseca, A. (2020). Be a Mom's Efficacy in Enhancing Positive Mental Health Among Postpartum Women Presenting Low Risk for Postpartum Depression: Results From a Pilot Randomized Trial. *International Journal of Environmental Research and Public Health*, *17*(13), 1–19. https://doi.org/10.3390/ijerph17134679
Novrianto, R., Marettih, A. K. E., & Wahyudi, H. (2019). Validitas Konstruk Instrumen. *Jurnal Psikologi*, *15* (1), 1–9. https://doi.org/http://dx.doi.org/10.24014/ jp.v14i2.6943
Ozyilmaz, A., Erdogan, B., & Karaeminogullari, A. (2018). Trust in Organization as a Moderator of the Relationship between Self-efficacy and Workplace Outcomes: A Social Cognitive Theory-based Examination. *Journal of Occupational and Organizational Psychology*, *91*(1), 181–204. https://doi.org/ https://doi.org/10.1111/joop.12189
Rabaglietti, E., Lattke, L. S., Tesauri, B., Settanni, M., & De Lorenzo, A. (2021). A Balancing Act During Covid-19: Teachers' Self-Efficacy, Perception of Stress in the Distance Learning Experience. In *Frontiers in Psychology* (Vol. 12). https://www.frontiersin.org/article/10.3389/fpsyg.2021.644108
Rosa, E. M., Fauzan, A., Makiyah, S. N. N., & Astuti, R. J. (2020). Sosialisasi Dan Assesment Tingkat Flourishing Dosen di Lingkungan Perguruan Tinggi Muhammadiyah (UMY dan UMP). *Prosiding Seminar Nasional PPM 2020*, 879–883. https://prosiding.umy.ac.id/semnasppm/index.php/psppm/article/ download/66/192/1359
Rothmann, S., Redelinghuys, K., & Botha, E. (2019). Workplace Flourishing: Measurement, Antecedents and Outcomes. *SA Journal of Industrial Psychology*, *45*(1), 1–11. https://doi.org/https://hdl.handle.net/10520/ EJC-13d251dc02
Said, A. A. (2020). *A Qualitative Study on Flourishing of Women From Different Ethnic Backgrounds in the Netherlands*. University of Twente.
Schotanus-Dijkstra, M., Pieterse, M. E., Drossaert, C. H. C., Westerhof, G. J., de Graaf, R., ten Have, M., Walburg, J. A., & Bohlmeijer, E. T. (2016). What Factors are Associated with Flourishing? Results from a Large Representative National Sample. *Journal of Happiness Studies*, *17*(4), 1351–1370. https://doi.org/ 10.1007/s10902-015-9647-3
Seligman, M. E. P. (2012). *Flourish: A Visionary New Understanding of Happiness and Well-being*. Simon and Schuster.
Seyranian, V., Madva, A., Duong, N., Abramzon, N., Tibbetts, Y., & Harackiewicz, J. M. (2018). The Longitudinal Effects of STEM Identity and Gender on Flourishing and Achievement in College Physics. *International Journal of STEM Education*, *5*(1), 1–14. https://doi.org/https://doi.org/10.1186/s40594-018-0137-0
Snyder, C.R. & Lopez, S.J. *Handbook of Positive Psychology*. New York, NY: Oxford University Press.
Whitehall, L., Rush, R., Górska, S., & Forsyth, K. (2021). The General Self-Efficacy of Older Adults Receiving Care: A Systematic Review and Meta-Analysis. *The Gerontologist*, *61*(6), e302–e317. https://doi. org/10.1093/geront/gnaa036

Families Mental Health and Challenges in the 21st Century – Arifin et al. (eds)

Training for peer counseling to overcome student adjustment problems at Gondanglegi Islamic Boarding School

Hanifa Timur Mawarizka* & Rr Siti Suminarti Fasikhah*
Department of Psychology, Muhammadiyah Malang University, Malang, Indonesia

ABSTRACT: Self-adjustment is a process of mental skills and behavior of a person in the face of demands, both from within himself and his environment. Peer counseling is a way for students (teenagers) to learn how to pay attention, help other teenagers, and apply it in everyday life. In the intervention method, the authors provide psychoeducation and *peer counseling* training to cadres. The results of this study indicate that there is an increase in the understanding of cadres regarding adjustment to adolescents, those who have carried out *peer counseling* have been able to master several counseling techniques to be applied when appointed as peer counselors. The conclusion is that the psychoeducational understanding related to the urgency of the problems that occur, namely an understanding of adolescent self-adjustment with N-Gain effectiveness of 100%, and counseling skills training obtained an effectiveness value of 73% which indicates are quite effective.

Keywords: Self-adjustment, psychoeducation, peer counseling

1 INTRODUCTION

Adolescence is a period of transition from children to adults, namely the developmental transition between childhood and adulthood which contains physical, cognitive, and psychosocial changes. Adolescence begins at the age of 12 years while late adolescence is at the age of twenties (Papalia & Feldman 2009). Adolescence is a critical period that can experience significant changes in themselves, both from within such as physical, cognitive, and psychological as well as from environmental factors such as pressure from the social environment (Azuji & Nwanna 2019). These transformations present opportunities for enhanced development; however, they also present challenges that can trigger vulnerabilities (Gerard & Booth 2015). They also must have five important psychosocial developmental challenges such as identity, sexuality, intimacy, autonomy, and achievement (Deighton-Smith & Bell 2017). Not only that, in Piaget's theory, it is stated that adolescents enter the formal operational stage. Adolescent understanding is no longer limited to actual or concrete experiences. They are able to manipulate events as if they actually happened to various situations that are still in the form of abstract proportion possibilities and try to reason logically. Adolescents tend to solve problems through trial and error, they begin to think and systematically try to find solutions (Santrock 2012).

Adolescent-reported life event measures capture salient stressors from their perspective. They typically include recent, negative appraised changes in the home, school, and community (e.g., parental separation, breakup with a boyfriend or girlfriend, exposure to violence, and crime that have occurred within the past six months to one year) (Flamm & Grolnick 2013). Adolescents are also inseparable from adjustment problems. According to

*Corresponding Authors: hanifa22596@gmail.com and suminarti@umm.ac.id

DOI: 10.1201/9781003402381-11

Schneiders (1960), self-adjustment is a process of one's mental skills and behavior in dealing with demands, both from within oneself and the environment. Schneiders also argues that there are characteristics of normal adjustment, namely: (1) the absence of excessive emotions, (2) the absence of psychological mechanisms, (3) the absence of personal frustration, (4) rational consideration and self-direction, (5) the ability to learn, (6) the use of past experiences, and (7) a realistic and objective attitude.

The ability to adjust has a positive effect on a person's various activities both inside and outside school. Students whose adjustment is normal can be identified by not finding excessive emotions. Individuals who respond to problems with calm and emotional control allow individuals to solve difficulties. Normal adjustment can be identified by learning continuously in solving problems that are full of conflict, frustration, or stress, students also use their experiences to be able to adjust. Students who have difficulty adjusting are prone to stress. Someone unable to cope with the demands of the environment will experience a depressed condition, both physically and psychologically. Stress is a situation that is triggered by internal and external demands that are dangerous for a person and are felt unable to deal with them (Nuryani 2020).

After conducting an assessment of the ustadzah and several students in Islamic boarding schools who were the sample of the author, it can be concluded that almost some of the students at the Islamic boarding school had problems with adjustment. This is because due to the non-attachment of the relationship with the guardian of the room, the incompatibility between friends, the strict rules in the cottage, and the daily schedule between formal school and cottage subjects which is quite dense, causing them to feel uncomfortable, wanting to get out of the cottage or running away from the cottage on purpose. Based on the explanation of the adjustment problem, peer counseling is needed.

Peer counseling can be defined as "people from similar groupings who are not professionals who help to clarify life problems and identify solutions by listening; clarifying; feeding back; summarizing; questioning and being positive, supportive and reassuring and then helping plan, organize and problem-solve" (Topping 2022). According to Carr (1981), peer counseling is a way for students (teenagers) to learn how to pay attention, help other teenagers, and apply it in everyday life. Tindall and Gray (1985) said peer counseling includes individual helping relationships (one-to-one helping relationships), group leadership, discussion leadership, giving consideration, tutorials, and all human interpersonal activities to help or help. In peer counseling, peer counselors are peers because of their abilities and strengths, they get debriefing to jointly help and assist in the process of solving problems faced by friends at school (Astiti 2019). Peer counselors are not intended to replace the role and function of professional counselors, but become a bridge between adolescents and professional psychological help so that in the future it is hoped that peer counselors can advise friends who need help to consult directly with professional counselors (Suwarjo 2008).

Therefore, with the explanation of the problem, the authors are interested in raising the title of "peer counseling for Ikatan Pelajar Muhammadiyah (IPM) cadres to overcome student adjustment problems in Gondanglegi Islamic boarding schools". The purpose of this paper is to provide education, and understanding to peers regarding self-adjustment, such as definitions, influencing factors, and the process of adjustment and to form peer counseling skills and abilities for students with the aim that these cadres can help or assist other students who have a psychological problem.

2 METHODS

The design of this research is one group pre-test and post-test. The pre-test is used when the material will be delivered with the aim of knowing the extent to which the material provided has been mastered by the individual, while the post-test is carried out at the end of the session with the aim of knowing the individual's understanding of the material given previously

(Magdalena *et al.* 2021). The pre-test and post-test psychoeducational adjustment and peer counseling consist of ten essay questions made based on theory which include an understanding of definitions, adolescent problems, adjustment, and factors that influence and impact on adolescent adjustment, while in the peer counseling questionnaire, the question items consist of understanding cadres related to peer counseling, who is involved in peer counseling, the principles of peer counseling, the goals of peer counseling and what techniques are carried out in peer counseling.

In the pre-test and post-test of peer counseling training, the assessment sheet consists of peer counseling techniques, including client acceptance, asking and initiating conversation skills, paraphrasing skills, empathy, concluding skills, and problem-solving skills (Permatasari 2021). The assessment sheet is given to cadres who act as counselees to assess cadres who act as counselors in the form of a checklist of five assessment options, namely very low, low, medium, high, and very high.

The assessment method used is interviews with ustadzah, room guardians, IPM member, and students. Interventions are psychoeducation related to adolescent self-adjustment and peer counseling training for Ikatan Pelajar Muhammadiyah (IPM) cadres. The number of individuals at the time of the assessment was eight people. Meanwhile, when the intervention was carried out on different days, the number of IPM cadres who were present on the first day to attend the psychoeducation session was twelve people and on the following day, only nine people attended the peer counseling training session. This peer counseling training is carried out in a roleplay manner with fellow cadres. The procedure is that the cadres are asked to pair up to act as counselors and as counselees. This is done alternately. They can express the problems that are happening to them.

3 RESULTS

The intervention carried out using the psychoeducational method of adjustment to members of the Ikatan Pelajar Muhammadiyah (IPM) at the Gondanglegi Islamic boarding school on the first day showed a change in the increase in scores related to the member's understanding of the adjustment material. The results of the intervention showed that there were changes before and after the intervention was given. These changes can be illustrated in Table 1 concerning N-Gain on SPSS.

Table 1. N-Gain score.

N-Gain Value	Category
$(g) > 0{,}7$	High
$0{,}3 < (g) \leq 0{,}7$	Medium
$(g) \leq 0{,}3$	Low

Table 2. Category of interpretation N-Gain.

Percentage (%)	Interpretation
< 40	Low
40–55	Less effective
56–75	Quite effective
>76	Effective

Source: Nashiroh *et al.* (2020)

Table 3. The output results of SPSS N-Gain (Psychoeducation adjustment and peer counseling).

	N	Minimum	Maximum	Mean	Std. Deviation
NGain_score	12	1.00	1.00	1.0000	.00000
Ngain_percentage	12	100.00	100.00	100.0000	.00000
Valid N (listwise)	12				

3.1 *Descriptive statistics*

Based on the data above, it can be seen that the N-Gain score is 1 (g>0,7). This means that the score is in the high category. While the percentage of psychoeducation effectiveness given is 100% (>76), where the psychoeducation provided is very effective for IPM cadres. This indicates that the cadres are good at understanding the variables of adjustment and peer ounselling after being given psychoeducation.

While the skills of peer ounselling cadres can be demonstrated through the output of N-Gain shown in Table 4.

Table 4. The output results of SPSS N-Gain peer counseling skills.

Descriptive Statistics	N	Minimum	Maximum	Mean	Std. Deviation
NGain_score	9	.25	1.00	.7390	.28997
NGain_percentage	9	25.00	100.00	73.8977	28.99731
Valid N (listwise)	9				

Results of peer counseling skills show that the skills of the cadres are 0.7 ($g \leq 0,7$), which means that the skills taught and then practiced by the cadres are in the medium category, and the effectiveness of the interventions shown at 73% (56–75) which indicates that the peer counseling skills applied by the cadres are quite effective and able to master several counseling techniques to be applied when appointed as peer counselors.

4 DISCUSSION

Providing psychoeducation to cadres as agents of change are defined as individuals or groups of people who act to influence the target group to be able to make changes (Qaddura 2021). Another definition of psychoeducation is an educational method that provides useful information and training to change an individual's mental or psychic understanding (Bhattacharjee *et al.* 2011). Psychoeducation methods can give an attractive impression to individuals and are more effective and fun (Zaini 2014). Psychoeducation can be done through training with methods of exploration, assessment, discussion, role-playing, and demonstration (Putra & Soetikno 2018). Psychoeducation can be carried out in various group or household settings. The process of providing this education also requires the involvement of participants as the key to the success of the intervention. Counselors as implementers can build a trusting relationship to carry out appropriate assessments and provide understanding to participants or cadres on how psychoeducation can provide benefits to them as agents of change (Amelia 2017).

In addition, the effectiveness of providing psychoeducation to cadres has also been investigated by Qaddura (2021) which proves that psychoeducation to agents of change implemented for religious leaders and community leaders can provide positive views. Setiani & Haryanto (2019) through their research on students revealed that there are differences in social adaptation abilities in new students before and after attending the psychoeducation session. Respondents felt that there was a better level of self-awareness than before so the increase was quite significant. Research related to psychoeducation was also conducted by Syifa & Oktavia (2021) on ten high school students and concluded that there was an effect of psychoeducation on reducing academic stress in students who received psychoeducational interventions as well as improving the quality of learning which was higher than the group that did not receive psychoeducation.

Then related peer counseling training that is implemented in cadres can be defined as groups consisting of children who have the same age, class, and social motivation. To be able to help good adjustment, peer counseling can be said to be quite effective if it is given by their peers because teenagers tend to personal fables, namely the belief that only they have unique experiences and are not adults. Therefore, strengthening through peer counseling is considered effective (Astiti 2019).

The effectiveness of peer counseling has been carried out by Ibandiyah & Hasanah (2021) with an experimental method for 9 students. Through this study, it was found that the average score of experimental students who received peer counseling experienced a significant increase compared to the control group. This means that peer counseling services are proven to be influential and effective in improving students' self-adjustment. The effectiveness of peer counseling training is also found in Syafitri & Rahmah's research showing that online peer counseling training is able to improve mental health literacy in participants peer counseling training through a significant increase in mental health literacy scores on the post-test, observation results of basic counseling skills and evaluation questionnaires after training (Syafitri & Rahmah 2021)

Another study conducted by Rusnawati Ellis (2020) on 40 students also showed that the peer counseling model was effective in reducing academic procrastination behavior in students. Parveen (2015) in her research on peer attachment also found that there was a significant correlation between girl's attachment with peers and adjustment behavior was found. Coefficient r was found to be .334 and thus it can be inferred from the coefficient r that with an increase in attachment with peers among girls there will be an increase in adjustment behavior too.

The results on psychoeducation obtained have a high level of effectiveness in terms of increasing the understanding of cadres related to self-adjustment and *peer counseling,* while in terms of peer counseling training, the results are quite effective and the abilities controlled by cadres have also begun to form. This is also supported by external factors from the cottage which strongly support the implementation of the event and the cooperative attitude of the cadres to follow each session to completion.

5 CONCLUSION

Based on the results of the interventions that have been carried out, these community problems can be overcome by providing interventions to IPM cadres who are agents of change at the Islamic boarding school through psychoeducational understanding related to the urgency of the problems that occur, namely an understanding of adolescent self-adjustment and peer counseling with N-Gain effectiveness of 100% (>76). Counseling skills training peers to assist, provide psychological support, and jointly overcome the problems experienced by peers obtained an effectiveness value of 73% (56–75), which are quite effective. It can be said that the cadres who did the training were good enough at providing peer counseling services.

Recommendations or suggestions to cadres can disseminate material to other cadres who do not participate and retrain each other's skills that have been done regularly to hone skills when they become peer counselors.

REFERENCES

Alpusari, M. (2013). Analisis Keterampilan Sains Siswa Dalam Pembelajaran IPA SD. *Jurnal Pendidikan Guru Sekolah Dasar*, *2*(1), 40–46. https://doi.org/http://dx.doi.org/10.33578/jpfkip.v2i1.2818

Amelia, D. R. (2017). Psikoedukasi Untuk Meningkatkan Pemahaman Tentang Pengasuhan Anak Pada Komunitas Ibu Anggota Posyandu. *Procedia – Social and Behavioral Sciences*, *5*(1), 12–17.

Astiti, S. P. (2019). Efektivitas Konseling Sebaya (Peer Counseling) dalam Menuntaskan Masalah Siswa. *IJIP: Indonesian Journal of Islamic Psychology*, *1*(2), 243–263. https://doi.org/10.18326/ijip.v1i2.243-263

Azuji, I. M., & Nwanna, U. C. (2019). *Elements of Adjustment Process in Adolescents: Suggestions for Counselling*. *III*(Xi), 456–462.

Bhattacharjee, D., Rai, A. K., Singh, N. K., Kumar, P., Munda, S. K., & Das, B. (2011). Psychoeducation: A Measure to Strengthen Psychiatric Treatment. *Delhi Psychiatry Journal*, *14*(1), 33–39.

Carr, R. (1981). *Theory and Practice of Peer Counseling*. Canada Employment and Immigration Commision.

Deighton-Smith, & Bell. (2017). Understanding Adolesents. *European Business Review*, *26*(3), 206–217. https://doi.org/10.1007/978-3-319-33450-9

Flamm, E. S., & Grolnick, W. S. (2013). Adolescent Adjustment in the Context of Life Change: The Supportive Role of Parental Structure Provision. *Journal of Adolescence*, *36*(5), 899–912. https://doi.org/10.1016/j.adolescence.2013.07.006

Gerard, J. M., & Booth, M. Z. (2015). Family and School Influences on Adolescents' Adjustment: The Moderating Role of Youth Hopefulness and Aspirations for the Future. *Journal of Adolescence*, *44*, 1–16. https://doi.org/10.1016/j.adolescence.2015.06.003

Ibandiyah, I. Z., & Hasanah, M. (2021). *Efektivitas Layanan Konseling Sebaya (Peer Counseling) Untuk Meningkatkan Penyesuaian Diri Santri*. *01*(02), 89–99.

Magdalena, I., Nurul Annisa, M., Ragin, G., & Ishaq, A. R. (2021). Analisis Penggunaan Teknik Pre-Test Dan Post-Test Pada Mata Pelajaran Matematika Dalam Keberhasilan Evaluasi Pembelajaran Di Sdn Bojong 04. *Jurnal Pendidikan Dan Ilmu Sosial*, *3*(2), 150–165. https://ejournal.stitpn.ac.id/index.php/nusantara

Nashiroh, P. K., Ekarini, F., & Ristanto, R. D. (2020). Efektivitas Penerapan Model Pembelajaran Kooperatif Tipe Jigsaw Berbatuan Mind Map terhadap Kemampuan Pedagogik Mahasiswa Mata Kuliah Pengembangan Program Diklat. *Jurnal Pendidikan Teknologi Dan Kejuruan*, *17*(1), 43. https://doi.org/10.23887/jptk-undiksha.v17i1.22906

Nuryani, (2020). Dampak Kesulitan Menyesuaikan Diri Pada Santri. *G-Couns: Jurnal Bimbingan Dan Konseling*, *4*(1), 174–179. https://doi.org/10.31316/g.couns.v4i1.469

Papalia, D.E. & dan Feldman, R. (2009). *Human Development, Edisi ke-10*. Salemba Humanika.

Parveen, A. & U. (2015). *Adolescents' Adjustment Behaviour and Self-image: The Role of Parent and Peer Attachment*.

Permatasari, N. *et al.* (2021). Bagaimana Remaja Menjadi Peer-Counselor di Masa Pandemi? *Buletin Konseling Inovatif*, *1*, 32–44.

Putra, A. S., & Soetikno, N. (2018). Pengaruh Intervensi Psikoedukasi Untuk Meningkatkan Achievement Goal Pada Kelompok Siswi Underachiever. *Jurnal Muara Ilmu Sosial, Humaniora, Dan Seni*, *2*(1), 254. https://doi.org/10.24912/jmishumsen.v2i1.1514

Qaddura, Z. H. (2021). Kekerasan Verbal pada Anak: Psikoedukasi pada Kader Posyandu Balita Puskesmas D Surabaya. *JIP (Jurnal Intervensi Psikologi)*, *13*(1), 21–30. https://journal.uii.ac.id/intervensipsikologi/article/download/15092/11284

Rusnawati Ellis, N. H. dan R. F. T. (2020). Efektivitas Model Konseling Teman Sebaya Untuk Mengurangi Perilaku Prokrastinasi Akademik Pada Mahasiswa. *Jurnal Bimbingan Dan Konseling Terapan*, *4*(1), 92–102.

Santrock, J. W. (2012). *Perkembangan Masa hidup Edisi Ketigabelas, Jilid 1*. Erlangga.

Schneiders, A. (1960). *Personal Adjustment and Mental Health*. Rinehart and Winston.

Setiani, T. P., & Haryanto, H. C. (2019). Efektivitas Psikoedukasi Terhadap Kemampuan Adaptasi Sosial Pada Mahasiswa Baru. *Psikoislamika: Jurnal Psikologi Dan Psikologi Islam*, *16*(1), 1. https://doi.org/10.18860/psi.v16i1.7531

Suwarjo. (2008). Pedoman Konseling Teman Sebaya untuk Pengembangan Resiliensi. *Yogyakarta*.

Syafitri, D. U., & Rahmah, L. (2021). Pelatihan Konselor Sebaya Daring Untuk Meningkatkan Literasi Kesehatan Mental Siswa di SMA Islam XY Semarang. *Gadjah Mada Journal of Professional Psychology (GamaJPP)*, *7*(1), 39. https://doi.org/10.22146/gamajpp.62299

Syifa, M., & Oktavia, C. (2021). Efektivitas Psikoedukasi Strategi Belajar dan Manajemen Stress pada Siswa SMA Negeri 1 Cerme, Gresik. *Temu Ilmiah Nasional (TEMILNAS XII), 2020* (Temilnas Xii), 203–207. https://ojs.unm.ac.id/Temilnas/article/view/20043

Tindall, J.D. and Gray, H. (1985). *Peer Counseling: In-Depth Look at Training Peer Helpers.* Accelerated Development Inc.

Topping, K. J. (2022). Peer Education and Peer Counselling for Health and Well-Being: A Review of Reviews. *International Journal of Environmental Research and Public Health, 19*(10), 6064. https://doi.org/10.3390/ijerph19106064

Zaini, M. (2014). Manajemen Perawatan Dengan Menggunakan Psikoedukasi Berbasis Komputer Terhadap Klien Depresi Ringan. *E-Journal WIDYA Kesehatan Dan Lingkungan, 1*(1), 68–74.

Families Mental Health and Challenges in the 21st Century – Arifin et al. (eds)
© 2024 The Author(s), ISBN 978-1-032-51442-0

The self-acceptance of the teenage who is living *married by accident*

Meriska Afrianti* & Andhita Dyorita Khoiryasdien*
Psychology S1 Study Program, University of 'Aisyiyah Yogyakarta, Jl. Siliwangi, Indonesia

ABSTRACT: Adolescence is the period between childhood and adulthood. Adolescents go through developmental tasks that must be completed successfully. Adolescents will become individuals who have lost their way if developmental tasks fail or are not completed correctly. This causes deviant behavior to emerge. Self-acceptance is the process of accepting or actualizing individuals with healthy personalities by adjusting and making themselves happy, which occurs between the ages of 11 and 22 years when the individual experiences change in mental, emotional, social, and physical maturity This research aims to determine the process of self-acceptance in adolescents who got married by chance. This study includes 2 (two) adolescent participants who were accidentally married. This was a qualitative data collection study that employed observation, interview, scale, and documentation techniques. The Berger Self-Acceptance Scale was used. This scale was a Likert scale with 36 statements about self-acceptance. This demonstrates that adolescents who are Married by Accident have self-acceptance based on consensual relationships that result in adultery. Subjects E and I both have fairly good self-acceptance, as evidenced by the highest aspect of the two subjects, namely confidence in their abilities; subject E receives a score of 2, which is in the range of scores X> 21, and subject I receives a score of 22, which is also in the range of scores X> 21.

Keywords: Adolescent, Self-Acceptance, Married by Accident

1 INTRODUCTION

Adolescence is a period of transition from childhood to adulthood. In the process of development, adolescents have developmental tasks that need to be passed on well. When developmental tasks fail or are not passed properly, adolescents will become individuals who lose their way. This triggers the emergence of deviant behavior. Juvenile delinquency cases are increasingly widespread and attract attention among the community and one of them is worrying about the freedom of sex carried out by teenagers today, often causing negative impacts and disturbing the community. Since the first, the Indonesian people have been recognized in the eyes of the world because of their morals, culture, civilization, and noble character and the majority of the people adhere to the Muslim religion, and free sex or what is known as pregnancy out of wedlock is considered taboo in the eyes of the community. Pregnancy out of wedlock is the aftermath of adultery which should be punished in Islam. As a result of adultery which results in pregnancy out of wedlock and there will be a disgrace in the family of the adulterer. For this reason, in order to reduce the sense of disgrace in the life of the Muslim community, the family should marry off the

*Corresponding Authors: meriskaafrianti@gmail.com and dyorita.kh@unisayogya.ac.id

DOI: 10.1201/9781003402381-12

child in order to protect the family from greater disgrace. It's a different thing when someone gets married but goes through illegal acts such as committing adultery. As a result of adultery which results in pregnancy out of wedlock and there will be a disgrace in the family of the adulterer. For this reason, in order to reduce the sense of disgrace in the life of the Muslim community, the family should marry off the child in order to protect the family from greater disgrace. It's a different thing when someone gets married but goes through illegal acts such as committing adultery As a result of adultery which results in pregnancy out of wedlock and there will be a disgrace in the family of the adulterer. For this reason, in order to reduce the sense of disgrace in the life of the Muslim community, the family should marry off the child in order to protect the family from greater disgrace. It's a different thing when someone gets married but goes through illegal acts such as committing adultery.

Pregnant out of wedlock or so-called *Married by Incident* happens among teenagers can be based on premarital sex (zina) and sexual intercourse by coercion. The factors that influence the cause of teenage pregnancy out of wedlock, according to (Rohmawati 2014), include individual factors, namely mental and spiritual weakness by 40% and environmental factors by 60% such as the role of parents and the development of science and technology which have a negative impact (Fuaji 2020). Environmental factors are also caused by the promiscuity of today's teenagers, the lack of supervision by parents, and the lack of knowledge of teenagers about the dangers of promiscuity. In addition, the lack of religious education is one of the causes of pregnancy outside marriage Married by accident (MBA). Nailaufar & Kristiana (2018) put forward adolescents who marry at a young age without thinking carefully, and only cared about loving each other and living together. Teenagers who have premarital sex usually begin with a sense of falling in love, then they immerse themselves in a bond they call "dating". Courtship behavior here tends to be interpreted as bad deeds. The attitude and model of today's teenage dating are already quite extreme, at least some of them have kissed or kissed. The behavior of sexual relations outside of marriage from year to year is increasing. Every year more than 600,000 million teenagers get pregnant worldwide, and 3 in 10 teens get pregnant before reaching the age of 20 (Malik et al. 2016).

As many as 42.5 percent of adolescents abroad aged 15–19 years have sexual relations outside of marriage as a result, many young women experience early pregnancy or unwanted pregnancies outside of marriage. Sexual behavior outside of marriage also done by E, E admitted that he had been dating since elementary school. After E sat on the bench junior high school 3rd grade, E began to dare to have sex with his girlfriend, and E admitted to doing it at his friend's boarding house. In addition, E said that the courtship period is a time when a person can love and be able to be loved by his lover, a form of expression of love or affection. Dear can be expressed in various ways, for example, giving presents, flowers, hugging, kissing, and even having sex. This is also as said by E, who has been in a dating relationship since elementary school and has often had sex with his girlfriend.

Teenagers who marry young because of MBA do not yet have the readiness to become parents and husband and wife. As a result of pregnancy out of wedlock (MBA) in addition to not being able to continue schooling and also having to become a mother at a very young age who does not have the provisions about parenting (Marufah & Sadewo 2019). As a result, most teenagers are still not fully able to deal with change. Teenagers want changes that can feel freedom but are afraid to accept the consequences to overcome the consequences that have been done. In this condition, self-acceptance is needed in adolescents to reduce the risk for adolescents and their prospective babies. As we know self-acceptance for teenagers in these times is very important (Fuaji 2020). Self-acceptance is an important factor in happiness, both toward oneself and being accepted in social life (Pradanistika 2021). Lack of self-acceptance is characterized by helplessness, incompetence, depression, self-blame, lack of motivation, low self-esteem, low positive behavior, and using difficulties for rehabilitation and adjustment (Fuaji 2020). Self-acceptance is a risk factor for depression and when individuals have low self-acceptance, they may experience depression. On the other hand, if the individual has high self-acceptance, he is not prone to depression (Princess 2012). Adolescents who have low self-acceptance will easily despair, always blame themselves, be ashamed, humble themselves, feel insignificant, feel jealous of other people's circumstances, will find it difficult to build positive relationships with others, and are unhappy (Pradanistika 2021). Based on the explanation above, the researcher wants to find out more about the self-acceptance of a teenager who is MBA.

2 METHODS

This research is a qualitative descriptive study with a case study method that produces descriptive data, in the form of written or spoken words from people and observable behavior. This qualitative research is natural. Researchers do not try to manipulate the circumstances or conditions of the research environment but conduct research on a situation in a situation where the situation does exist. This research deliberately sees and allows the conditions under study to be in their actual state. This research method is a case study. This case study is an in-depth study of individuals and is relatively long-term, continuous and uses a single object, meaning that the case is experienced by one person. In this case study, the researcher collects data about the subject himself from his past, present and surrounding environment. The researcher conducts a case study with a theoretical basis as a reference when the researcher will explore something related to the subject. The individual who is the subject of the research is a 16-year-old teenager with the initials E. His last formal education was high school. He was married on October 18, 2018, and has been in a marriage for three years and works as a household mother.

3 RESULTS

Based on the auto anamnesis interview, the following findings were obtained: subject E during the interview, the subject's voice sounded low. The subject's facial expression looked cheerful and occasionally smiled. The subject does not seem to hesitate to tell his experience from the beginning to the present. The subject sat in a fairly comfortable position on the sofa chair while facing the researcher. The first subject stopped for a moment while doing the interview because his son called the subject. Subject E actively communicated with the researcher and was friendly and cheerful, one of the activities carried out by chatting casually while laughing and asking how each family was. Subject E, when he started the interview, showed a cheerful character to the people around him, even the subject laughed and smiled a few times when telling his life story.

Subject E shows that subject E has an equal feeling dimension, this is evident from the results of the interview submitted by subject E, namely, the subject sees himself as a strong woman. With this incident, the subject feels grown up and Alhamdulillah, he has survived until now. Significant others also said that subject E often gathers and socializes with the people in his complex. Subject E has a dimension of confidence in self-ability and responsibility, which can be seen from the results of the interview with subject E, namely, believe in yourself because you can get through all the trials so far you feel enough and have to stay strong through it because Alhamdulillah, he feels that God is good and also feels responsible because until this moment subject E can go through and survive subject E also feels he has a responsibility who grew up taking care of her husband and children. Subject E has a stand that is evident from the results of the interview submitted by subject E: the principle of life is brave to act, and dares to be responsible. The subject says that everything that is done there must be a consequence. Subject E accepts humanity by being grateful by worshiping Allah by getting closer to asking forgiveness for the sins that have been committed and the subject doing something of his own volition.

Significant Others also said that the subject was able to accept himself in the circle of friends and society because the subject had often gathered and socialized with the community in the complex because the subject did not feel sorry but significant others said that subject E's parents required them to work and live independently because they were already married. ladder.

Based on the self-acceptance scale, the subject is in the medium category with a total score of 90 which is in the score range of $72 < X$ 108. The highest aspect, namely self-confidence, gets a score of 2 which is in the range of $X > 21$. So it can be concluded that subject E has high self-confidence. Furthermore, based on the self-acceptance scale, it shows that subject E in the aspect of feeling equal is in the moderate category with a total score of 14 in a score range of $10 < X$ 15. While the responsible aspect has a total score of 1 in the medium category, namely in the range of scores of $2 < X$ 3. Then the aspect of opinion has a total score of 8 in the medium category, namely in the range of scores $8 < 6 < X$ 9. Although subject E has high self-confidence, the subject also has feelings of equality, responsibility, opinion, and acceptance of good humanity with himself and his environment

4 DISCUSSION

Based on the results of interviews conducted by researchers, it was found that subject E had equal feelings, believed in his abilities, was responsible, held an opinion, and accepted humanity. Subject E has an equal feeling as seen from the results of the subject's interview which said that the subject saw himself as a strong woman. With this incident, the subject also felt that he was grown up and Alhamdulillah, the subject was able to survive until now. Based on the results of this study, the subject already had an equal feeling which was quite good in undergoing MBA.

The equal feeling is the individual considers himself valuable as a human being who is equal to others so that the individual does not feel like a special person or deviate from others. Individuals feel that they have weaknesses and strengths as well as other people. In other words, individuals who have self-esteem are able to see themselves as individuals who have the same rights as others.

Confidence in self-ability: The researcher found that subject E had confidence in their ability to be able to face life. This can be seen from the results of the interview with subject E which found that subject E believes in himself because he can get through all the trials so far he feels enough and has to stay strong to live it because Alhamdulillah he feels that God is good and also feels that he is responsible because until this moment the subject being able to go through and survive the subject also feels a great responsibility to take care of her children and husband. Based on the results of this study, the subject has very good self-confidence in self-acceptance as someone who underwent a MBA. Self-confidence is an individual who has the ability to face life. This can be seen from the attitude of the individual who is confident, prefers to develop his good attitude and eliminate the bad, rather than wanting to be someone else, the individual will be satisfied to be himself. In other words, individuals have confidence in their abilities to be able to face life.

Responsible: The researcher found that subject E was willing to take responsibility for their behavior. This can be seen from the results of the interview with subject E which found that subject E felt that he had a great responsibility to take care of his children and husband. Significant others also said that subject E was able to take care of his children and husband and was responsible for his marriage. Based on the results of this study, the subject has a fairly good sense of responsibility in accepting himself as someone who undergoes MBA. Responsible is the individual who dares to take responsibility for his behavior. This trait can be seen in the behavior of individuals who accept criticism and make it a valuable input for self-development.

The researcher's stance found that subject E had a life stance, which can be seen from the results of the interview with subject E. He stated that he had the principle of being brave to act, and daring to be responsible, emphasizing that everything he does has consequences, the subject always follows his heart and his husband always reminds him about their children and their struggles so far have not been easy. Based on the results of this study that the subject has a stand or principle of life. Standing: Individuals prefer to follow their own standards rather than follow social pressures. Individuals who have attitudes and confidence in their actions rather than following the standards of others and have ideas, aspirations, and self-expectations. In other words, individuals have principles,

Accepting human nature, the researcher found that subject E had gratitude or was able to accept every situation he had, was open about every feeling he had, told the truth about his feelings, and had positive thoughts. This can be seen from the results of the interview with subject E, which found that Subject E was grateful to Allah by asking forgiveness for their sins and acting on their own accord. Based on the results of this study, the subject was able to accept the nature of humanity in accepting himself as someone who underwent a Married by accident (MBA). Accepting the nature of humanity is the individual does not deny the emotions he has and does not feel guilty excessively.

Based on the aspect of self-acceptance, Sheerer (Rahayu 2019) explains about five, namely, aspects of feeling equal, believing in one's abilities, being responsible, taking a stand, and accepting humanity. Subject E can be said to have self-acceptance because it fulfills these five aspects. In addition, according to the results of the self-acceptance scale, subject E has a total score of 90 and has the highest score in the aspect of self-confidence, getting a score of 2 which is in the range of scores X> 21. So it can be concluded that subject E has confidence in self-efficacy. tall. This is also in accordance with the conclusion of the interview which shows that subject E believes in himself because he can get through all the trials so far he feels enough and must remain strong through it because Alhamdulillah he feels that Allah is good and also feels responsible because until now subject E being able to go through, live and survive subject E also feels that she has a big responsibility to take care of her children and husband. So it can be concluded that subject E has good self-acceptance. In general, adolescents who experience pregnancy and then give birth without getting married have a self-acceptance process consisting of five stages, namely, denial, anger, bargaining, depression, and acceptance. This process can be experienced differently by each individual in achieving self-acceptance. The subject can be said to have achieved self-acceptance because it has characteristics at that stage completely. Based on the results of the study, the stages that most often appear are denial, anger, depression, and acceptance.

This study has a weakness, namely the difficulty in finding subjects. This research must be done online and offline so there are still many shortcomings. Difficulty in choosing the time of the interview and the difficulty of meeting directly with the subject

5 CONCLUSION AND IMPLICATION

5.1 *Conclusion*

Based on the results of the analysis of the self-acceptance research of adolescents who underwent MBA in Morowali, especially in the village of Bente, it can be concluded that subjects who are pregnant out of wedlock or what are called MBA are based on consensual relationships that cause premarital sex (zina). Aspects of self-acceptance are aspects of feeling equal, believing in one's own abilities, being responsible, taking a stand, and accepting human nature. Based on the results of this study, the subject has a very good self-confidence in his self-acceptance as someone who underwent an MBA, Subject E already has fairly good self-acceptance, this can be seen from the highest aspect of the subject, namely believing in the subject's self-ability. Subject E gets a score of 2 which is in the range of X scores > 21. The subject's self-acceptance process is based on research results, the stages that most often appear are denial, anger, bargaining, depression, and acceptance. Based on the results of the discussion, it can be seen that the impact experienced by adolescents who experience an MBA has different forms, such as not being well received by the teenager in the community, getting social discrimination from various parties, and mental damage due to bullying, hampering self-confidence.

5.2 *Suggestion*

For teenagers, the community and especially parents of teenagers who are pregnant outside of marriage are expected to be able to specifically assess which ones will be the big impact in the event of a MBA, and if the child or teenager is already pregnant out of wedlock. The surrounding area is expected to assist adolescents in accepting their pregnancy by providing material and motivational support to their children, as well as providing guidance to children so as not to repeat the wrong actions and also for parents of teenagers who are expected to pay more attention to and supervise their children so that avoid free sex that can lead to pregnancy out of wedlock. Things like MBA really need attention.

REFERENCES

Adyartasari, VP (2018). Relationship Between Climateric Symptoms and Middle Adult Women's Self-Acceptance. Thesis.

Afifah, LA (2017). *Youth in Review From the Perspective of Islamic Education* (Case Study on Adolescent Dropouts in Jambu District, Semarang Regency). Thesis.

Amin, R. D. (2020). Menikahi Wanita Hamil Diluar Nikah Menurut Pandangan Dewan Hisbah Persis dan Majelis Tarjih Muhammadiyah (Doctoral dissertation, UIN Sunan Gunung Djati Bandung).

Apsari, A. R., & Purnamasari, S. E. (2017). Hubungan Antara Konformitas Dengan Perilaku Seksual Pranikah Pada Remaja. Insight: Jurnal Ilmiah Psikologi, *19*(1), 1–12.

Ardianti, F., Fakhrurrozi, M., & Marissa, A. (2017). Psychological Well-being Pada Remaja Akhir Yang Hamil Di Luar Nikah. Jurnal Psikologi, *9*(1).

Auliya, DQB (2018). Formation of the Sakinah Family by Married by Accident (MBA) Couples.

Azzahra. F. (2020). Marriage Due to Premarital Pregnancy in the View of Islamic Jurisprudence and the Compilation of Islamic Law (KHI). Thesis.

Fuaji, L. (2020). *Penerimaan Diri pada Remaja yang Hamil Diluar Nikah* (Doctoral dissertation, Universitas Muhammadiyah Surakarta).

Haidar, G., Apsari, NC, Studies, P., Welfare, I., & Fisip, S. (2020). *Phornography in Adolescents.* Undergraduate Study Program in Social Welfare, Padjadjaran University. 136–143.

Hurlock, EB (2014). Child Development. Sixth Edition volume *1*. Jakarta: Erlangga https://doi.org/10.1078/0944-2006-00091

Hurlock, EB (2019). Developmental Psychology: A Lifespan Approach. Fifth edition Jakarta: Erlangga.

Khoiryasdien, A. D., & Warastri, A. (2020). Pelatihan Berpikir Positif Untuk Meningkatkan Penerimaan Diri Survivor Bipolar Di Yogyakarta. *Insight:* Jurnal Pemikiran Dan Penelitian Psikologi, *16*(2), 317–335. https://doi.org/10.32528/ins.v

Malik, D., Astuti, AB, & Yulianti, NR (2016). (*Study of Phenomenology in Desa Baru*, District of Ibu, West Halmahera). 259–286.

Marufah, W. N., & Sadewo, F. X. S. (2019). Pengalaman Kekerasan Seksual Pelajar Putri di Jombang. Paradigma, 7(2).

Nailaufar, U., & Kristiana, I. F. (2018). Pengalaman Menjalani Kehidupan Berkeluarga Bagi Individu Yang Menikah Di Usia Remaja. Jurnal Empati, *6*(3), 233–244.

Noviani, LP (2017). Adolescent Self-Acceptance Ability Level.

Pradanistika, S. (2021). Self-Acceptance of Students Who Get Pregnant Out of Wedlock.

Putri, A. K. (2012). Hubungan Antara Penerimaan Diri Dengan Depresi Pada Wanita Perimenopause (Doctoral dissertation, Universitas Airlangga).

Rahayu, F. S. (2019). Konseling Rasional Emotif Perilaku untuk Meningkatkan Penerimaan Diri pada Remaja Hamil Diluar Pernikahan. Prophetic: Professional, Empathy, Islamic Counseling Journal, *2*(1), 143–152.

Rofiatin, I. (2016). The Relationship of Self-Acceptance with Depression Levels in the Elderly.

Rohmawati, EDA (2014). Factors that Occur in Pregnancy Before Marriage in the Area of the Religious Affairs Office, Tegalsari, Surabaya. 04.

Sasotya, JA (2021). Self-Acceptance Process in Adolescent Girls who are Pregnant and Give Birth Out of Wedlock. Thesis.

Triwiyarto, U. (2015). *Case Study on the Causes of Juvenile Delinquency*.

Wangge, B. D. R. & Hartini, N. (2014). *Hubungan Antara Penerimaan Diri Dengan Harga Diri Pada Remaja Pasca Perceraian Orangtua* (Doctoral dissertation, Universitas Airlangga). 2(1).

Families Mental Health and Challenges in the 21st Century – Arifin et al. (eds)
© 2024 The Author(s), ISBN 978-1-032-51442-0

The role of family functioning toward internalizing problems and externalizing problems of adolescents

Dian Caesaria Widyasari & Nandy Agustin Syakarofath*
Fakultas Psikologi, Universitas Muhammadiyah Malang, Malang, Jawa Timur, Indonesia

ABSTRACT: Internalizing and externalizing problems are common mental health problems experienced by adolescents. Internalizing problems are when individuals control pressure on themselves, causing anxiety and depression. In contrast, externalizing problems attempt to project pressure on the external environment resulting in problematic behaviors that harm themselves and others. Family functions are a vital factor assumed to determine adolescents' internalizing and externalizing problems, as a family is the closest environment affecting adolescents' development. This study aimed to identify the contribution of family functioning to the internalizing and externalizing problems in adolescents. The research instruments used are the Family Assessment Device (FAD) scale and the Strengths and Difficulties Questionnaire (SDQ). A total of 170 adolescents living in East Kutai Regency, Kalimantan, aged 16–18, participated in this study based on multistage random sampling. The multiple linear regression analysis showed that family function could predict adolescents' internalizing and externalizing problems. The research findings implied that optimal family functioning contributes to the lower possibility of adolescents experiencing internalizing and externalizing problems.

Keywords: adolescent, externalizing problems, family function, internalizing problems

1 INTRODUCTION

One of the psychological problems commonly experienced by adolescents is internalizing problems (Graber 2013) and externalizing problems (Georgiou & Symeou 2018). Internalizing and externalizing problems are constructs that include adolescents' emotional and behavioral problems.

Internalizing problem is a tendency where adolescents direct the problems face into themselves so that various complaints arise, such as symptoms of stress, anxiety, and depression (Kwan *et al.* 2021). Meanwhile, externalizing problems are tendencies that direct the problems faced by adolescents to the outside of themselves so that various behavioral problems arise, such as delinquency, aggressiveness (fighting, vandalism, stealing, lying), and other rule-breaking behaviors (Georgiou & Symeou 2018; Soto-Sanz *et al.* 2019). Internalizing problems are closely related to psychological distress while externalizing problems are more toward problems or conflicts related to others and the adolescent's social environment.

In Indonesia, no global data can describe a comprehensive case of internalizing problems and externalizing problems in the child and adolescent population. However, basic health research data related to internalizing problems in the general population shows that as many as 5.7% of individuals experience emotional disorders in the form of anxiety and depression

*Corresponding Author: nandysyakarofath@umm.ac.id

DOI: 10.1201/9781003402381-13

(Health Research and Development Agency of the Ministry of the Republic of Indonesia 2013). This figure continued to increase to around 9.8% in 2018 when people reported experiencing symptoms of anxiety and depression (Health Research and Development Agency of the Ministry of The Republic of Indonesia 2018). Furthermore, data on externalizing problems can be seen from the number of crimes reported by the Bureau of Operation Control of the Police Headquarters from 2011–2013, showing 1,030,848 cases that occurred throughout Indonesia (Central Agency for Criminal Statistics 2014). The forms of criminality are acts of violence and persecution, immoral crimes, theft and destruction of goods, and substance abuse. Data on internalizing problems and externalizing problems in the general population are manifestations of mental health problems whose early onset of symptoms can be traced and prevented as early as possible in adolescence (Hiscock *et al.* 2012). However, if early tracing and prevention efforts in adolescence are less effective, it will cause mental health problems such as those that arise in the general population.

The negative consequences of internalizing problems and externalizing problems as mental health problems commonly experienced by adolescents need to be the focus of attention from various practitioners and researchers from various settings and cultures to identify the causes, prevention and treatment as early as possible. According to previous studies, family and parental factors are consistently important indicators that significantly determine the level of internalizing and externalizing problems in children and adolescents from the point of view of developmental psychology (Masten *et al.* 2005). The family environment is the closest environment for adolescents to interact, grow and develop throughout the stages of human development (Bronfenbrenner & Morris 2006). Furthermore, the family is considered a system capable of influencing various subsystems in the family, namely the child and the parents (Cox *et al.* 2001). Thus, any changes in the family environment will affect adolescents as members of the family and have an impact on other subsystems, including the structure and functioning of the family, such as parent–child relationships and relationships with the entire family (Olson 2000).

Syakarofath and Subandi (2019) explained that family is a protective factor and a risk factor that can prevent and cause negative mental health problems in adolescents. However, individual vulnerabilities caused by genetic or biological factors are also less likely to appear when they do not interact with stress or unpleasant experiences derived from family factors such as negative parenting and how parents treat their children. The toxic family stress model presented by Peris and Miklowitz (2015) reinforces this assumption that the increase in disease turnover and the recurrence rate experienced by children are also due to children's interactions with parents, both father and mother.

Based on the description above, this study aims to test the influence of family functioning on internalizing problems and externalizing problems experienced by adolescents. This research can describe and provide an understanding of the extent of the contribution of family functioning to psychological problems experienced by adolescents, both internalizing and externalizing problems. Furthermore, the information obtained from this study can be a source of mapping possible solutions that can be done from a family perspective. The hypothesis proposed in this study is the functioning of the family being able to predict internalizing problems and externalizing problems in adolescents.

2 METHOD

2.1 *Research participants and procedures*

A total of 170 adolescents in East Kutai Regency, Kalimantan, aged 15–17 years, participated in this study based on multistage random sampling with the help of situsrandom.

org. A total of 94 adolescent girls (55.295%) and 76 adolescent boys (44.705%) participated in the study. The sample selection process is carried out in stages, combining stratified random and cluster sampling. The application of this combination of sampling techniques ensures that all members of the population have the same opportunity to be involved as a sample so that it is hoped that the data obtained can represent the population (Duli 2019).

The application of multistage random sampling is carried out through several stages; the first stage determines the Primary Sampling Unit (PSU) in the form of a geographical area. The PSUs in this study were four East Kutai Regency sub-districts randomly selected from 18 existing sub-districts. The four districts are Bengalon, Muara Ancalong, Sangatta North, and Sangatta South. Furthermore, the second stage is a process of randomizing smaller PSUs in the form of equivalent high schools located in four districts that become PSUs. Four equivalent high schools were selected to represent each sub-district, namely SMAN 1 Sangatta Utara, SMAN 1 Sangatta Selatan, SMAN 1 Bengalon, and SMAN 1 Muara Ancalong.

Participation in this study is voluntary, where the researcher asks for research permission from the school and asks students to fill in informed consent before the data collection process according to their willingness. In addition, researchers seek to ensure accuracy in filling in data and avoid double filling by manually checking personal data fields such as initials, name, age, gender, regional origin, school origin, and contacting by email.

2.2 *Data research and analysis instruments*

This study has three main variables: family functioning as an independent variable, internalizing problems, and externalizing problems as dependent variables. The three instruments were distributed online using a Google form link and shared with the help of the school, which is the place of research through the medium of sending messages, such as WhatsApp.

The Family Assessment Device (FAD) is used to measure the perception of family functioning, where the measuring instrument consists of 60 items (Epstein *et al.* 1983). This instrument measures general functioning and six dimensions of family functioning: problem-solving, communication, roles, affective responsiveness, affective involvement, and behavioral control. Measurements use four answer choices, namely SS (Very Appropriate), S (Appropriate), TS (Non-conforming), and STS (Highly Incompatible). In this instrument, there is a favorable item where a score of 1 is given for the STS answer choice to 4 for the SS answer. As for the unfavorable items, use a scoring system of 1 for SS to 4 for STS. Fad instruments have a reliability index of $\alpha = 0.96$ which indicates that the instrument has good internal consistency.

Furthermore, the Strengths and Difficulties Questionnaire (SDQ) is to measure emotional problems (5 items) and peer problems (5 items) which measure subscale internalizing problems, as well as conduct problems (5 items) and hyperactivity/inactivity (5 items), which measure subscale externalizing problems (Goodman *et al.* 2010). This instrument has good internal consistency with a Cronbach Alpha value of $\alpha = 0.72$ and is already available in Indonesian with the use of research and practice in the clinical field as a screening instrument (Wiguna & Hestyanti 2012). This scale has three answer choices, namely Incorrect (TB), Somewhat True (AB), and True (B), with an answer score range ranging from 0–2, whereas a rather correct (AB) answer will always be worth 1 while Incorrect (TB) and True (B) have varying values depending on the item. Each subscale ranges from 0– 20 with normal, borderline, and abnormal interpretations.

Data analysis uses simple linear regression techniques with the help of the Statistical Program for Social Science (SPSS) for Windows version 22 to test research hypotheses. Previously, it tested the assumption of normal ($p>0.05$) and linear distributed data.

Table 1. Subject demographic data.

Demographic Data	Males N = 131 (43.7%)	Females N = 94 (56.3%)
Age		
16 years old	5 (6.5%)	9 (9.5%)
17 years old	44 (57.9%)	56 (59.5%)
18 years old	27 (35.5%)	29 (30.8%)
Subdistrict Origin		
Bengalon	18 (23.6%)	23 (24.4%)
Muara Ancalong	20 (26.3%)	22 (23.3%)
Sangatta Utara	26 (34.1%)	32 (33.9%)
Sangatta Selatan	12 (15.7%)	17% (18%)
Father's Education		
Magister Degree	1 (1.3)	
Bachelor Degree	26 (34.1%)	7 (7.44%)
Diploma	18 (23.6)	33 (35.1%)
Senior High School	29 (38.1)	11 (11.7%)
Junior High School	0 (-)	41 (43.6%)
Others	2 (2.6%)	2 (2.1%)
Mother's Education		
Magister Degree	2 (2.6%)	0 (%)
Bachelor Degree	3 (3.9%)	12 (12.7%)
Diploma	6 (7.8%)	12 (12.7%)
Senior High School	30 (39.4%)	26 (27.6%)
Junior High School	18 (23.6%)	12 (12.7%)
Others	17 (23.6%)	31 (32.9%)
Elementary School	0 (-)	1 (1%)
Marital Status Parents		
Divorced live	2 (2.6%)	2 (2.1%)
Married	1 (1.3%)	1 (1%)
Divorced dead	73 (96%)	91 (96 %)
Birth Order		
1st child	28 (36.8%)	6 (6.3%)
2nd child	34 (44.7%)	35 (37.2%)
3rd child	10 (13.11%)	30 (31.9%)
4th child	2 (2.6%)	13 (13.8%)
5th child	1 (1.3%)	7 (7.4%)
6th child	1 (1.3%)	3 (3.1%)

3 RESULT

The demographic data of the subjects in Table 1 showed that the subjects who actively participated in this study were aged 16 to 18 years, with the majority being women spread across Bungalon District, Muara Ancalong District, Sangatta Utara District, and Sangatta Selatan District. Dad's educational background is mostly high school, diploma, and bachelor's degree, while the mother's educational background is mostly middle and high school. Furthermore, the majority of the birth sequences of the study subjects were the first and second children.

Table 2 shows that there is no difference between male and female subjects regarding internalizing and externalizing problems. Teenage boys and girls have an equal chance of experiencing one or both of the child's and adolescent's typical mental health problems.

Table 2. Overview of internalizing problems and externalizing problems by gender.

	Internalizing Problems					Externalizing Problems				
Variables	N	Mean	SD	t	*p*	N	Mean	SD	t	*p*
Male	76	7.18	2.813	−0.490	0.558	76	8.43	2.087	−1.894	0.351
Female	94	7.40	2.992			94	9.06	2.209		

Table 3 shows the results of a simple linear regression test that family functioning can predict internalizing problems (adjusted R2 = 0.118; F(23.676) = 169, $p < 0.001$) and externalizing problems (adjusted R2 = 0.026; F(5.551) = 169, p = 0.020). This means that children's chances of developing internalizing problems and externalizing problems are reduced if their families fully function.

Table 3. The role of family functioning on internalizing problems and externalizing problems using simple linear regression.

	Internalizing Problems				Externalizing Problems			
Factors	B	Adjusted R Square	F	*p*	B	Adjusted R Square	F	*p*
Family functioning	−0.264	0.118	23.676	0.000^b	−0.100	0.026	5.551	0.020^b

Predictors: (Constant), family functioning
Dependent Variable: Internalizing problems, externalizing problems

4 DISCUSSION

The results of this study show the role of family functioning in internalizing problems and externalizing problems in adolescents in the East Kutai Regency. The more optimal the functioning of the family, the lower the chances of adolescents developing internalizing problems and externalizing problems. This is in line with the opinion of previous studies, which stated that family functioning is the main predictor of mental disorders faced by children as family members (White *et al.* 2014; Hughes & Gullone 2008). Furthermore, family functioning is also a protective factor that can prevent children from experiencing mental health problems and become an opportunity for a source of social support for children (Baena *et al.* 2021). Therefore, these findings prove that the research hypothesis is acceptable.

Six family indicators function optimally, namely problem-solving, communication, the role of family members (role), the ability to respond well to the feelings of others according to conditions (affective responsiveness), the ability to pay attention to the feelings of others by showing appropriate expressions and attitudes (affective involvement), and behavioral control (Epstein *et al.* 1983). Adolescents' ability to solve problems, communicate, respond, and express negative emotions well can also be learned through their attachment to parents in a family that they feel is safe and full of warmth (Stefan & Avram 2017). In the relationship between parents and children, the role of parents is vital to accompany adolescents to face various developmental tasks, especially in the emotional aspect, so that adolescents can achieve optimal self-potential without significant mental health problems.

In addition, family conflicts also can interfere with adolescents' mental health because they will interfere with family functions. The conflicts that occur are generally around the disagreement between adolescent opinions and parents' opinions (Laursen & Collins 2009). The disagreement is often related to daily activities such as duties and responsibilities and their respective roles at home (Branje 2020). Such duties, roles, and responsibilities are closely related to the functioning of the family. The conflict between parents and children can cause various psychological problems in both parties, such as increased stress, anxiety, and depression (Low 2021) as a manifestation of internalizing problems which can also be an indication of externalizing problems (Foley *et al.* 2021). There are common forms of externalizing problems that arise, such as oppositional defiant disorder (ODD) and conduct disorder (CD) (Syakarofath & Subandi 2019).

Conflicts between father and mother in the family context are also correlated with the possibility of the child experiencing internalizing and externalizing problems. Research conducted by Hosokawa and Katsura (2019) on the population of parents who have adolescent children in Japan shows several important findings about family relationships affect the likelihood of adolescents experiencing internalizing and externalizing problems. For example, parents of adolescents who solve husband and wife relational problems constructively will prevent their teenage boys and girls from experiencing internalizing and externalizing problems. Conversely, teenage parents who use destructive means to solve husband and wife relational problems will increase the likelihood of teenage boys experiencing externalizing problems and adolescent girls experiencing internalizing problems.

Internalizing and externalizing problems are mental health problems commonly encountered when entering adolescence that need to be treated immediately so as not to develop until adulthood. Various strategies can prevent these mental health problems from worsening, namely improving the quality of the relationship between parents and children so that they become warmer and more prosperous (Thomas *et al.* 2017). The quality of family relationships, including social support (for example, providing love, advice, and attention) and relationship tensions (for example, differences of opinion, critical attitudes, and the abundance of demands), can help children understand and express the emotions they feel more healthily.

5 CONCLUSION AND IMPLICATION

Family is a determining factor for adolescents in developing internalizing problems and externalizing problems experienced. Prevention of mental health problems commonly experienced by adolescents can be prevented through various strategies, namely by maintaining the quality of the relationship between children and parents. This is true for both adolescent boys and girls.

The research implies that the family factors are very broad, so to be more detailed and on target in designing an intervention for prevention, it is expected for subsequent researchers to choose more specific variable components related to the family.

FUNDING

This research was funded by the Directorate of Research and Community Service, University of Muhammadiyah Malang (DPPM-UMM) through the P2I scheme.

ACKNOWLEDGMENTS

We thank Andi Azisyah Azzahra, who has assisted in the data collection process in this study.

REFERENCES

Alenazi, S. F., Hammad, S. M., & Mohamed, A. E. (2019). Prevalence of Depression, Anxiety and Stress Among Male Secondary School Students in Arar city, Saudi Arabia, During the School Year 2018. *Electronic Physician, 11*(2).

Badan Penelitian dan Pengembangan Kesehatan Kementrian Republik Indonesia. (2013). *Riset Kesehatan Dasar. Riset Kesehatan Dasar.* Retrieved from http://www.litbang.depkes.co.id/sites/download/rkd2013/Laporan_Riskesdas2013

Badan Penelitian dan Pengembangan Kesehatan Kementrian Republik Indonesia. (2018). *Riset Kesehatan Dasar. Riset Kesehatan Dasar.* Retrieved from https://www.depkes.go.id/resources/download/info-terkini/hasil-riskesdas-2018.pdf

Baena, S., Jiménez, L., Lorence, B., & Hidalgo, M.V. (2021) Family Functioning in Families of Adolescents with Mental Health Disorders: The Role of Parenting Alliance. *Children, 8*, 222. https://doi.org/ 10.3390/children8030222

Bifftu, B. B., Takele, W. W., Guracho, Y. D., & Yehualashet, F. A. (2018). Depression and its Help Seeking Behaviors: A Systematic Review and Meta-analysis of Community Survey in Ethiopia. *Depression Research and Treatment*, 1–11. https://doi.org/10.1155/2018/1592596

Branje, S. (2020). Parent Conflict Resolution. *The Encyclopedia of Child and Adolescent Development, 1–10.* https://doi.org/10.1002/9781119171492.wecad412

Bronfenbrenner, U., & Morris, P. A. (2006). The Bioecological Model of Human Development. In Richard M. Lerner (Ed.), *Handbook of child psychology, Volume 1: theoretical models of human development.* 6th ed (pp. 793–828). Hoboken, NJ: John Wiley & Sons

Cox, M. J., Paley, B., & Harter, K. (2001). Interparental Conflict and Parent–child Relationships. In J. H. Grych & F. D. Fincham (Eds), *Interparental conflict and child development* (pp. 249–272). Cambridge: Cambridge University Press. https://doi.org/10.1017/ CBO9780511527838.011.

Epstein-Lubow G.P., Beevers C.G., Bishop D.S., & Miller, I.W. (2009). Family Functioning is Associated with Depressive Symptoms in Caregivers of Acute Stroke Survivors. *Archives of Physical Medicine and Rehabilitation, 90*(6), 947–55. https://doi.org/10.1016/j.apmr.2008.12.014

Foley, S., Badinlou, F., Brocki, K. C., Frick, M. A., Ronchi, L., & Hughes, C. (2021). Family Function and Child Adjustment Difficulties in the Covid-19 Pandemic: An International Study. *International Journal of Environmental Research and Public Health, 18*(21), 11136. http://dx.doi.org/10.3390/ijerph182111136

Georgiou, S. N., & Symeou, M. (2018). Parenting Practices and the Development of Internalizing/Externalizing Problems in Adolescence. In *Parenting—empirical advances and intervention resources.* InTech. https://doi.org/10.5772/66985.

Goodman A, Lamping DL, Ploubidis GB. (2010). When to Use Broader Internalizing and Externalizing Subscales Instead of the Hypothesized Five Subscales on the Strengths and Difficulties Questionnaire (SDQ): data from British Parents, Teachers and Children. *Journal of Abnormal Child Psychology, 38*, 1179–1191.

Goodman, R. (1997). The Strengths and Difficulties Questionnaire: A Research Note. *Journal of Child Psychology and Psychiatry, 38*(5), 581–586. https://doi.org/10.1111/j.1469-7610.1997.tb01545.x.

Goodman, R., Meltzer, H., & Bailey, V. (1998). The Strengths and Difficulties Questionnaire: A Pilot Study on the Validity of the Self-report Version. *European Child & Adolescent Psychiatry, 7*(3), 125–130. https://doi.org/10.1007/s007870050057.

Graber, J. A. (2013). Internalizing Problems During Adolescence. In *Handbook of adolescent psychology* (pp. 587–626). Hoboken, NJ, USA: John Wiley & Sons, Inc. https://doi.org/10.1002/ 9780471726746.ch19

Hammarström, A., Korhonen, J., Blomqvist, I., & Hägglöf, B. (2017). Increase of Internalized Mental Health Symptoms Over the Last 33 Years Among Adolescents. *European Journal of Public Health, 27* (suppl_3), 1–11.

Hiscock, H., Bayer, J.K., Lycett, K. *et al.* (2012). Preventing Mental Health Problems in Children: The Families in Mind Population-based Cluster Randomised Controlled Trial. *BMC Public Health 12*, 420. https://doi.org/10.1186/1471-2458-12-420

Hosokawa, & Katsura, T. (2019). Exposure to Marital Conflict: Gender Differences in Internalizing and Externalizing Problems Among Children. *PloS One*, *14*(9), e0222021–e0222021. https://doi.org/10.1371/journal.pone.0222021

Hughes, E.K. & Gullone, E. (2008). Internalizing Symptoms and Disorders in Families of Adolescents: A Review of Family Systems Literature. *Clin. Psychol. Rev. 28*, 92–117.

Kwan, M.Y.W., Ceccacci, A., Paolucci, N. *et al.* (2021). Physical Activity and Internalizing Symptoms During the Transition from Adolescence to Emerging Adulthood: A Systematic Review of Prospective and Longitudinal Studies. *Adolescent Res Rev*, *6*, 75–89 https://doi.org/10.1007/s40894-020-00132-3

Liu J. (2004). Prenatal and Perinatal Complications as Predispositions to Externalizing Behavior. *Journal of Prenatal and Perinatal Psychology and Health*, *18*, 301–311

Low, Y.T.A. (2021). Family Conflicts, Anxiety and Depressive Symptoms, and Suicidal Ideation of Chinese Adolescents in Hong Kong. *Applied Research Quality Life*, *16*, 2457–2473. https://doi.org/10.1007/s11482-021-09925-7

Madigan, S., Atkinson, L., Laurin, K., & Benoit, D. (2012). Attachment and Internalizing Behavior in Early Childhood: A Meta-analysis. *Developmental Psychology*, *49*(4), 672–689. https://doi.org/10.1037/a0028793.

Olson, D. H. (2000). Circumplex Model of Marital and Family Systems. *Journal of Family Therapy*, *22*(2), 144–167. https://doi.org/10.1111/1467-6427.00144.

Soto-Sanz, V., Castellví, P., Piqueras, J. A., Rodríguez-Marín, J., Rodríguez-Jiménez, T., Miranda-Mendizábal, A., ... & Alonso, J. (2019). Internalizing and Externalizing Symptoms and Suicidal Behaviour in Young People: A Systematic Review and Meta-analysis of Longitudinal Studies. *Acta Psychiatrica Scandinavica*, *140*(1), 5–19, https://repositori.upf.edu/

Ştefan, & Avram, J. (2017). Investigating Direct and Indirect Effects of Attachment on Internalizing and Externalizing Problems through Emotion Regulation in a Cross-Sectional Study. *Journal of Child and Family Studies*, *26*(8), 2311–2323. https://doi.org/10.1007/s10826-017-0723-7

Syakarofath, N. A., & Subandi, S. Faktor Ayah Dan Ibu Yang Berkontribusi Terhadap Munculnya Gejala Perilaku Disruptif Remaja. *Jurnal Psikologi*, *18*(2), 230–244. https://doi.org/10.14710/jp.18.2.230-244

Thomas, P.A., Liu, H., & Umberson, D. (2017). Family Relationships and Well-Being, *Innovation in Aging*, *1* (3), 1–11. https://doi.org/10.1093/geroni/igx025

White, J., Shelton, K.H., & Elgar, F.J. (2014). Prospective Associations Between the Family Environment, Family Cohesion, and Psychiatric Symptoms Among Adolescent Girls. *Child Psychiatry Hum. Dev.*, *45*, 544–554.

Families Mental Health and Challenges in the 21st Century – Arifin et al. (eds)
© 2024 The Author(s), ISBN 978-1-032-51442-0

Am I a real, ideal, or false self on Social Networking Site (SNS)? The differences in online self-presentation among people in Indonesia

Nandy Agustin Syakarofath*
Faculty of Psychology, Universitas of Muhammadiyah Malang, Malang, Indonesia

Nurul Aiyuda
Psychology Study Programme, Universitas Abdurrab Pekanbaru, Pekanbaru, Indonesia

ABSTRACT: The use of technology in society has an impact on the lives of individuals, including the use of increasingly massive social media. Many people demand social media because it provides various facilities, such as facilitating access to long-distance communication and can provide opportunities for individuals to present themselves according to what they want. This study mainly investigates the differences among Malayanan, Madurese, and Javanese people as one of the largest ethnic groups in Indonesia when they present themselves on social networking sites (SNS). A total of 123 subjects, 36 males and 87 females, were actively involved as the research participants. The Self-presentation on Facebook Questionnaire was used to measure the participants' self-presentation. The data were analyzed using a two-way ANOVA. The results showed no differences between Malayan, Madurese, and Javanese people in presenting themselves on SNS ($F = 1.464$; db = 14, $p = .235$; $\eta^2 = 0{,}235$).

Keywords: ethnicity, self-presentation, SNS

1 INTRODUCTION

The reputation of social networking sites (SNSs) over the past few years has frequently become widespread and increased public recognition. Based on a survey conducted by Statista (2019), users of SNSs currently reach 2.77 billion, and their use is increasing year by year namely, in the previous year, 2018, there were several 2.62 billion users, in 2017, several 2.45 billion users and in 2016, there were 2.28 billion users. Other data also mentions that around 4,545 billion more users of SNSs spend an average of 2 hours 43 seconds every day to access the platform, a figure that is quite fantastic to explain the extent to which people are familiar with using them (Alassiri 2014).

According to Zarghooni (2017) and Jain (2017), the SNS is a website that provides a forum for users to interact with their surrounding environment, both family, friends, and colleagues at work by sharing interests, topics and increasing their circle of acquaintances. Another opinion states that the SNS is a web that allows a person to create a public or semi-public profile by registering in a limited system and can connect with other people both individually and in groups so that they can share information daily without being limited by space and time (Boyd & Ellison 2008; Liu & Ying 2010). Using a SNS can connect with a

*Corresponding Author: nandysyakarofath@umm.ac.id

DOI: 10.1201/9781003402381-14

broader network (both already known and not) by sharing information daily such as personal information, social issues, politics, health, sports, work, or academics.

The use of SNSs has a significant impact on people's lives. It is why they are interested in spending time accessing it. For the general public, the utilization of SNSs can be an amusement media and a means to develop new relationships (Jain 2017), and for academics, it can be used as a medium to share information on academic tasks, find information on lecture assignments, and job info (Rashmi & Neetu 2012). Therefore, this is why SNS is an essential element in life because its use can be felt by the community in critical fields such as education, politics, and economics, so its use cannot be ignored (Zairian 2012).

In addition to having a positive impact, using SNSs also has a negative impact. Abdulahi, Samadi, and Gharleghi (2014) mentioned that the impact of using SNSs is an addiction and privacy issue. Other problems arising from the use of SNSs are negligence of duties and responsibilities, addiction, kidnapping, murder, robbery, loss of productive time, unhealthy social relations, social lies or hoaxes, privacy, and misuse of images, videos, or other content and identity uncertainty (Siddiqui & Singh 2016).

Studies investigating SNSs show that about 12% of users experience addiction (Alabi 2012; Wolniczak *et al.* 2013; Wu *et al.* 2013). The addiction's symptoms can manifest in physical, cognition, mood, emotional reactions, and interpersonal and psychological problems (Tang *et al.* 2016). In addition, this condition impacts mental health problems such as stress, anxiety, and depression and is negatively associated with long-term well-being (Marino *et al.* 2017).

Another negative impact of using SNSs is making individuals more involved in various harmful behaviors that harm others, like criminal acts. For example, about 28% of bullying cases, and many are unaware of the severe impact, although, in some countries, it can lead to suicide (Tarigan *et al.* 2018). Common types of bullying are flaming, harassment, denigration, impersonation, outing, trickery, exclusion, and cyberstalking (Rahayu 2013). Furthermore, the use of SNSs is also attached to fraud and misuse. Most fraud types are identity manipulation and diffusion of misinformation (Apte *et al.* 2018). These negative impacts drive the urgency of research related to the use of SNSs further so that the causes and impacts and actions can be taken to mitigate the negative impact can be identified.

Although it has a negative impact, individuals actively use SNSs for self-existence. According to Mehdizadeh (2010), people who tend to be narcissistic and have low self-esteem are associated with more intense online activities and often promote themselves to the public. Narcissism itself is a widespread pattern of selfishness, and it requires admiration from others and the sense that the self is essential that is excessive (Oltmanns *et al.* 2006) or how they over-evaluate and judge themselves by considering themselves perfect (Weiten 2004). It means that individuals who have a negative evaluation of themselves tend to take action to present themselves so that their needs for attention and recognition from others are fulfilled.

Efforts to fit the needs of attention and acceptance from others encourage individuals to make massive self-presentations online. According to de Lamater and Myers (2007), self-presentation named an effort to control oneself by the desired image in establishing social interaction. Its behavior appears to fulfill general assumptions, and the aim is to reach an agreement, avoid rejection because of the universal orientation of behavioral norms, and answer the expectations of others towards us (Peleckis 2013). The aim is to get an evaluation and a comfortable and acceptable impression of the surrounding environment (Kusumasari & Hidayati 2014). However, sometimes self-presentation carried out online is far from reality and has a specific purpose so that his image is constructed in such a way, like good or bad. Toma and Hancock (2011) state that individuals tend to require that they present an authentic self-profile (picture or self-information) but based on the measurement and objective assessment is not the case.

Mayawati (2016) explained that two main components form online self-presentation, namely the motives and appearance of oneself. The motive aspect consists of personal motives (self-expression, self-construction, nurturance, popularity, power, and social motives (affiliation, social acceptance, and social support). In self-appearance, consists of identifying themselves with the real world (real self), the new self (self), the self that is shown selectively (selective self), and the ideal self (ideal self). That is, someone has particular criteria to present themselves on the social networking site (SNS) to make it looks enjoyable and exciting so that there is a public lie.

There is various effort made to get favorable judgment and impression. According to Goffman *et al.* (1978), someone does impression management by forming perceptions of information, organizing, and integrating it so that it creates a consistent look from other people, for example, in terms of personality and interpersonal attitudes and primarily visual appearance (e.g., hairstyles, clothing). SNSs such as Facebook, Twitter, Instagram, and WhatsApp are becoming popular because users can present themselves ideally following the impressions expected by others through postings in the form of photos, statuses, or comments when surfing on their social media accounts.

Previous research states that race is related to individual self-presentation. For example, Siibak (2009) showed that women in Estonia tend to select pictures or photos of themselves that are considered the most interesting to post. Kapidzic and Herring (2014) also mention gender and racial differences related to how to represent themselves, that images are considered to have a unique attraction for someone to interact online. It means someone tries to present himself based on an appropriate or acceptable evaluation to the local community. The indication of the relationship between race and online self-presentation is a compelling reason to be researched to add scientific insight. Besides that, there is various race in the world, especially in Indonesia, which has cultural diversity from various tribes that are different.

The formulation of the problem in this paper is that there are differences in individual self-presentation based on the tribe in Indonesia. The study was initiated based on Kodiran's opinion (2004), which states that certain cultures have a capital personality structure, some character traits owned by most citizens of a society. Therefore, Indonesia, as a country rich in ethnic and cultural diversity, is the right choice to be further researched from various aspects, including sociocultural and personality, in the context of current developments. The tribe studied are Malayan, Madurese, and Javanese people, one of the largest ethnic groups in Indonesia (Pitoyo & Triwahyudi 2017). The process of tribal selection to be researched is focused on random sampling techniques that will be explained in detail in the research method. This study aimed to test the differences in self-presentation in the tribes mentioned before.

2 METHODS

A total of 123 subjects, composed of 36 males and 87 females, were collected using random sampling with the main criteria actively using SNSs and came from Malayan, Madurese, and Javanese people as one of the largest ethnic groups in Indonesia. The determination of three ethnicities is chosen using random.org. Self-presentation on Facebook Questionnaire (SPFQ) is used to collect data (Michikyan *et al.* 2014). SPFQ has 17 items, and participants answered each item on a scale of 1 to 5 (1—strongly disagree; 5—strongly agree). The scale consists of 3 aspects, real self-presentation (α = 0,638), ideal self-presentation (α = 0,715), and false self-presentation (α = 0.850). For statistical analysis of data, two-way ANOVA and Variations Homogeneity Test are used through SPSS for Windows version 22.0. The homogeneity test in this study uses Levene's test, with sig 0.436 (the self-presentation), sig 0.342 (real self), sig 0.460 (ideal self), and sig 0.401 (false with). Based on the homogeneity test results, each meets the requirements to be continued in the analysis of the difference test (p> 0.05).

In Table 1, there is an inequality in the number of subjects, so this study does not have a balanced number of conducting different tests, but according to Widiarsho (2013), it can be overcome by a homogeneity test. Therefore, the research data met the requirements of the homogeneity test before proceeding to the analysis of the difference test.

The ANOVA test presents data that there is no difference in self-presentation on social media based on ethnicity, the Javanese tribe (*M* = 49.7, *SD* = 10.2), Malayan (*M* = 46.0, *SD* = 11.9) and Madurese (*M* = 46.5, *SD* = 8.3) with a significance of 0.235 (*F* = 1.464). However, if analyzed separately, it was found that there were differences in the aspects of false self-presentation in the part of improper self-exploration with a significance level of 0.048 (p <0.05). The Javanese occupy the highest position (*M* = 8.6, *SD* = 2.6) in the false self-exploration compared to the other two tribes namely Malayan (*M* = 7.5 *SD* = 2.6) and Madurese (*M* = 7.1, *SD* = 2,6). Whereas in the false section, self (compare/impress) there is also a difference if seen based on ethnic origin with a significance level of 0.044 (p <0.05). The Javanese (*M* = 8.16, *SD* = 3.06) has the next highest false self compared to the Malayan tribe (*M* = 8.07 *SD* = 3.9) and Madurese (*M* = 6.25, *SD* = 2.2). This shows the uniqueness of each tribe with all the diversity it has.

3 DISCUSSION

As described in Table 1, the findings concerned that there were no significant differences between the three tribes in presenting themselves using SNSs. This is because the three tribes studied are one unit derived from the same culture. The Javanese, Malayan, and Madurese are a tribe originating from Indonesia, where the culture owned by the majority is the collective culture. The visible characteristic of collective culture is that individuals are very

Table 1. The differences of online self-presentation based on ethnicity using two-way ANOVA.

		N	Mean	SD	F	sig
Self-presentation	Javanese	88	49.77	10.293	1.464	.235
	Malayan	15	46.00	11.988		
	Madurese	20	46.50	8.382		
Real self	Javanese	88	17.84	3.487	1.419	.246
	Malayan	15	16.27	3.555		
	Madurese	20	18.20	4.312		
Ideal self	Javanese	88	6.34	1.767	1.865	.159
	Malayan	15	5.67	2.093		
	Madurese	20	5.60	2.010		
False self	Javanese	88	25.59	7.159	1.505	.226
	Malayan	15	24.07	7.923		
	Madurese	20	22.70	5.648		
false self (deception)	Javanese	88	8.80	2.792	.408	.666
	Malayan	15	8.47	3.204		
	Madurese	20	9.30	2.658		
false self (exploration)	Javanese	88	8.64	2.692	3.125	.048*
	Malayan	15	7.53	2.696		
	Madurese	20	7.15	2.641		
False self (compare/impress)	Javanese	88	8.16	3.062	3.200	.044*
	Malayan	15	8.07	3.973		
	Madurese	20	6.25	2.221		

attached to social groups, including how they present themselves (Saliyo 2012). Indonesia, a country with high collectivism and low individualism, is described as a society with a strong desire to be close to one another and to be aware of any factors that can hinder or minimize the quality of their togetherness (Nadar 2007). It is this collectivist culture that shapes the way they behave, behave, and their self-concept.

The formation of self-concept is influenced by the individual's belief in himself based on the opinions of others towards him. Like a mirror, what appears to reflect what is in front of it (looking glass self). Wayne (1992) mentions that looking glass self consists of three elements, the first, imagining how the individual self looks before others, the second, the way how to interpret the response of others to him, and the third is the development of the self-concept of the evaluation and interpretation of other people's responses to the individual. Because it comes from the same culture, Javanese, Malayan, and Madurese develop the same self-concept so that the impact on each other's general characteristics is not much different, including in presenting themselves.

However, if analyzed separately in false self-presentation, there are differences, especially in false self-exploration and false self compare. Although a collective culture of self-public nature influences the general individual characteristics of the three tribes, it does not deny that individuals also develop their private selves in smaller groups. It means Indonesia, as a country that adheres to a collective culture, does have a common characteristic of its name, but in each region, it can also develop the peculiarities of its characteristics. The flow of how self and culture, when experiencing acculturation, influence the self-presentation can see in Figure 1.

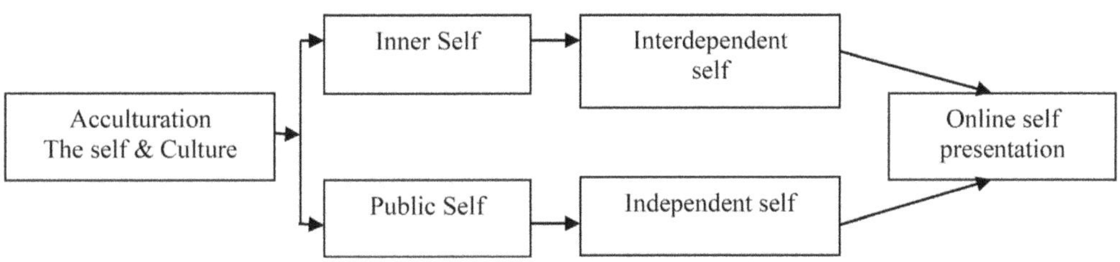

Figure 1. The model of self-acculturation and culture towards the self-presentation of the community.

In the false self-presentation (aspect exploration and compare/impress), Javanese occupy the highest position compared to the other tribes, Malayan and Madurese. Susetyo, Widiyatmadi, and Sudiantara (2014) mention the concept of self and appreciation of the Javanese self rooted in the implementation of the principle of "harmony and respect," which is strictly characterized by Javanese personality. So that in displaying themselves, Javanese people tend to display behavior that is more suitable and acceptable to society at large because it avoids conflict (Muluk & Muniarti 2007). One way is to form an impression (impression management) through false self-exploration, namely by exploring themselves through a comparison to the outside world called the false self to compare, unlike the other two tribes who tend to be more open and assertive.

The Malayan tribe showed moderate self-presentation. Malayans generally have a loose and open cultural condition, making it possible to accommodate changes from different cultural elements (Reginasari 2018b). The same thing was also conveyed by Sunandar (2015) that Malayan society has flexible characteristics in amending the external culture, including accepting technological changes, such as social media. On the other hand, individuals are also asked to hold the mandate and maintain shame in knowing the limits of using technology. The same thing was also conveyed in Effendy (2004) that Malayan people have inherited characteristics of teaching and learning with self-knowledge. Individuals as Malayan tribes must have "self-knowledge," which means personal self-awareness of the nature and meaning of life, the end of life, and the rights and obligations that must be fulfilled. This reason is what causes the self-presentations of Malayan Tribes to the level of moderate.

The level of moderate self-presentation Malayan tribe can describe by the rules and point of teaching Malayan as forming the character of Malayan people. According to Effendy (2004), through Malayan teaching points, individuals are expected to have an open nature as a reflection of honesty and responsibility, which can display in self-presentation on social media. This statement was strengthened by Sunandar (2015), who confirmed that Malayan society has a loose social structure so that it quickly affiliated with something new, including technology. Hasbullah (2014) added that the characteristic of Malayanness itself is openness but is still based on living together with differences.

In accepting good and superior external influences, Malayan people tend to be tolerant and open, even innovating themselves on the technology received (Thamrin 2015). In the Malayan tribe, parents are required to be able to encourage children as creative and dynamic people, but on the one hand, they also hold the mandate and keep shame (Reginasari 2018a). Creative and dynamic conditions are illustrated by how individuals of Malayan ethnicity can be open and creative by presenting themselves through surfing experiences that utilize technology such as social media. However, the openness of the Malayan community certainly has limits too. In carrying out the daily life of Malayan people, they must not separate from Malayan guidelines identical to Islamic teachings (Hasbullah 2014; Sunandar 2015; Thamrin 2015). Individuals in Malayan ethnic groups are advised not to go beyond religious boundaries in their daily lives and have simplicity (Thamrin 2015).

Reginasari (2018a) mentioned several Malayan cultures handed down from generation to generation to be a force to face the challenges of globalization, including technological developments. However, on the other hand, this can also be a boomerang for Malayan tribes if it is not well adapted.

Furthermore, the Madurese tend to display their real self but display false self in the form of deception at certain moments. According to Ali (2010), Madurese people have a distinctive character and stand out; that is what they are. He tends to be expressive, spontaneous, and open, where the nature possessed manifests itself in responding to everything he faces, especially if it is related to the treatment of others himself. With such characteristics, the Madurese tend to display their real self as the findings obtained. However, on the other hand, unique Madurese people can also commit false self-deception (not real self by committing fraud). The purpose of why they make the impression is to get the right image. It is because self-esteem for Madurese people is essential to keep it necessary to build a self-image (self-image) that can be accepted by all people (Tsabit 2008). Maintaining self-esteem for the Madurese community is even the main thing in order to avoid "*malo*" (feeling of shame), especially for male Madurese (Bustami 2002).

In general, although there were no contrasting differences between the three tribes studied related to how they presented themselves on SNSs, each region still had distinctive characteristics that highlighted the cultural and socio-demographic characteristics of their respective regions.

Indonesia is famous for its easy-to-acculturate country and even assimilates to new cultures even though its people are known for their values and upholding local culture, each of which has its repertoire. SNSs influence the difference in how to present the self. Nasution (2017) revealed that the main trigger for the acculturation of Western culture toward natives was the rapid development of technology and communication. Individuals began to see technology development as something that must utilize in Indonesia and even one of the highest internet users in Asia.

The limitations in this study are the number of research subjects that are less balanced in each tribe, considering that not all people from each tribe studied use SNSs; this is only limited to young people whose self-formation is still proceeding, so there is the possibility of change when entering advanced adulthood.

4 CONCLUSION

Based on the explanation above, it can conclude that: (1) There is no difference individual self-presentation on SNSs based on ethnicity, third in Javanese, Malayan, and Madurese people where the self-presentation of the three tribes is lower than it should be. It means that Javanese, Malayan, and Madurese peoples are less flexible in presenting themselves on SNSs because they tend to limit themselves to changes according to their characteristics based on the prevailing collective culture locally, (2) Each tribe has a different self-presentation influenced by local culture, where if analyzed separately, the Javanese tend to display their ideal self and false self-presentation in the form of exploration and compare/impression compared to the other two tribes. It means that even though globally, the self-presentation of the three tribes studied was not found to be a contrasting difference but still had a distinctiveness that distinguishes them from other tribes. The culture and local wisdom influence this in each region where the tribe is located.

The implication for further research is to explore local wisdom and local culture further to enrich the study of literature so that the peculiarities of each region can in detail explain its influence on the formation of the self in individuals not limited to Javanese, Malayans, and Madurese, given that Indonesia has many other tribes.

REFERENCES

Abdulahi, A., Jalil, B., & Lumpur, K. (2014). A Study on the Negative Effects of Social Networking Sites Such as Facebook among Asia Pacific University Scholars in Malayansia. *International Journal of Bussiness and Social Science*, *5*(10), 133–145.

Alabi, O. F. (2012). A survey of Facebook Addiction Level Among Selected Nigerian University Undergraduates. *New Media and Mass Communication*, *10*, 70–80.

Alassiri, A. A., Munda, M. B., & Ghazali, R. B. (2014). Usage of Social Networking Sites and Technological Impact on the Interaction-enabling Features. *International Journal of Humanities Ad Social Science*, *4*(4), 46–61.

Apte, M., Palshikar, G. K., & Baskaran, S. (2018). *Frauds in Online Social Networks: A Review. Social Networks and Surveillance for Society*, *1–18*. doi:10.1007/978-3-319-78256-0_1

Baron, A. R., & Branscombe, R. N. (2012). *Social Psychology* (13th Edition). New York: Pearson.

Boyd, D. M., & Ellison, B. N. (2008). Social Networking Sites: Definition, History and Scholarship. *Journal of Computer-mediated Communication*, *13*(1), 210–230. http://doi.org/10.1111/j.1083-6101.2007.00393.x

Bustami, A.L. (2002). Carok: Konflik Kekerasan dan Harga Diri Orang Madurese. LkiS, Yogyakarta

Counts, S., & Stecher, K. (2009). Self-presentation of Personality During Online Profile Creation. *Proceedings of the Third International ICWSM Conference*. 191–194.

De Graff. (2011). The Relationship Between Adolescents' Personality Characteristics and Online Self-presentation. Retrieved from https://pdfs.semanticscholar.org/233e/61174a122b1e8c0428e646b001e5ace249d9.pdf. (accessed 19 Juni 2019)

Effendy, T. (2004). *Tunjuk Ajar Malayan; Butir-butir budaya Malayan Riau*. Pekanbaru: Balai Kajian dan Pengambangan Budaya Malayan Berkerjasama Dengan Penerbit AdiCita.

Goffman, Er. (1959). *The Presentation of Self in Everyday Life*. United States: Anchor Books

Hanifah, J., Aiyuda, N., Ramadhan, N. H., Kusumawati, I., Sinda Suadana, Muis, I. A., ... Nasution, I. N. (2019, 20 December). *Tunjuk Ajar Malayan: Shame and Guilt of Malayan Population in Riau*. Presented at the ICIPSY (International Conferences Islamic Psychology), Yogjakarta.

Hasbullah. (2014). Islam dalam Bingkai Budaya Lokal (Kajian Tentang Integrasi Islam Dalam Budaya Malayan Riau). *TOLERANSI: Media Komunikasi Umat Bergama*, *6*(1), 1–15.

Hofstede, G. (2010). *Cultures and Organizations: Software of the Mind* (3rd Edition). McGraw-Hill Education.

Hongladarom, S. (2011). Personal Identity and the Self in the Online and Offline. *Minds & Machines*, *21*, 533–548. http://doi.org/10.1007/s11023-011-9255-x

Jain, D. (2017). Information Sharing on Social Networking Sites (SNS): An Empirical Study. *Annual Research Journal of SCMS, Pune*, *5*(March), 38–60.

Kapidzic, S. & Herring, S.C. (2014). Race, Gender and Self-presentation in Teen Chat Profiles. *New Media & Society*. Retrieved from https://journals.sagepub.com/doi/abs/10.1177/1461444813520301?journalCode=nmsa (accessed 16 Juni 2019)

Liu, D., & Baumeister, R. F. (2016). Social Networking Online and Personality of Self-worth: A Meta-analysis. *Journal of Research in Personality*, *64*(July), 79–89. http://doi.org/10.1016/j.jrp.2016.06.024

Liu & Ying (2010). A Review of Social Network Sites: Definition, Experience and Applications. *Scientific Research*, 749–752. Retrieved from http://file.scirp.org/pdf/18-2.1.31.pdf (accessed 15 May 19)

Manago, A. M., Graham, M. B., Greenfield, P. M., & Salimkhan, G. (2008). Self-presentation and Gender on MySpace. *Journal of Applied Developmental Psychology*, *29*(6), 446–458. http://doi.org/10. 1016/j. appdev.2008.07.001

Marino, C., Finos, L., Vieno, A., Lenzi, M., & Spada, M. M. (2017). Objective Facebook Behaviour: Differences Between Problematic and Non-problematic Users. *Computers in Human Behavior*, *73*, 541–546. https://doi.org/10.1016/j.chb.2017.04.015

Mayawati, E.H. (2016). Presentasi Diri Online Pada Remaja Pengguna Jejaring Sosial. Tesis. Fakultas Psikologi, Universitas Gadjah Mada. Tidak di Publikasikan.

Mehdizadeh, S. (2010). Self-Presentation 2.0: Narcissism and Self-esteem on facebook. *Cyberpsychology, Behavior, and Social Networking*. *13*(4). Retrieved from https://www.liebertpub.com/doi/full/10.1089/cyber.2009.0257 (accessed 15 May 19).

Michikyan, M., Dennis, J., & Subrahmanyam, K. (2014). Emerging Adulthood can you Guess Who I am? Real, Ideal and False Self-presentation on Facebook among Emerging Adults. *Emerging Adulthood*, *3*(1). http://doi.org/10.1177/2167696814532442

Michikyan, M., Subrahmanyam, K., & Dennis, J. (2014). Computers in Human Behaviour can you tell who I am? Neuroticism, Extraversion, and Online Self-presentation among Young Adults. *Computers in Human Behavior*, *33*, 179–183. http://doi.org/10.1016/j.chb.2014.01.010

Muluk, H., & Murniati, J. (2007). Konsep Kesehatan Mental Menurut Masyarakat Etnik Javanese dan Minangkabau. *Jurnal Konsep Sehat Mental*, *13*(2), 167–181.

Nadar. (2007). The Prominent Characteristics of Javanese Culture and their Reflections in Language Use. *Humaniora*, *19*(2), 168–174.

Nasution, R. D. (2017). Pengaruh Perkembangan Teknologi Informasi Komunikasi terhadap eksistensi Budaya Lokal. *Jurnal Penelitian Komunikasi dan Opini Publik*, *21*(1), 30–42.

Oltmanns FT, Emery ER, Taylor S (2006) *Abnormal psychology*. Toronto: Pearson Education, Canada.

Paliszkiewicz, J., & Magdalena M.-S. (2012). Impression Management in Social Media: The Example of LinkedIn. *Management*, *11*(3), 203–212.

Pitoyo, A.J & Triwahyudi, H. (2017). Dinamika Perkembangan Etnis di Indonesia Dalam Konteks Persatuan Negara. *Populasi*, *25*(1), 64–81.

Rahayu, Flourensia Sapty. (2013). Cyberbullying Sebagai Dampak Negatif Penggunaan Teknologi Informasi. *Jurnal Sistem Informasi*, *8*(1), 22. https://doi.org/10.21609/jsi.v8i1.321

Rashmi, R., & Neetu, S. (2014). The Use of Social Networking in Academics: Benefit and Opportunity. *International Journal of Science and Research (IJSR)*, *3*(6).

Reginasari, A. (2018a). *Tunjuk Ajar Malayan; Dalam Bingkai Psikologi*. Indie Book Corner.

Reginasari, A. (2018b). Tunjuk Ajar; Pengasuhan ala Malayan Riau Sebagai Optimalisasi Fungsi Keluarga Pada Tahap Perkembangan Keluarga Dengan Anak Usia Prasekolah. In *Psikologi Perkawinan dan Keluarga*. Yogjakarta: Kanisius.

Saliyo. Konsep diri dalam Budaya Javanese. (2012). *Buletin Psikologi*. *20*(1–2), 26–35. http://doi.org/10.22146/bpsi.11946

Shiraev, E., & Levy, D. A. (2010). *Cross-cultural Psychology: Critical Thinking and Contemporary Applications* (4th ed). Boston: Allyn & Bacon.

Siddiqui, S. & Singh, T. (2016). Social Media its Impact on Positive and Negative Aspects. *International Journal of Computer Applications Technology and Research*, *5*(2), 71–75.

Siibak, A. (2009). Constructing the Self through the Photo Selection – Visual Impression Management on Social Networking Websites. Cyberpsychology: *Journal of Psychosocial Research on Cyberspace*, *3*(1), Retrieved from: https://www.researchgate.net/publication/283986315_Constructing_the_Self_through_the_Photo_Selection_Visual_Impression_Management_on_Social_Networking_Websites [accessed 17 July 2019].

Statista. (2019). Number of Social Media Users Worldwide from 2010 to 2021 (in billions). Retrieved from https://www.statista.com/statistics/278414/number-of-worldwide-social-network-users/. (accessed 16 June 2019).

Sunandar. (2015). Malayan Dalam Tantangan Globalisasi: Refleksi Sejarah Dan Berubahnya Sistem Referensi Budaya. *Jurnal Khatulistiwa – Journal of Islamic Studies*, Volume *5* (Nomor 1), 60–73.

Supradewi, R dan Setyaningsih, R. (2013). Self Construal Pada Remaja Etnis Javanese. *Prosiding Seminar Nasional*. Semarang: Fakultas Psikologi Universitas Islam Sultan Agung.

Susetyo, D.P. B., Widiyatmadi, H.M.E., & Sudiantara, Y. (2014). Konsep self dan Penghayatan Self Orang Javanese. *Psikodimensia*, *13*(1), 47–59. https://doi.org/10.24167/psiko.v13i1.277

Tang, J. H., Chen, M. C., Yang, C. Y., Chung, T. Y., & Lee, Y. A. (2016). Personality Traits, Interpersonal Relationships, Online Social Support, and Facebook Addiction. *Telematics and Informatics*, *33*, 102–108. https://doi.org/10.1016/j.tele.2015.06.003

Tarigan, V.C.T., Hasibuan, L.R., Purba, r. b., Irawan, Sari, P.B., ... Yossie Rossanty, Muhammad Dharma Tuah Putra Nasution, Cybercrime Case on Social Media in Indonesia, *International Journal of Civil Engineering and Technology*, *9*(7), 2018, pp. 783–788. http://www.iaeme.com/IJCIET/issues.asp?JType=IJCIET&VType=9&IType=7

Taylor, S. E., Peplau, L. A., & Sears, D. O. (2006). *Social Psychology* (12th Edition). London: Pearson.

Tazghini, S. & Siedlecki, K. (2013). A Mixed Approach to Examining Facebook Use and its Relationship to Self-esteem. *Computers in Human Behavior*, *29*, 827–832.

Thamrin, H. (2015). Enkulturasi Dalam Kebudayaan Malayan. *Al-Fikra: Jurnal Ilmiah Keislaman*, *14*(1), 54.

Tsabit, M. (2008). Perilaku agresi Masyarakat Madurese (Studi fenomenologi tentang carok di desa Kalebengan Kecamatan Ruberu Kabupaten Sumenep). Skripsi. Universitas Islam Negeri Malang.

Wayne, E. (1992). Looking Glass and Penetration Social Penetration. the Annual Speech Communication Association Convention. Retrieved from https://files.eric.ed.gov/fulltext/ED351724.pdf

Weiten, W. (2004). *Psychology themes and variations*. Belmont, CA: Wadsworth-Thomson Learning

Widhiarso, W. (2010). Membuat Kategori Skor Hasil Pengukuran Dari Skala. Fakultas Psikologi. Universitas Gadjah Mada. Yogyakarta.

Widiarsho, W. (2013). UJI Hipotesis Komparatif (Uji-t). In *wahyupsy.blog.ugm.ac.id* (Issue 1, pp. 1–4). http://wahyupsy.blog.ugm.ac.id/tag/uji-t/

Wolniczak, I., Caceres-DelAguila, J. A., Palma-Ardiles, G., Arroyo, K. J., Solís-Visscher, R., Paredes-Yauri, S., & Bernabe-Ortiz, A. (2013). Association Between Facebook Dependence and Poor Sleep Quality: A Study in a Sample of Undergraduate Students in Peru. *PLoS One*, *8*, e59087. https://doi.org/10.1371/journal.pone.0059087

Wu, A. M., Cheung, V. I., Ku, L., & Hung, E. P. (2013). Psychological Risk Factors of Addiction to Social Networking Sites Among Chinese Smartphone Users. *Journal of Behavioral Addictions*, *2*, 160–166. https://doi.org/10.1556/JBA.2.2013.006

Zarghooni, S. (2007). *A Study of Self-Presentation in Light of Facebook. Institute of Psychology*, University of Oslo

Zhao, S., Grasmuck, S., & Martin, J. (2008). Identity construction on facebook: Digital empowerment in anchored relationships. *Computers in Human Behavior*, *24*(5), 1816–1836. http://doi.org/10.1016/j.chb.2008.02.01

Families Mental Health and Challenges in the 21st Century – Arifin et al. (eds)
© 2024 The Author(s), ISBN 978-1-032-51442-0

Validity and reliability of quality of life scale for parents of autistic children in Indonesia

Ni'matuzahroh* & Nurreza Nimas Putri*
Department of Psychology, University of Muhammadiyah Malang, Indonesia

Mein-Woei Suen
Department of Psychology, Asia University, Taiwan
Taiwan Applied Psychology Association, Taiwan
Gender Equality Education and Research Center, Asia University, Taiwan
Department of Medical Research, Asia University Hospital, Asia University, Taiwan
Department of Medical Research, China Medical University Hospital, China Medical, Taiwan

ABSTRACT: This study aims to determine the validity and reliability of the quality of life measurement tool for autistic children which has been adapted by Baghdadli from the measuring instrument developed by Berdaux. Thus, it is hoped that this research can be a reference for further researchers in measuring the quality of life of parents with autistic children. This psychometric research uses the Par-DD-Qol measuring instrument which has been translated into Indonesian. The research sample amounted to 446 parents with the criteria of having children aged 2–18 years who were taken using a non-probability purposive sampling technique. The research analysis method used is the Confirmatory Factor Analysis method using Lisrel version 8.8. Based on the first stage of the CFA test, it was found that there were 2 invalid items and continued with model modification which showed the results that the emotional aspects and daily disturbances were valid (0.32–0.79) and quite reliable (0.84) with 16 items and dropped 2 items on aspects of quality of life in general.

Keywords: Quality of Life, Autism, Parents, Measuring Instruments, Validity, reliability

1 INTRODUCTION

Autism spectrum disorder (ASD) is a neurodevelopmental disorder characterized by disturbances in socialization and communication that are very severe and last for a long time when compared to other developmental disorders (Baghdadli *et al.* 2014). According to Cappe, Pedoux, Poirier, Downes, and Nader-Grosbois (2019), ASD is considered a neurodevelopmental disorder that affects approximately 1% of the world's population which can be defined as a persistent disturbance in mutual communication and social interaction, behavior, interests, or restricted and repetitive activities. So it can be said that ASD is a disorder that occurs by communicating and interacting. According to Vasilopoulou and Nisbet (2016) raising autistic children can lead to a low quality of life for parents and mental health risks and stress levels.

Quality of life is an individual's perception of their existence in society in the context of the existing culture and value system and is related to their goals, expectations, standards, and views (GROUP 1998). According to Vasilopoulou and Nisbet (2016), the factors that influence the quality of life of parents who have children with autism are parental factors

*Corresponding Authors: zahroh@umm.ac.id and nurrezanimas17@webmail.umm.ac.id

DOI: 10.1201/9781003402381-15

which include gender and mental health, child-related factors which include symptoms and behavior of children who keep repeating themselves, contextual factors which include income, job title, social support and participation.

In Indonesia, the measuring tool for the quality of life of parents with special needs with intellectual disabilities and parents with autism that has been tested in Indonesia is the WHOQol-BREF (Sukmadi *et al.* 2020). The WHOQOL-BREF scale consists of 27 items consisting of 2 questions for aspects of life, 4 questions for physical aspects, 8 questions for psychological aspects, 5 questions for interpersonal relationships, and 6 questions for relationships in the environment. However, the WHOQOL-BREF scale in a study conducted by S.M. Skevington, Lotfy, and O'Connell (2004) shows that there is a deficiency on the WHOQOL-BREF scale, namely that there are individual items or questions that do not fit the domain, such as physical safety items from the environmental domain and energy items from the physical health domain which have more correlation stronger with the domain of psychological health than the domain of origin.

Other measuring tools that specifically measure the quality of life of parents with autistic children are the Parental-developmental disorder-quality of life scale or commonly abbreviated as Par-DD-QoL. In contrast to the WHOQOL-BREF scale, which has four domains to measure, namely physical, psychological, social, and environmental health. The Par-DD-QoL scale focuses on three dimensions, namely emotional, daily disturbances, and global quality of life (Baghdadli *et al.* 2014). Dimensions on the Par-DD-QoL scale are aligned with aspects of quality of life which include feelings of well-being, positive social interactions, and opportunities to achieve personal potential (Robert L. Schalock *et al.* 2002).

The Par-DD-QoL scale was first developed by Baghdadli *et al.* (2014) in French and is an adaptation of the Par-ENT-QoL scale of G. Berdeaux, C. Hervié, Smajda, and Marquis (1998). The Par-DD-QoL scale has been adapted in Arabic and has been tested for the validity of the scale by Achachra (2016) showing the results that the validity value can be said to be of good psychometric quality and can be used by professionals in the Algerian environment. In addition, the Par-DD-QoL scale is considered concise and can be used in clinical practice which can also make it possible to independently assess each person's quality of life (Achachra 2016).

The Par-DD-QoL scale has been adapted into Indonesian by Hakim (2021) and Latifah (2021). However, there has been no research that has carried out further validation and reliability on the Indonesian Par-DD-QoL adaptation scale. This study aims to measure the validity and reliability of the developed quality-of-life measurement tool for autistic children, namely the Par-DD-QoL (Baghdadli *et al.* 2014; Berdeaux G. *et al.* 1998). Therefore, it is hoped that this research can be a reference for further researchers in measuring the quality of life of parents with autistic children.

2 METHODS

This study uses a psychometric study where the researcher aims to test the validity and reliability of the measuring instrument of the variables to be tested, namely the quality of life (Budiastuti & Bandur 2018) with research subjects totaling 446 parents who have autistic children with an age range of 2–18 years, and the subject-taking technique used is a non-probability purposive sampling technique. The instrument used was Baghdadli's quality of life scale, namely Par-DD-QoL which was adapted from the Par-ENT-QoL questionnaire. This scale contains 18 questions with three components, namely: 1) Emotional related to deviant behaviors of autistic children that affect the psychological well-being of parents (stress, anxiety, sleep problems) contains 10 items, 2) Daily disturbances related to the adaptive skills of autistic children and some environmental characteristics such as scholarships, financial assistance, and protective measures contain 6 items, and 3) Quality of life, in general, contains 2 items (Baghdadli *et al.* 2014; Berdeaux G. *et al.* 1998). This scale is in the form of a 5-point Likert scale which will be distributed online using Google Forms and distributed directly to autistic institutions or communities in Indonesia. Data analysis was carried out in this study using the Confirmatory Factor Analysis method using the Lisrel 8.8 program.

3 RESULT

The results of the study show the standardized loading factor value of each item in the Quality of Life measuring instrument. A standardized loading factor is used for the validity of a measuring instrument. Items to measure emotional aspects can be seen from 10 items with EMO code. From the standardized loading factor value, all items on the emotional aspect can be said to be valid because all items have a standardized loading factor value > 0.3. In addition to referring to the standardized loading factor value to determine the validity of an item, it can be seen by comparing the calculated R-value with the R table. If the calculated R-value is greater than the R table, the item can be said to be valid. The R table value for a significance level of 5% with a research sample of more than 400 people is 0.098. Items to measure the emotional aspect with the EMO code were declared valid with 10 items. Then, the item to measure the daily distraction aspect with the GAHA code was declared valid with 6 items. Meanwhile, items to measure aspects of quality of life, in general, can be seen from 2 items with QOYU code. From the standardized loading factor value, it can be seen that one of the items in the quality of life aspect in general has a standard loading factor value of < 0.3 (QOYU 2) so the item is declared invalid. Furthermore, the calculated R values for each item on the aspect of quality of life, in general, are −0.24 and 0.57. Item QOYU 1 has a calculated R-value that is < from the table R-value, namely −0.24, which is declared invalid. So, for the aspect of quality of life in general it can be said to be invalid. The reliability test is done by looking at the value of Construct

Table 1. Validity and reliability test on the quality of life scale.

Validity and Reliability Variable	Standardized loading	Measurement Error	CR	VE	R Hitung
Emotional			0.85	0.37	
EMO 1	0.51	0.74			0.56
EMO 2	0.67	0.55			0.68
EMO 3	0.63	0.60			0.58
EMO 4	0.74	0.45			0.68
EMO 5	0.66	0.57			0.64
EMO 6	0.59	0.66			0.64
EMO 7	0.62	0.61			0.65
EMO 8	0.63	0.61			0.64
EMO 9	0.66	0.57			0.66
EMO 10	0.30	0.91			0.39
Problem			0.81	0.43	
GAHA 1	0.57	0.68			0.55
GAHA 2	0.68	0.53			0.62
GAHA 3	0.75	0.44			0.72
GAHA 4	0.75	0.44			0.70
GAHA 5	0.71	0.50			0.66
GAHA 6	0.39	0.85			0.46
Quality of Life in General			0.03	0.22	
QOYU 1	0.35	0.88			−0.24
QOYU 2	−0.57	0.67			0.57

Reliability (CR) and Variance Extracted (VE). A construct can be said to be reliable if it fulfills one of the reliable requirements, namely, the value of CR 0.7 and the value of VE 0.5. The CR and VE values can be seen from the SLF and Error values in Figure 1. Where it shows that there are 2 invalid items (validity value is around −0.57–0.75).

In the overall model fit test, to achieve a good model fit, several conditions must be met, namely the Root Mean Square Error of Approximation (RMSEA) value 0.08, Standardized Root Mean Square Residual (Standardized RMR) of 0.05, and Normed Fit Index (NFI), Non-Normed Fit Index (NNFI), Relative Fit Index (RFI), Incremental Fit Index (IFI), Comparative Fit Index (CFI), Goodness of Fit Index (GFI), and Adjusted Goodness of Fit

Table 2. Goodness of fit statistics results on quality of life scale (Par-DD-QoL).

GOFI	Nilai Hasil Hitung	Nilai standar untuk kecocokan baik	Conclusion
RMSEA	0.111	$\leq$ 0.08	*Marginal fit.*
NFI	0.91	$\geq$ 0.9	*Good Fit*
NNFI	0.91	$\geq$ 0.9	*Good Fit*
CFI	0.92	$\geq$ 0.9	*Good Fit*
IFI	0.92	$\geq$ 0.9	*Good Fit*
RFI	0.89	$\geq$ 0.9	*Marginal fit.*
Standardized RMR	0.072	$\leq$ 0.05	*Marginal fit.*
GFI	0.82	$\geq$ 0.9	*Marginal fit.*
AGFI	0.77	$\geq$ 0.9	*Marginal fit.*

Index (AGFI) 0.9. The results of Goodness of Fit Statistics can be seen in Table 2.

In Table 2, it is known that the RMSEA, RFI, Standardized RMR, GFI, and AGFI values are below the standard values for a good fit so they are categorized into a marginal fit. While the value of NFI, NNFI, CFI, and IFI shows a value of > 0.9 so it can be said to have a good model fit (good fit). Where these results can be concluded that this measuring instrument already has a fairly good model fit with 18 items and 3 aspects. In order to produce better results of Goodness of Fit Statistics than the previous results, modifications were made to the model by excluding invalid items and referring to Modification indices.

In this model modification recommendation, it can be seen that there are many modification suggestions given by the program to improve the fit of the model, but in this study, the researcher will only make modifications by excluding aspects of the general quality of life (QOYU) because it is invalid and adding an error covariance between 2 the same aspects, namely error covariance on items EMO2 with EMO1, items GAHA5 with GAHA3, items EMO4 with EMO3, items GAHA2 with GAHA1, and items EMO9 with EMO8. The results of the modification showed that 10 items on the emotional aspect were declared valid with a standardized loading factor value of around 0.33–0.70 and 6 items on the daily disturbance aspect were declared valid with a standardized loading factor value of around 0.39–0.79.

After the modification of the validity test model and the reliability test were carried out again to see the standardized loading value, CR and VE values in each aspect of the measuring instrument. The results of the validity test showed that the 16 items on the quality of life measuring instrument were declared valid and the reliability test showed that after the modification of the model, there was not much change in the CR and VE values. In addition, it can also be seen that the value of VE < 0.5 which does not meet the requirements to be said to be reliable. Therefore, the aspect where the VE value is < 0.5

but the CR is 0.7 is said to be "reliable enough" because it only fulfills one of the requirements to be said to be reliable.

4 DISCUSSION

Items to measure aspects of quality of life, in general, can be seen from 2 items with the code QOYU. From the standardized loading factor value, it can be seen that one of the items in the quality of life aspect in general has a standard loading factor value of < 0.3 (QOYU 2) so the item is declared invalid. Furthermore, the calculated R values for each item on the aspect of quality of life, in general, are −0.24 and 0.57. Item QOYU 1 has a calculated R-value that is < from the table R-value, namely −0.24, which is declared invalid. So, for the aspect of quality of life in general it can be said to be invalid. Aspects of quality of life are generally invalid, which can be caused by differences in the calculation method for each aspect where in the research conducted by Achachra (2016), Achachera (2014), and Baghdadli *et al.* (2014) it is said that the score on the quality of life aspect is the sum of the scores of the emotional aspect and the daily disturbance aspect. In the aspect of quality of life there are 2 items, the first item or item 17 contains the question "Are you satisfied with the treatment offered to your child?" and the second item or item 18 contains the question "Regarding your child's autism disorder, how is your current quality of life?".

The reliability test is done by looking at the value of CR and VE. A construct can be said to be reliable if it fulfills one of the reliable requirements, namely, the value of CR 0.7 and the value of VE 0.5. The CR and VE values can be seen from the SLF and Error values in Figure 1. Where it shows that there are 2 invalid items (validity value is around −0.57–0.75). Then from the calculation results of CR values 0.03 – 0.82 and VE 0.22 – 0.43, then every aspect can be said to be reliable with the fulfillment of one of the reliable conditions. Meanwhile, in a study conducted by Hakim (2021) and Latifah (2021), it was

Table 3. Test the validity and reliability of the quality of life scale after modification.

Validitas dan Relianilitas					
Variable	Standardized loading	Measurement Error	CR	VE	R Hitung
Emotional			0.84	0.35	
EMO 1	0.44	0.81			0.56
EMO 2	0.62	0.62			0.68
EMO 3	0.57	0.68			0.58
EMO 4	0.70	0.51			0.68
EMO 5	0.67	0.55			0.64
EMO 6	0.62	0.61			0.64
EMO 7	0.64	0.59			0.65
EMO 8	0.63	0.60			0.64
EMO 9	0.63	0.60			0.66
EMO 10	0.32	0.90			0.39
Ganguan harian			0.84	0.48	
GAHA 1	0.54	0.71			0.55
GAHA 2	0.66	0.57			0.62
GAHA 3	0.79	0.38			0.72
GAHA 4	0.74	−0.1345			0.70
GAHA 5	0.75	0.44			0.66
GAHA 6	0.38	0.85			0.46

stated that the validity value of the quality of life scale (Par-DD-QoL) was 0.223–0.692 and the reliability value was 0.903 which stated that all items on this quality of life scale valid and reliable.

5 CONCLUSION AND IMPLICATION

From this study, it can be concluded that the quality of life scale tested on the research sample of parents who have autistic children aged between 2 and 18 years has good validity and reliability if they abort 2 items on aspects of quality of life in general. This can be seen from the validity value of each item in the number 0.32–0.79 and the reliability value of 0.84 with a total of 16 items. Some of the advantages of this study are, the research sample used is specific, namely, parents who have autistic children aged 2–18 years in Indonesia. The limitation of this study is the difference in the score calculation process where, for the aspect of quality of life in general, it is calculated by adding up the scores of emotional aspects and aspects of daily disturbances. And, in the process of spreading the scale of the research, it is in a COVID-19 pandemic situation and the number of samples collected is feared not to be able to represent every parent of autistic children scattered in Indonesia. The spread of the scale using the Google form in addition to having more value there is a weakness, namely the research sample cannot fully understand the meaning of the measuring instrument because there is no direct explanation from the researcher. This makes it difficult for the research sample to understand each item. Therefore, it is hoped that future researchers, especially researchers who will test the validity and reliability of the quality of life measurement tools, can use a more interesting dissemination platform and can reach a wider research sample. Meanwhile, for further research that wants to measure the level, relationship or influence, it can eliminate items on the quality of life aspect in general if they have the same sample criteria, namely parents with autistic children aged between 2 and 18 years.

It is hoped that future researchers will use a quality of life measurement tool (Par-DD-QoL) as a measuring tool for quality of life research to identify quality of life in parents of children with autism. Also, the results of this study are used as a reference to develop psychological interventions, especially for autistic parents.

REFERENCES

Achachera, A. (2014). Impact de l'autisme sur la Qualité de vie des Parents. *le Bulletin scientifique de l'arapi*, (33), 85–89.

Achachra, A. (2016). Adaptation d'un Questionnaire à l'environnement Algérien Évaluant l'impact du Handicap de l'enfant sur la Qualité de vie(QDV) des parents « Le par-qol». *2*(2), 163–180.

Baghdadli, A., Pry, R., Michelon, C., & Rattaz, C. (2014). Impact of Autism in Adolescents on Parental Quality of Life. *Qual Life Res*, *23*(6), 1859–1868. doi:10.1007/s11136-014-0635-6

Berdeaux G., Herviè C., Smajda, C., & Marquis, P. (1998). Parental Quality of Life and Recurrent ENT Infections in their Children: Development of a Questionnaire. *Quality of Life Research*, *7*, 501–512.

Budiastuti, D., & Bandur, A. (2018). *Validitas dan Reliabilitas Penelitian* Jakarta Mitra Wacana Media

Cappe, É., Pedoux, A., Poirier, N., Downes, N., & Nader-Grosbois, N. (2019). Adaptation and Quality of Life of Parents with a Child with Autism Spectrum Disorder: A Comparative Exploratory Study between France, French-Speaking Belgium and Quebec. *Psychologie Française*, *65*(2), 141–155. doi:10.1016/j.psfr.2018.11.002

GROUP, T. W. (1998). Development of the World Health Organization WHOQOL-BREF Quality of Life Assessment. *Psychological Medicine*, *28*, 551–558. Doi:10.1017/S0033291798006667

Hakim, M. L. (2021). *Hubungan Antara Self-Compassion Dan Kualitas Hidup Orang Tua Yang Memiliki Anak Autism Spectrum Disorder (Asd)*. Universitas Muhammadiyah Malang UMM Institutional Repository.

Hambatan Intelektual Dan Orang Tua Yang Memiliki Anak Dengan Hambatan Autism Di Skh Madina Kota Serang-Banten). *Prosiding Seminar Nasional Pendidikan FKIP*, *3*(1), 470–484.

Latifah, T. N. (2021). *Hubungan Antara Mindfulness Trait Dan Kualitas Hidup Pada Orang Tua Yang Memiliki Anak Autis*. (S1), Universitas Muhammadiyah Malang, Umm Institutional Repository.

Robert L. Schalock, Ivan Brown, R. B., Robert A. Cummins, David Felce, Leena Matikka, Kenneth D. Keith, & Parmenter, T. (2002). Conceptualization, Measurement, and Application of Quality of Life For Persons With Intellectual Disabilities: Report Of An International Panel of Experts. *Mental Retardation*, *40*(6), 457–470. Doi:10.1352/0047-6765(2002)040<0457:Cmaaoq>2.0.Co;2

S.M. Skevington, Lotfy, M., & O'connell, K. A. (2004). The World Health Organization's Whoqol-Bref Quality of Life Assessment: Psychometric Properties and Results of the International Field Trial a Report from the Whoqol Group. *Quality of Life Research*, *13*(2), 299–310.

Sukmadi, M. R., Sidik, S. A., & Mulia, D. (2020). Kualitas Hidup Orang Tua Yang Memiliki Anak Berkebutuhan Khusus (Studi Kasus Pada Orang Tua Yang Memiliki Anak Berkebutuhan Khusus Dengan.

Vasilopoulou, E., & Nisbet, J. (2016). The Quality of Life of Parents of Children With Autism Spectrum Disorder: A systematic review. *Research in Autism Spectrum Disorders*, *23*, 36–49. doi:10.1016/j.rasd.2015.11.008

Families Mental Health and Challenges in the 21st Century – Arifin et al. (eds)

Belief in conspiracy and chronic hatred toward China

Yuni Nurhamida*, Chairunnisak & Fath Mashuri
Fakultas Psikologi, Universitas Muhammadiyah Malang, Malang, Indonesia

ABSTRACT: The study aims to reveal the effect of belief in a conspiracy on hatred toward the Chinese in Indonesia. The sampling technique in this study used convenience sampling with 362 research participants. They are Indonesian citizens aged at least 18 years. The instruments in this study are the Scale of Beliefs in Jewish Conspiracy Theory that modified to belief in Chinese Conspiracy and the scale of Group-Based Hatred. Meanwhile, the data retrieval method used in this study is a Likert scale with a simple linear regression test. Results of studies show that beliefs in conspiracy theories have a significant influence on hatred in ethnic Chinese ($r = 0.465$). According to the explanation that has been described, it can be concluded that the emergence of hatred towards ethnic Chinese is due to the belief in conspiracies which then encourage individuals to have prejudices, then high prejudices against other people or groups will cause hatred.

Keywords: Ethnic Chinese, Beliefs in Conspiracies, Hatred

1 INTRODUCTION

Hate is a very strong negative emotion and a strong dislike in a person. Individuals or groups who have hatred toward an object will perceive the object as something bad, moral, dangerous, and must be avoided (Staub 2003). Meanwhile, Sternberg (2003; 2005) stated that hatred is a mixture of several negative emotions, including anger, fear, disgust, and condescension. Based on research conducted by Halperin (2008), 83.3% said that they wanted something bad to happen to the hated group.

The hate group that is the focus of this research is the indigenous hatred for China. Therefore, the researcher will explain some of the issues that often occur between the two groups. The Center for Strategic and International Studies (CSIS) (in voa Indonesia 2021) found 9,545 hate speeches against Shia, Ahmadiyah, and Chinese groups on Twitter from January 1, 2019, to July 31, 2021. Based on the survey conducted that hate speech received the most attention, and the response from Warganet was hate speech against China which reached 28,933 likes.

The phenomenon of hatred between indigenous and ethnic Chinese can be traced in the history of relations between natives and non-natives. It can be traced back to the Dutch colonial period, when the Chinese in Indonesia became business partners of the VOC so that they were treated preferentially compared to the natives. At that time, the source of indigenous suffering was because the VOC put China in the same position. Then not long after that the Sultans of Java, the Netherlands, and England made the Chinese as tax collectors in the areas they controlled. This incident sparked hatred toward China and became one of the causes of the emergence of the stereotype of Chinese as mercenaries and extortionists (Dhani

*Corresponding Author: yuninurhamida@umm.ac.id

DOI: 10.1201/9781003402381-16

2016). Another example is the tragedy of the May 1998 riots, when President Suharto stepped down, which is an example of the riots, which were colored by the news spread to the public that people of Chinese descent caused the economic crisis in Indonesia (Himawan 2020). Even though it has been more than two decades since the tragedy occurred, discrimination and hatred against ethnic Chinese still exist today. Recently, the COVID-19 pandemic has hit the whole world, including Indonesia. The COVID-19 pandemic, which started in Wuhan, China, has become a new ammunition to re-create negative emotions towards ethnic Chinese in Indonesia.

Jusuf Kalla when serving as vice president once stated that the rich people in Indonesia are of Chinese descent and are mostly Christians and Confucians, while it is the indigenous Muslims who are poor. Therefore, hatred and discriminatory views against ethnic Chinese by people who identify themselves as part of the lower class who feel that the rights and power of their group have been usurped by Chinese-Indonesians (Setijadi 2017).

Several factors cause hatred between groups: first is the fact that hate speeches are carried out by individuals which are then discussed and spread among ingroups. Thus, ingroup members will also have hatred for the outgroup, not because of what they do but about who they are. The second factor is the presence of past memories that result in individuals having an assessment of future events. From the ingroup's knowledge about violent behavior from the outgroup, it affects future behavior, thus concluding that the outgroup is the bad guy. The last factor is the lack of direct interaction to strengthen hatred between groups (Fischer et al. 2018).

Based on the first factor, the cause of hatred between groups is hate speech made by individuals which are then spread among group members. Hate speech spread within the group then encourages people with higher roles to create conspiracies that can influence ingroup hatred in outgroups. People who believe in the conspiracy are discussed in psychology with the concept of belief in conspiracy theory.

Conspiracy theory is defined as a false narrative that explains the main cause of an event that is composed of more than one bad actor cooperating with each other. Conspiracy content is often spread with dangerous ideas or understandings that contain hate speech that is often used by hate-spreading groups (Faisal & Sadida 2021). Individual beliefs about conspiracy theories can be defined using the concept of prejudice. Watson (1984) states that prejudice is a negative attitude that is not tolerated.

2 METHODS

The research design used in this study is correlational. The correlational design examines the relationship between variables in a real social context, which aims to reveal the relationship between variables that naturally have occurred (Adelina et al. 2017).

In this study, the target population is indigenous people, aged at least 18 years. The indigenous criteria are because in this study we want to see the hatred of the Indigenous group in the Chinese group. Then the criteria for a minimum age of 18 years are determined by considering the complexity of the scale that requires adequate cognitive understanding, and the authority to provide the willingness to fill in the data independently as an adult.

Sampling in this study used a convenience sampling technique, with a number of 362 participants involved. The subjects in this study were Indonesian citizens who had a minimum age of 18 years and came from various ethnic groups in Indonesia. The total number of subjects in this study was 362 people.

It is known that there are 167 male subjects with a percentage of 46.1% and 195 female subjects with a percentage of 53.9%. Most of the respondents were Javanese with a total of 124 respondents or 34.4%, then Madurese with 44 respondents or 12.2%, Sundanese with 33 respondents or 9.1%, and other ethnicities with 161 respondents or 44.4%.

3 VARIABLES AND RESEARCH INSTRUMENTS

Conspiracy beliefs are beliefs or beliefs of an individual or group in conspiracy theories that are actually different from the facts around them. The collective scale used for this variable is the Belief in Jewish Conspiracy Scale compiled by Swami (2012) and then adapted into Indonesian by Zein, Rikardi, & Arinda (2020) with 12 favorable items. The items are then adjusted to the needs of the researcher so that they use 10 items from this scale.

The hatred used in this study is group-based. This research is focused on knowing the hatred between the natives and the Chinese group. The group-based hate scale instrument is referred to on the hate scale compiled by Halperin, Canetti, & Kimhi (2012) which was then contextualized by Nurhamida (dissertation, on progress, unpublished). This instrument has 5 items of chronic hatred. Chronic/mild hatred is a persistent emotional attitude, which absolutely rejects all out-group members which involve a number of negative feelings, as well as a stable cognitive perception that the hated group member causes violations against the in-group in a severe, repeated, unfair, and intentional. This is the result of a crime rooted in the character of a hated group member. In order to prevent future painful offenses, the hate group wants absolute separation from other group members.

The scale model used is the Likert scale, each scale has 6 points (1 'Strongly Disagree' and 6 'Strongly Agree'). There are 2 scales used in this study, namely the Belief in Jewish Conspiracy Scale (10 items) and the Chronic Hatred Scale (5 items).

Furthermore, the research scale instrument was transferred in the form of a Google form, then a tryout was carried out on 60 subjects who were distributed through social media (Instagram, WhatsApp, and Twitter). Then to analyze the validity and reliability of the researcher using SPSS 23, from the implementation of statistical tests, the results for the hate scale that have been tested are declared valid with a validity index of 0.712–0.876 with a reliability score of = 0.960. And on the Belief in Jewish Conspiracy Scale which has been tested and adjusted for items by doing a tryout and it is stated that all items are valid with a validity index of 0.617–0.809 with a reliability score of = 0.936. Based on the results obtained, it is stated that both measuring instruments have valid items and have good reliability so both scales can be used in this study.

4 RESULT

The results of descriptive statistical analysis showed that 68 (18.8%) respondents had low conspiracy beliefs, 225 (62.2%) respondents were in the medium category, and in the high category there were 69 (19.1%) respondents. While on the hate variable, there are 57 (15.7%) respondents who have a low perception of hatred, and 241 (66.6%) respondents in the medium category, then in the high category 64 (17.7%) respondents. Of the 362 respondents who have participated in this study, it shows that the level of hatred and belief in conspiracies is at a moderate level.

Regression analysis

A simple linear regression test was used to measure how much influence the belief variable on conspiracy (X) had on the hate variable (Y).

Table 1. Simple linear regression test belief in conspiracy and chronic hatred.

	Unstandardized coefficients				
Model	B	Standard Error	R Square	F	Sig.
(Constant)	.360	1.864			
Belief in Conspiracy	.902	.051	.465	312.860	.000

Based on the presentation of the results of the regression test, the value of significance of 0.000 (< 0.05) means that the belief in conspiracy variables has a significant effect on the hate variable. The value of the coefficient of determination (R^2), which is 0.465, explains that the belief in the conspiracy has a 46.5% influence on chronic hatred.

5 DISCUSSION

This study was conducted with the aim of finding out the influence that exists between belief in conspiracy and hatred of China. Through the results of research conducted on 362 Indonesian citizens aged 18 years and over and based on the results of statistical analysis, it shows that there is a significant influence on belief in conspiracy with hatred against China. If a person has an increased belief in the conspiracy then the hatred experienced will also increase, and vice versa. The findings in this study provide empirical evidence that the indigenous belief in conspiracies triggers the emergence of hatred towards China. Then 362 respondents who participated in this study indicated that the level of hatred and belief in conspiracies is at a moderate level.

Individuals with a belief in conspiracies are in fact individuals who choose shortcuts to simplify complex and difficult things to understand. Results of research that has been carried out (March & Springer 2019) show that individuals who have strange beliefs/magical thoughts, manipulative, dominant, and heartless are more likely to believe in conspiracy theories. As has been found in several studies that feelings of helplessness tend to make individuals believe conspiracy theories to be a defense for them because by believing in conspiracy theories individuals feel they can blame others.

As explained by Haider (1958) in Douglas, Sutton, & Cichocka (2017), three motives can encourage individuals to believe in conspiracy theories, namely, epistemic motives, existential motives, and social motives. Next, the researcher will discuss one by one how the three motives can affect beliefs in conspiracies to their effect on hatred. First, epistemic motives, epistemic, namely making the world more understandable and predictable, because it provides a clear and simple causal explanation. Basically, every individual has a need for cognitive closure, so to fulfill his knowledge needs and to cover up feelings of uncertainty, individuals will easily believe in conspiracy theories and tend to make instant decisions (Winata 2021). Van Prooijen and Jostmann (2013) found through their experiments that uncertainty increases belief in conspiracy theories. One example of conspiracy theories is the constellation about China recently. It is believed by most natives that the Covid-19 virus created in the Chinese Communist Party (CCP) military laboratory (Kominfo 2022). Although this theory is not known to be true but not a few believe it because of an epistemic need. So that individuals who believe in this conspiracy theory tend to have hatred for China.

The second is existential motives, people believe in conspiracy theories because of the desire to have control and control over their lives (Muhammad 2022). Individuals with conspiracy beliefs tend to show self-confidence, and a sense of security and seem to have control over what happens to the individual or social group (Aminulloh et al. 2021). This is relevant to the recent Covid-19 vaccination issue. It is known that many people refuse to comply with health protocols and are reluctant to receive the Covid-19 vaccine. People's refusal to obey this order is a form of impact from their belief in conspiracy theories. The public believes that Covid-19 is not a natural disease but is engineered by the elite and the Jews (Arumsari et al. 2021). Indeed, individuals who have existential motives believe in conspiracy theories solely to maintain their existence or control in threatened situations. To maintain their existence, individuals with this understanding will then blame certain parties who are considered to have caused the existing problems. Therefore, the indigenous belief in the conspiracy theories that exist can make them have hatred for China.

The last motive, namely the social motive, refers to the individual's need to maintain a positive image of himself and his group (Douglas et al. 2017). In order to maintain self and group image, individuals will prefer to believe in conspiracy theories in order to cast blame or a bad image on other groups. From this motive, it is clear that conspiracy theories are very interesting to believe for people who are in a position of being disadvantaged and deprived of their rights (Prooijen & Douglas 2018). As the item on the confidence gauges against conspiracy used in this study states, "The Chinese have caused an economic crisis in this country due to their ambition to achieve the interests of their own people".

This theory is certainly very suitable for indigenous people who have been affected by the economic crisis. So that from this condition the status of the indigenous group became increasingly inferior, with this inferiority leading the natives to hatred towards China because of the difference in status in which the natives were the disadvantaged party.

After discussing the three motives that can encourage individuals to believe in the conspiracy, it can be concluded that these three motives can lead individuals to hate ethnic Chinese. However, this does not just happen but goes through a long process so that prejudice can be formed which can then lead to hatred. Individuals with conspiracy beliefs can have prejudice if these beliefs are followed by social norms in the surrounding environment. In addition, based on factors that can cause prejudice, namely the desire to find a scapegoat for mistakes that have been made. This is relevant to the social motives of conspiracy theories, which both want to maintain a positive image of themselves and their groups. Then prejudice is also influenced by past experiences and stereotypes (Mahardika et al. 2019).

6 CONCLUSION AND IMPLICATION

The results of this study indicate that belief in the conspiracy has an effect on hatred toward the Chinese in Indonesia. It means that the higher the individual's belief in conspiracy theories, the higher the individual's hatred for China. Because belief in the conspiracy theory has the effect of being able to trigger hatred in other groups, it is necessary to make efforts to avoid believing in conspiracy theories.

Some programs can be made to prevent believing in conspiracy theories including an awareness to evaluate ingroup, and an awareness that our groups can also make mistakes. Therefore, cooperation from various parties (educational institutions, government, and public figures) is needed to prevent the development of conspiracy theories that can trigger hatred between groups. One of the efforts that can be applied is the discussion of inclusiveness on a regular basis so that it is not easy to believe in conspiracies, especially regarding Chinese citizens. To build good relations between the indigenous and ethnic Chinese, special activities can be held that involve both parties. As much as possible we can fight conspiracy theories that can foster prejudice or hatred between groups. Suggestions for further research are that researchers can continue to develop other variables that affect hatred, and expand the subject used in the study so that it does not only involve certain areas and it can be a reference to be applied in society.

REFERENCES

Adelina, F., Hanurawan, F., & Suhanti, I. Y. (2017). Hubungan Antara Prasangka Sosial dan Intensi Melakukan Diskriminasi Mahasiswa Etnis Jawa Terhadap Mahasiswa yang Berasal dari Nusa Tenggara Timur. *Jurnal Sains Psikologi*, 1–8.

Aminulloh, A., Artaria, M. D., Fianto, L., & Setiamandani, E. D. (2021). Propaganda dan Teori Konspirasi: Wacana Masyarakat Terhadap Covid-19 di Indonesia. *Jurnal Nomosleca.*

Arumsari, W., Desty, R. T., & Kusumo, W. E. (2021). Gambaran Penerimaan Vaksin Covid-19 di Kota Semarang. *Indonesian Journal of Health Community*, 35–45.

Dhani, A. (2016). Sejarah Kebencian Terhadap Etnis Tionghoa. *tirto.id.*
Douglas, K. M., Sutton, R. M., & Cichocka, A. (2017). The Psychology of Conspiracy Theoris. *Current Directions in Psychological Science.*
Faisal, F. F., & Sadida, N. (2021). Hati-hati Dengan Berita Kosnpirasi: Studi Pengaruh Terpapar Berita Konspirasi Covid-19 Terhadap Persepsi Risiko dan Kepatuhan Terhadap Prtokol Kesehatan. *Mind Set*, 51–62.
Fischer, A., Halperin, E., Canetti, D., & Jasini, A. (2018). Why We Hate. *Emotion Review*, 10(4).
Halperin, E. (2008). Group-based Hatred in Intractable Conflict in Israel. *Journal of Conflict Resolution*, 52 (5), 713–736.
Halperin, E., Canetti, D., & Kimhi, S. (2012). In Love With Hatred: Rethinking the Role Hatred Plays in Shaping Political Behavior. *Journal of Applied Social Psychology*, 2231–2256.
Himawan, Eunika, M. (2020, Mei 20). 22 Tahun Setelah Kerusuhan Anti-Cina Mei 1998, Riset Ungkap Prasangka dan Trauma Masih Ada. [Unggahan blog]. *Diakses dari* https://theconversation.com/22-tahun-setelah-kerusuhan-anti-cina-mei-1998-riset-ungkap-prasangka-dan-trauma-masih-ada-138962
Kominfo. (2022). (HOAKS) Virus Corona Covi-19 Dibuat di Lab Militer Partai Komunis China.
Mahardika, P., Rosalinda, S. Y., Asyiva, N., Hilmawan, H. S., & Rianingrum, D. (2019). *Prasangka.*
Muhammad. (2022). Peningkatan Korban Kejahatan Siber Selama Pandemi Covid 19. *NUSANTARA: Jurnal Ilmu Pengetahuan Sosial*, 2043–2054.
Prooijen, J.-W. V., & Douglas, K. M. (2018). Belief in Conspiracy Theories: Basic Principles of an Emerging Research Domain. *European Journal of Social Psychology*, 897–908.
Setijadi, C. (2017). *Chinese Indonesians in the Eyes of the Pribumi Public.* ISEAS Yosuf Ishak Institute, 73.
Staub. (2005). History, Vulnerability, and the Frustration of Basic Human Needs. In Sternberg, R. (Ed.). *The psychology of hate.* Washington, DC: American Psychological Association.
Sternberg, R. J. (2003). A Duplex Theory of Hate: Development and Application to Terrorism, Massacres, and Genocide. *Review of General Psychology.*
Sternberg, R.J. (2005). Understanding and Combating Hate. In Sternberg, R.J. *The Psychology of Hate.* Washington, DC: American Psychological Association.
Swami, V. (2012). Social Psychological Origins of Conspiracy Theories: The Case of the Jewish Conspiracy Theory in Malaysia. *Fromtiera in Psychology.*
Zein, R. A., Rikardi, A. A., & Arinda, M. F. (2020). The Enemy of 'Ummah: Belief in Jewish Conspiracy Theories Indirectly Affected Vaccination Decisions.
Zulkarnain. (2020). Ujaran Kebencian (hate speech) di Masyarakat Dalam Kajian Teologi. *Studia Sosia Religia.*

Families Mental Health and Challenges in the 21st Century – Arifin et al. (eds)
© 2024 The Author(s), ISBN 978-1-032-51442-0

Parental acceptance and family support for *psychological well-being* in parents of children with special needs

Susanti Prasetyaningrum*, Cahyani Syafitri & Rizky Susanti

ABSTRACT: Parents who have children with special needs will experience more challenges when compared to parents who have normal children in general. The limitations that children have can cause parents to have low psychological well-being. Psychological well-being can be influenced by several things, including parental acceptance and family support. The purpose of this study is to determine the effect of parental acceptance and family support on the psychological well-being of parents of children with special needs. The approach in this research is correlational quantitative. This study uses the Parental Acceptance-Rejection Questionnaire (PARQ), The MOS Social Support Survey, and Ryff's Scale which are distributed offline. A sampling of data using a purposive sampling technique with a total of 130 respondents was carried out. Based on the results of multiple linear regression analysis, it is known that parental acceptance and family support affect the psychological well-being of parents of children with special needs ($p<0.05$ and $R^2 = 0.594$).

Keywords: parental acceptance, family support, psychological well-being

1 INTRODUCTION

Children with special needs are children who have physical, cognitive, or emotional limitations that cause children's delays in development, including learning difficulties, Attention deficit hyperactivity disorder (ADHD), mental retardation, physical disorders, sensory, speech and language disorders, autism, as well as emotional and behavioral disorders (Ni'matuzahroh *et al.* 2021). Based on data from the Central Statistics Agency (BPS), it is known that the number of children with special needs in Indonesia reaches 1.6 million (Kemdikbud 2019). Meanwhile, in the data presented by the Data and Information Technology Center (2020), it is known that as many as 26,448 children are enrolled in extraordinary schools in East Java Province. When having children with special needs, indirectly parents will experience greater challenges when compared to parents who have normal children in general. Children's imperfections can indeed have a negative impact on some parents, where parents feel disappointed because reality does not match what is expected.

Kubicek (2013) found that parents with children with special needs often face behavioral challenges and experience life pressures due to delays or inability of children in their development. When parents take care of children with special needs, it is necessary to be mentally and physically prepared, it also requires greater attention and affection than parents in general. In addition, parents with children with special needs need more effort than parents with normal children, because they mostly do not yet understand how to deal with their children with needs. Bougher-Muckiana (2015) explained that children's limitations in communicating cause parents with children with special needs to feel confused about building a good relationship with their child.

*Corresponding Author: susanti_p@umm.ac.id

DOI: 10.1201/9781003402381-17

Parents with children with special needs tend to have higher levels of stress and lower levels of *psychological well-being* compared to parents who have normal children (Kuhn *et al.* 2018). This is in accordance with research conducted by Cox *et al.* (2015), which explains that some mothers who have autistic children have a low level of *psychological well-being* due to anxiety about their child's future. Parents with ADHD children in particular a mother will experience higher psychological distress than other parents because they take more responsibility in parenting (Moen *et al.* 2014). Meanwhile, children with special needs will need the existence of healthy parents both physically and psychically in their care during the development period. Nikmatunasikah (2020) explained that parents who have high *psychological well-being* will foster parental involvement in the lives of children with special needs, and from this, they will achieve good psychomotor, cognitive, affective, social and academic development in children with special needs.

Psychological well-being of parents with children with special needs will arise if there is acceptance from parents of their child's condition. One of the studies conducted by Ghoniyah and Savira (2015) showed that parents with *Down syndrome* children have a *psychological well-being* picture that is superior to the dimension of self-acceptance, where parents are able to accept their child's condition sincerely without considering their child a burden in their lives. Martin *et al.* (2019) explained that gratitude and hope can affect *psychological well-being*, depression, and anxiety in parents of children with developmental disorders. Listiyandini *et al.* (2015) explained that gratitude is shown by the ability to appreciate others and the existence of God, being able to create positive emotions toward his life and being able to express his feelings and appreciation.

One form of realizing *the psychological well-being* of parents with children with special needs is the acceptance of parents related to the needs of the child. However, it is undeniable that many parents with children with special needs have not been able to accept their children. As for their development, such children are in dire need of acceptance from others, especially acceptance from parents. This is in line with the statement by Yavuzer (2010) who explains that a positive relationship between the child and the mother can be formed due to the acceptance by the mother of the child. In addition, the acceptance of parents will also cause a sense of love, affection, and love for their child so that parents can participate in their child's activities to support the development and growth of the child. This is in accordance with research conducted by Neff and Fuso (2014) where the results of the study showed that the acceptance and rejection of parents and attitudes in raising children with mental disabilities have a positive relationship, this means that if parents are able to accept their children, then parents will have a good attitude when nurturing and raising their children who have mental disabilities. Not only does it have an impact on children, but this can also have an impact on parents. The inability of parents to accept the imperfections of their children will cause psychological distress in parents (Petcharat & Liehr 2017).

This can also be seen from the results of the researcher's assessment of parents who attended school at one of the special schools in Malang. Based on the results of interviews conducted on December 22, 2021, with 3 parents of students. It is found that parents who have not been able to accept their child's situation because they feel ashamed when facing other people's negative responses regarding their child, they feel that their child has no limitations, they feel that their child is the same as normal children in general. This is in accordance with research by Kosasih and Virlia (2016) which explains that there is a low parental acceptance of children with mental retardation. Tarnoto (2016) found that there were 42.27% of parents who were less concerned about children with special needs, while 41.21% of parents had less understanding related to children with special needs so that it had an impact on the growth and development of children.

Parental acceptance is a condition where parents can accept a reality that exists, this is evidenced by the existence of affection, affection, attention, and the emergence of comfort and encouragement from parents toward their children regardless of the limitations possessed by the child (Khaleque & Rohner 2011). The process of accepting parents to children with special needs often takes a fairly long time. Kania and Yanuvianti (2018) found that the acceptance of parents of children with special needs will appear if parents receive emotional support, instrumental support, reward support, and information support from those around them.

In addition to parental acceptance, the *psychological well-being* of parents with children with special needs will be more perfect if there is support from the closest family. This is reinforced by research that has been carried out by Kusnadi *et al.* (2021) which states that with family support, the *psychological well-being* of parents with children with disabilities is getting better. Olianda and Rizal (2020) found that social support has an effect on the stress of parenting carried out in mothers who have children with mental disabilities. Puspita (2018) explained that there is a relationship between social ethnicity and the acceptance of parents who have children with *Down syndrome*.

Previous research has examined a lot related to parental acceptance, family support, and *psychological well-being*. Research conducted by Carrasco *et al.*, (2019), Winarsih *et al.*, (2020), Budiarti and Hanoum (2019), Adha (2018), and Rusdiana (2018) examined the relationship between parental acceptance and family support with *psychological well-being* separately. However, so far, there has not been a single conceptualization between parental acceptance, family support and *psychological well-being* in parents of children with special needs. When knowing how these three variables are related, it will give an idea of what parents need to pay attention to improve *psychological well-being*.

Based on this phenomenon and see how important this problem is. So researchers are interested in examining the effect of parental acceptance and family support on *psychological well-being* in parents of children with special needs. The purpose of this study is to determine the effect of parental acceptance and family support on *psychological well-being* of parents of children with special needs. Meanwhile, the benefits of this research can be used as a scientific contribution for counselors, teachers and parents with children with special needs in overcoming problems related to *psychological well-being*, parental acceptance and family support.

2 METHODS

The research used is non-experimental quantitative research. Quantitative research is a structured research that qualifies data to be generalized (Kurniawan & Puspitaningtyas 2016). This study uses a correlational study approach which is then analyzed by multiple linear regression analysis. Multiple linear regression analysis was performed with the aim of testing the influence of two or more free variables on one bound variable (Janie 2012).

This study used *a non-probability purposive sampling* technique. The *non-probability purposive sampling* technique is a process of selecting samples based on certain desired criteria (Latipun 2017). The criteria used in this study were (1) Having children with special needs; (2) Domiciled in the Greater Malang area (Batu City, Malang City, and Malang Regency).

There are 3 variables in this study, namely parental acceptance, family support, and *psychological well-being*. Parental acceptance is a condition in which parents who have children with special needs can accept the situation of their child by showing affection and care for children with special needs. The scale used in this study was the *Parental Acceptance Rejection Questionnaire* (PARQ) developed by Rohner *et al.* (2005) with a validity index of 0.316–0.628 and a reliability of 0.893. The PARQ scale consists of 26 items using *the Likert* scale. Meanwhile, family support is the attitude, actions, and acceptance of each family member aimed at children with special needs. This study used *The MOS Social Support Survey* measuring instrument developed by Sherbourne and Stewart (1991) with a validity index of 0.416 – 0.811 and reliability of 0.944. The scale consists of 19 items. There are 4 answer choices that use *the Likert* scale to measure the PARQ scale and the MOS Social Support, namely 1 "*strongly disagree*", 2 "*disagree*", 3 "*agree*" and 4 "*strongly agree*". An example of a statement item in The MOS Social Support is "*I have someone to talk to about my child's development*".

Next is the bound variable (Y) is *psychological well-being* which is defined as a condition where parents who have children with special needs are able to live their lives well because they have life goals, have a positive attitude towards themselves, and can develop their potential so that they are able to manage their lives independently. This study used *the Ryff's Scale* (2006) measuring instrument with a validity of 0.308–0.708 and reliability results of 0.933. The instrument consists of 42 items using *a Likert* scale with answers of 1 "*strongly disagree*", 2 "*disagree*", 3 "*agree*" and 4 "*strongly agree*".

Scale distribution in respondents in 5 special schools located in the Greater Malang area. Each SLB researcher gets 20–30 respondents. The deployment will be carried out from May 22, 2022, to June 15, 2022. The respondents obtained in this study were 136 parents who had children with special needs. However, a total of 6 respondents were eliminated because they did not meet the criteria in this study, so this study only used 130 respondents.

Furthermore, conducting data analysis starts with scoring, then validity and reliability tests, assumption tests and multiple linear regression tests using *SPSS for windows* version 26 software. Sugiyono (2011) explained that multiple linear tests were carried out to predict the magnitude of the influence of more than two free variables (X1, X2) on bound variables (Y).

3 RESULT

This study used multiple linear regression analysis. Before conducting multiple linear regression analysis, it is required to perform tests of normality, linearity, multicollinearity, and homoskedasticity. Based on the residual normality test, it is known that the data is distributed normally if the significance value > 0.05. Based on the normality test that has been carried out, it can be seen that the data is distributed normally with a significant value of 0.2 ($p > 0.05$). Meanwhile, the linearity test results on the variables of parental acceptance and family support were 0.009 and 0.021. This shows that there is no linear relationship between the variables of parental acceptance and family support *with psychological well-being*. In addition to this, researchers also conducted a multicollinearity test on *a free* variable (independent) which showed a VIF value of 1,516 ($p < 10$). Based on these results, it can be seen that there is no multicollinearity in *the independent* variables in this study. Then the researchers conducted a homoskedasticity test with a significant value of 0.001 ($p < 0.05$), which showed that the data were homoskedasticity.

The results of hypothesis tests that have been carried out through multiple linear regression shows that $F = 93.012$ and a significant value of 0.00 ($p < 0.05$). Based on this, it can be seen that there is an influence of parental acceptance and family support together on *psychological well-being* in parents of children with special needs.

Furthermore, a partial test was carried out to determine the influence of *independent* variables on *partial dependent* variables. Based on the table above, it shows that the parental acceptance variable has a significant effect on *the psychological well-being* of parents of children with special needs with a value of 0.00 ($p < 0.05$) and the family support variable affects *the psychological well-being* of parents of children with special needs with a value of 0.00 ($p < 0.05$). The parental admission variable has an *unstandardized beta* value of 0.982. While the family support variable has a value of 0.413. This shows that the parental acceptance variable has a greater influence on the *psychological well-being* variable when compared to the family support variable.

It can be known if the *value of R Square* obtained is 0.594. The regression equation in this study is $Y = (0.982*PO) + (0.413*DK) + 17.60$. Thus, it can be said that parental acceptance and family support have a role of 59.4% in *psychological well-being*, while 40.6% *of psychological well-being is* influenced by variables that are not studied. Therefore, it can be concluded that parental acceptance and family support affect *psychological well-being* in parents of children with special needs.

4 DISCUSSION

The results showed that parental acceptance and family support influenced the *psychological well-being* of parents of children with special needs in a positive direction. This means that parental acceptance and family support have a positive influence on *psychological well-being*. This is in accordance with research conducted by Budiarti and Hanoum (2019) and Prabowo (2017) showing that parental acceptance and family support have an influence on *psychological well-being* in parents who have children with special needs. If parents are able to accept the condition of children with special needs and get support from the family, then parents of children with special needs will have high *psychological well-being*, and vice versa.

The results of the analysis showed that parental acceptance has a greater influence on *psychological well-being* in parents of children with special needs when compared to family support. This means that the internal factor of parents, namely acceptance of the condition of the child who has special needs, gives rise to positive feelings and attitudes of parents who are able to direct *psychological well-being*. This external factor in the form of family support as a support for parents has positive emotions. Research conducted by Rahmahdianti and Rusli (2020) found that working mothers who have children with mental disabilities have high *psychological well-being*. Aurelia *et al.* (2022) explained that parents who have children with special needs have the most prominent *psychological well-being* picture in the dimension of self-acceptance, where parents are able to accept and try to make peace with the circumstances and conditions of their children who have limitations.

The individual can achieve self-actualization, functioning optimally and maturely when he is able to accept well the advantages and limitations possessed (Rachmayani & Ramdani 2014). When parents are able to accept the fact that their child has limitations, then parents will carry out maximum care for the child, and from this, parents will feel a meaningful life and have a purpose in life. On the other hand, if the individual is unable to accept the limitations he has, it will cause a negative perception of himself to cause feelings of disappointment and unable to achieve life satisfaction (Ryff & Singer 2006). Individuals who have gratitude in themselves will realize and always take positive things so as to be able to perceive themselves that they receive a lot of goodness and gifts from both God and people in the surrounding environment so as to be able to increase their motivation to act well and repay the kindness to others in the form of deeds, words, and feelings and will eventually be able to create positive relationships with others, independent, and able to function fully in a social environment (Prabowo 2017).

In addition to parental acceptance, *psychological well-being can be* influenced by family support. The results showed that family support *affects psychological well-being* in parents of children with special needs. This is supported by research conducted by Hikmah (2020) which states that the *psychological well-being of mothers* with children with special needs will increase if they get support from their families. Balcells-Balcells *et al.* (2019) also found that family support is an important factor in the quality of life of parents with children with special needs. This means that the higher a parent with a child with special needs receives support from his family, the more *psychological well-being* will increase.

With the existence of support from the family, parents with children with special needs will be lighter when carrying out their roles as parents, they are not burdened in meeting the needs of themselves and their children because there is someone who will help them (Winarsih *et al.* 2020). A similar study conducted by Zahira (2020) states that the psychological pressure that occurs on parents of children with special needs can have a negative impact on the psychological well-being felt by parents. Therefore, it takes some support provided by family members, both physically, psychically, and materially. If there are parents with children with special needs who do not get support from their families properly, then over time parents with children with special needs will feel feelings of anxiety, stress to depression, this will also affect the care of children with special needs. Individuals who have difficulty in establishing positive relationships with others will certainly have an impact on the lack of support received. Distina and Kumail (2019) explained that the lack of support received by individuals can worsen their mental health conditions.

According to Sherbourne and Stewart (1991), family support consists of several aspects, namely emotional support, informational support, real support, affection support, and positive social interaction. In this study, it was found that information support is a support that is widely received by parents with children with special needs. This is in accordance with research conducted by Kyzar *et al.* (2012), where parents of children with special needs get information support in the form of knowledge about the care of children with special needs, communication strategies, teaching, how to recognize the talents and interests of children with special needs and relax for parents of children with special needs. The challenges and difficulties faced by parents with children with special needs will hinder the development of potential, take up parents' time to work, and carry out activities necessary for psychological well-being (Jesslin & Kurniawati 2020). Kyzar *et al.* (2012) explained that with emotional, physical, material, and informational support, it can reduce the burden of parenting felt by parents of children with special needs.

Based on the results of the analysis, it can be seen that the variables of parental acceptance and family support have an influence of 59.4% on *psychological well-being* in parents of children with special needs. Meanwhile, 40.6% were influenced by other variables that were not studied in this study. Some variables that affect *psychological well-being* in parents of children with special needs are *parenting self-efficacy* and optimism (Pasyola *et al.* 2021), *death concerns* (Cox *et al.* 2015), resilience (Purwanti & Kustanti 2018), coping stress (Budiarti & Hanoum 2019), gratitude and personality forms (Nurarini 2016) and other variables.

The limitation of this study is the scope of the study which was only carried out on parents who have children with special needs in the Greater Malang Region (Malang City, Malang Regency, and Batu City). In addition, researchers only focused on the role of parental acceptance and family support in building good psychological well-being in parents and under-examined in more detail aspects of parental acceptance and family support. While the advantage of this study is that researchers do not use specific subjects, researchers use the subjects of parents of children with special needs in general which is expected to provide wider benefits. In addition, there have not been many other studies that examine the topics of parental acceptance, family support, and psychological well-being simultaneously.

5 CONCLUSION AND IMPLICATION

Based on the results of the analysis and discussion, it can be concluded that the three hypotheses in this study are accepted. Parental acceptance and family support have a shared influence on the psychological well-being in parents of children with special needs. Psychological well-being of parents who have children with special needs is also influenced by the support of the family.

Through research that has been carried out, it is hoped that it can provide an explanation of the importance of parental acceptance, family support, and psychological well-being in parents who have children with special needs. So that related parties, such as parents, counselors, teachers, and families can make considerations in providing appropriate solutions and treatment when experiencing problems related to parental acceptance, family support, and psychological well-being. In addition, interventions can be given in the form of psychoeducation which contains how to accept children with special needs and get family support so that good psychological well-being can be created when caring for children with special needs.

REFERENCES

Aurelia, C., Siregar, M., Felicia., Chandra., & Marpaung, W. (2022). Gambaran Psychological Well-being Orang Tua Yang Memiliki Anak Autistic Spectrum Disorder (ASD). *Syntax Literate: Jurnal Ilmiah Indonesia*, *7*(4). http://dx.doi.org/10.36418/syntax-literate.v7i4.6664

Balcells-Balcells, A., Giné, C., Guàrdia-Olmos, J., Summers, J. A., & Mas, J. M. (2019). Impact of Supports and Partnership on Family Quality of Life. *Research in Developmental Disabilities*, *85*, 50–60. doi: 10.1016/j.ridd.2018.10.006.

Bougher-Muckiana, H. R., Roota, A. E., Coogleb, C. G., & Floydb, K. K. (2015). The Importance of Emotions: The Socialization of Emption in Parents of Children with Autism Spectrum Disorder. *Early Child Development and Care*. http://dx.doi.org/10.1080/03004430.2015.1112799

Budiarti, E., & Hanoum, M. (2019), Koping Stress Dan Dukungan Keluarga Terhadap Kesejahteraan Psikologis Orang Tua Yang Memiliki Anak Berkebutuhan Khusus. *Jurnal Ilmiah Psikologi*, *11*(1).

Carrasco, M. A., Delgado, B., & Holgado-Tello, F. D. (2019). Parental Acceptance and Children's Psychological Adjustment: The Moderating Effects of Interpersonal Power and Prestige Across Age. *PLOS One*, *14*(4). https://doi.org/10.1371/journal.pone.0215325.

Cox, C. R., Eaton, S., Ekas, N. V., Van Enkevort, E. A. (2015). Death Concerns and Psychological Well-being in Mothers of Children with Autism Spectrum Disorder. *Research in Developmental Disabilities*, *45–46*, 229–238. https://doi.org.10.1016/j.ridd.2015.07.029

Distina, P. P., & Kumail, M. H. (2019). Pengembangan Dimensi Psychological Well-being Untuk Pengurangan Risiko Gangguan Depresi. *Jurnal Dakwah dan Pengembangan Sosial Kemanusiaan*, *10*(1), 39–59. https://doi.org/10.32923/maw.v10i1.768.

Ghoniyah, Z., & Savira, S. I. (2015). Gambaran Psychological Well Being Pada Perempuan Yang Memiliki Anak Down Syndrome. *Character*, *3*(2).

Hikmah, N. K. (2020). *Peran Resiliensi Sebagai Mediator Antara Dukungan Keluarga dan Kesejahteraan Psikologis Pada ibu Dengan Anak Berkebutuhan Khusus*. Skripsi, Universitas Islam Indonesia, Yogyakarta.

Janie, D. N. A. (2012). *Statistik deskriptif & Regresi Linier Berganda Dengan spss*. Semarang: Semarang University Press.

Jesslin., & Kurniawati, F. (2020). Perspektif Orang Tua Terhadap Anak Berkebutuhan Khusus di Sekolah Inklusi. *Jurnal Pendidikan Inklusi*, *3*(2), 072–091.

Kania, P. Z., & Yanuvianti, M. (2018). Hubungan Dukungan Sosial Dengan Penerimaan Pada ibu Yang Memiliki Anak Berkebutuhan Khusus di SLB X Bandung. *Prosiding Psikologi*, *4*(1), 103–107.

Khaleque, A., & Rohner, R. P. (2011). Pancultural Associations Between Perceived Parental Acceptance and Psychological Adjustment of Children And Adults: A Meta-analytic Review of Worldwide Research. *Journal of Cross-Cultural Psychology*, *43*(5), 784–800. https://doi.org/10.1177/0022022111406120.

Kosasih, H., & Virlia, S. (2016). Gambaran Penerimaan Orang Tua Dengan Anak Retardasi Mental di Sekolah Luar Biasa (SLB) C "DG" dan SLB C "SJ". *Jurnal Psibernetika*, *9*(1).

Kubicek, L. F., Riley, K., Coleman, J., Miller, G., & Linder, T. (2013). Assessing the Emotional Quality of Parent–Child Relationships Involving Young Children with Special Needs: Applying the Constructs of Emotional Availability and Expressed Emotion. *Infant Mental Health Journal*, *34*(3), 242–256. doi:10.1002/imhj.21384.

Kuhn, J., Ford, K., & Dawalt, L. S. (2018). Brief report: Mapping Systems of Support and Psychological Well-being of Mothers of Adolescents with Autism Spectrum Disorders. *Journal of Autism and Developmental Disorders*, *48*(3), 940–946. https://doi.org/10.1007/s10803-017-3381-0.

Kusnadi, S. K., Irmayanti, N., Anggoro, H., & Agustina, K. S. B. (2021). Hubungan Antara Dukungan Keluarga Dengan Psychological Well-being Pada Orang Tua Yang Memiliki Anak Tunagrahita Sedang. *Jurnal Psikologi Insight*, *5*(1), 79–86.

Kyzar, K. B., Turnbull. A. P., & Summers, J. A. (2012). The Relationship of Family Support to Family Outcomes: A Synthesis of Key Findings from Research on Severe Disability. *Research and Practice for Persons with Severe Disabilities*, *37*(1), 31–44. https://doi/org/10.2511/027494812800903247.

Latipun. (2017). *Psikologi eksperimen*. Malang: UMM Press.

Listiyandini, R.A., Nathania, A., Syahniar, D., Sonia, L., & Nadya, R. (2015). Mengukur Syukur: Pengembangan Model Awal Skala Bersyukur Versi Indonesia. *Jurnal Psikologi Ulayat*, *2*(2), 473–496. https://doi.org/10.24854/jpu22015-41

Martin, F., Clyne, W., Pearce, G., & Turner, A. (2019). Self-management Support Intervention for Parents of Children with Developmental Disorders: The Role of Gratitude and Hope. *Journal od Child and Family Studies*, *28*, 980–992. https://doi.org/10.1007/s10826-018-01308-1

Moen, Ø. L., Hedelin, B., & Hall-Lord, M. L. (2014). Public Health Nurses' Conceptions of Their Role Related to Families with a Child Having Attention-Deficit/Hyperactivity Disorder. *Scandinavian Journal of Caring Sciences*, *28*, 515–522. https://doi.org/10.1111/scs.12076

Neff, K. D., & Faso, D. J. (2014). Self-compassion and Well-being in Parents of Children with Autism. *Mindfulness*, *6*(4), 938–947. https://doi.org/10.1007/s12671-014-0359-2

Ni'matuzahroh., Yuliani, S. R., & Mein-Woei, S. (2021). *Psikologi Intervensi Pendidikan: Anak Berkebutuhan Khusus*. Malang: UMM Press.

Nikmatunasikah, K. (2020). *Psychological well-being dan Keterlibatan Orang Tua Yang Memiliki Anak Berkebutuhan Khusus*. Skripsi, Universitas Muhammadiyah Malang, Malang.

Nurarini, F. (2016). *Pengaruh Rasa Syukur dan Kepribadian Terhadap Psychological Well-being Orang Tua Yang Memiliki Anak Berkebutuhan Khusus*. Skripsi. Universitas Islam Negeri Syarif Hidayatullah, Jakarta.

Olianda, R. A., & Rizal, G. L. (2020). Hubungan Antara Hardiness dan Dukungan Sosial Terhadap Stress Pengasuhan Pada ibu Yang Memiliki Anak Tunagrahita. *Attadib: Journal of Elementary Education*, *4*(2).

Pasyola, N. E., Abdullah, A. M., & Puspasari, D. (2021). Peran Parenting Self-efficacy dan Optimism Terhadap Psychological Well-being Yang Memiliki Anak Intellectual Disability. *PSYMPATHIC: Jurnal Ilmiah Psikologi*, *8*(1), 131–142. https://doi.org/10.15575/psy.v8i1.12645

Petcharat, M., & Liehr, P. (2017). Mindfulness Training for Parent of Children with Special Needs: Guidance for Nurses in Mental Health Practice. *Journal of Child and Adolescent Psychiatric Nursing*, *30*(1), 35–46. https://doi.org/10.1111/jcap.12169.

Prabowo, A. (2017). Gratitude dan Psychological Well-being Pada Remaja. *Jurnal Ilmiah Psikologi Terapan*. *5*(2).

Purwanti, D. A., & Kustanti, E. R. (2018). Hubungan Antara Resiliensi Dengan Psychological Well-being Pada ibu Yang Memiliki Anak Dengan Gangguan Autis. *Jurnal Empati*, *7*(1), 283–287. https://doi.org/10.14710/empati.2018.20219

Puspita, S. D., (2018). *Hubungan Antara Dukungan Sosial dan Penerimaan Orang Tua (ibu) Yang Memiliki Anak Down Syndrome*. Skripsi, Universitas Islam Indonesia, Yogyakarta.

Rachmayani, D., & Ramdani, N. (2014). Adaptasi Bahasa dan Budaya Skala Psychological Well-being. *Proceeding: Seminar Nasional Psikometri*, 253–268.

Rahmahdianti., & Rusli, D. (2020). Gratitude dan Psychological Well-being Pada ibu Bekerja Yang Memiliki Anak Tuna Grahita. *Jurnal Riset Psikologi*, *2020*(1). http://dx.doi.org/10.24036/jrp.v2020i1.

Rohner, R. P., & Ali, S. (2016). Parental Acceptance-rejection Questionnaire (PARQ). *Encyclopedia of Personality and Individual Differences*, 1–4. https://doi.org/10.1007/978-3-319-28099-8_56-1.

Rohner, R. P., Khaleque, A., & Cournoyer, D. E. (2005). Parental Acceptance-rejection: Theory, Methods, Cross-cultural Evidence, and Implications. *Ethos*, *33*(3), 299–334. https://doi.org/10.1525/eth.2005.33.3.299.

Ryff, C. D., & Singer, B. H. (2006). Know Thyself and Become what you are: A Eudaimonic Approach to Psychological Well-being. *Journal of Happiness Studies*, *9*(1), 13–39. https://doi.org/10.1007/s10902-006-9019-0.

Sherbourne, C. D., & Stewart, A. L. (1991). The MOS Social Support Survey. *Social Science & Medicine*, *32* (6), 705–714. https://doi.org/10.1016/0277-9536(91)90150-b.

Tarnoto, N. (2016). Permasalahan-permasalahan Yang Dihadapi Sekolah Penyelenggara Pendidikan Inklusi Pada Tingkat SD. *Humanitas Indonesian Psychological Journal*, *13*(1). DOI:10.26555/humanitas.v13i1.3843.

Winarsih, M., Nasution, E. S., Ori, D. (2020). Hubungan Dukungan Keluarga Dengan Penerimaan Diri Orang tua Yang Memiliki abk di SLB Cahaya Pertiwi Kota Bekasi. *Jurnal IKRA-ITH Humaniora*, *4*(2).

Yavuzer, H. (2010). *Parents and the Child* (21st ed.). Istanbul: Remzi.

Zahira, G. R. (2020). *Family Support dan Gratitude Terhadap Family Quality of Life Pada Orang Tua Dengan Anak Berkebutuhan Khusus*. Skripsi, Universitas Islam Negeri Sultan Syarif Kasim Riau, Pekanbaru.

Families Mental Health and Challenges in the 21st Century – Arifin et al. (eds)
© 2024 The Author(s), ISBN 978-1-032-51442-0

The effect of work–life balance toward performance employee of PT. X Branch Malang

Ulinnuha Aqilah Ramadhani* & Nida Hasanati*
Faculty of Psychology, University of Muhammadiyah Malang, Malang, Indonesia

ABSTRACT: Performance is a potential that must be owned by each individual to be able to carry out each task and responsibility given. Good employee performance will have a positive impact on the company so that a common achievement can be realized. One of the things that can do to respond to the condition of each employee is Work life balance. Employees who have a good work–life balance (WBL) will be able to improve the quality of employee performance in a company. The purpose of this study is to determine the effect of WBL on Employee Performance. This type of research uses quantitative with a correlation design. The population in this study is employees at PT. X Malang Branch. The number of subjects used as many as 142 people using a sampling technique that is simple random sampling. The scale used in this research is the Individual Work Performance Questionnaire (IWPQ) and WBL. The results showed that there was a significant effect between WBL and Employee Performance with a total percentage of 26%. Of the four dimensions of WBL, there are two influential dimensions, namely Personal Life Interference with Work and Work Enhancement of Personal Life. The better the WBL owned by each employee, the better the level of quality of employee performance, and vice versa.

Keywords: Employee Performance, work–life balance, Retail Employee

1 INTRODUCTION

Business growth in the retail sector is increasing day by day so the competition between retail companies is also getting tougher. Therefore, the contribution of human resources owned by retail companies is the main component that must be empowered appropriately to be able to produce competent human resources in their fields and also be able to compete with other competing companies. Concerning human resources, cases that often occur today are regarding the decline in the quality of employee performance.

This is supported by the results of a survey conducted by Hasbi (2020) that there was an increase in the level of employee absenteeism from PT. Sumber Alfaria Trijaya Tbk Talang Kelapa branch in 2018 was 1.06% with a total absenteeism rate of 0.41%, 1.14% in 2019 with a total absenteeism rate of 0.43%. In 2019, the absenteeism rate for truancy increased by 0.02% compared to the previous year and, in 2020, it became 1.17% with a total absentee rate of truancy of work as much as 0.35%. In 2020, the absentee rate for skipping work decreased by 0.08% compared to the previous year, but the company only tolerated the attendance rate of 0.31%. This can be used as an indicator that can explain that the company experienced a decrease in the quality of employee performance due to

*Corresponding Authors: ulinnar@webmail.umm.ac.id and nida@umm.ac.id

DOI: 10.1201/9781003402381-18

the high level of employee absenteeism caused by skipping work or employees being absent.

Based on the explanation above, this can be supported by research conducted by Flippo (2001) explaining the problems related to staffing, including high absenteeism and high tardiness. If the absentee level is high, it will have an impact on low employee performance and the expected targets are difficult to achieve. This can also occur if the high absenteeism rate reaches more than 10% of the total number of employees, it can result in many activities being hampered and affect the overall performance of employees.

Regarding employee performance, several companies have started to act quickly to be able to respond to the conditions of their employees, one of the crucial things for a company to pay attention to is the balance of personal life with work life (work–life balance; WBL).

Several previous studies explained that WBL could affect employee performance, namely research conducted by Mendis & Weerakody (2014) found that there is a strong relationship and positive influence between the balance between personal life and employee performance. A good balance between personal life can benefit employees, so employees feel benefit by feeling happier at work and having relatively less stress levels. The results of research from Obiageli, Uzochukwu, & Ngozi (2015) also found that WBL has a significant positive correlation with employee performance at commercial banks in Lagos.

This research was conducted to determine the effect of WBL on the performance of employees of PT. X Malang branch. The purpose of this study is to determine the effect of WBL on the performance of employees of PT. X. The benefits of this research are that it can enrich and expand the study of industrial and organizational psychology (PIO) and can be used as a reference for related companies in managing existing human resources.

2 REVIEW LITERATURE

Campbell (1990) defines performance as behavior or action given by the company to its employees in accordance with the company's goals. Performance is not only an action or behavior but also an evaluation and assessment. According to Koopmans et al. (2014), there are three dimensions that affect employee performance, namely, first task performance in which task performance dimension contains an individual skill to be able to perform or do a given task. Second, contextual performance, in the contextual performance dimension, is defined as individual behavior in supporting the environment in the organization both psychologically and socially so that the organization can carry out its functions and duties properly. And lastly, counterproductive work behavior, in the dimension of counterproductive work behavior, is defined as an action or behavior that can harm the company and its shareholders.

Kalliath & Brough (2008) define WBL as an individual's perception that work activities with non-work must be commensurate to encourage growth according to the priorities needed. This balance leads to growth and development in the world of work such as gaining recognition and career potential, as well as non-career in the form of traveling, taking care of children and personal life. Fisher, Bulger, & Smith (2009) say that there are four dimensions that can form WBL, namely: Work Interference With Personal Life (WIPL), which refers to the extent to which work can interfere with activities or personal lives carried out by individuals; Personal Life Interference With Work (PLIW), which refers to the extent to which personal life interferes with an individual's work life; PLEW, which refers to the extent to which a person's personal life can increase productivity and individual performance in the world of work; Work Enhancement of Personal Life (WEPL), which refers to the extent to which work can improve the quality of an individual's personal life. These dimensions will make an individual will feel happy in his work life and personal life.

3 METHODS

The subjects in this study were office employees at PT. X Malang branch. This study uses a sampling technique that is simple random sampling by determining the number of samples to be taken and later processed through the Statistical Package for Social Science (SPSS) and will produce samples according to the number selected. In this study, two variables were used: the independent variable (X) and the dependent variable (Y). The independent variable (X) is WBL, and the dependent variable (Y) is employee performance. The scale used in this study uses five answer options, namely Strongly Agree (SS), Agree (S), Slightly Agree (US), Disagree (TS), and Strongly Disagree (STS). The items in this study have a value of SS = 5, S = 4, AS = 3, TS = 2, and STS = 1.

In the WLB variable, a research scale from Fisher, Bulger, & Smith (2009) is used, which consists of 17 items translated by Gunawan, Nugraha, Sulastiana, & Harding (2019). This item is a derivative of the dimensions forming WLB, namely work interference personal life, personal life interference work, PLEW, and WEPL.

The employee performance variable uses a research scale from Koopmans et al. (2014) with three indicators, namely task performance, contextual performance, and counter-productive work behavior, translated by Widyastuti & Hidayat (2018) with a total of 18 items. On the employee performance scale, the level of reliability is 0.78. While on the WLB scale on the WIPL dimension, the level of reliability is 0.83, on the PLIW dimension the reliability level is 0.77, on the PLEW dimension the reliability level is 0.62, and on the WEPL dimension the reliability level is 0.62.

4 RESULTS

Table 1. Multiple regression test.

	R2	F	Sig.
Regression	0.260	12.032	0.000b
Residual			
Total			

Based on the results of multiple linear regression analysis in the table, the results of the Fcount 12,032 show that WBL affects employee performance. Then there is a significant value of $p = 0.000$ ($p < 0.01$) indicating that the hypothesis is accepted. WBL affects employee performance. The value of R Square (R2) is 0.260, this means that 26% of the WBL variable affects employee performance and other factors influence the rest.

Results of the partial significance test value shows the influence of two dimensions of WBL on employee performance. WBL dimensions affect employee performance if the significance value is < 0.05.

Table 2. Partial significance test.

Work life balance	β	Sig.
WIPL	−0.115	0.214
PLIW	0.215	0.023
PLEW	0.133	0.090
WEPL	0.445	0.000

The dimension of WIPL shows a significance value of 0.214 > 0.05, which means that this dimension has no significant effect on employee performance. The coefficient (β) shows a value of −0.115 indicating a negative effect of WIPL on employee performance.

The dimension of PLIW shows a significance value of 0.023 < 0.05, which means that this dimension significantly influences employee performance. The regression coefficient (β) shows a value of 0.215 indicating a positive influence on PLIW with employee performance. So that it can improve employee performance.

In the last dimension, PLEW shows a significance value of 0.090 > 0.05, which means that this dimension has no significant effect on employee performance. The regression coefficient (β) shows a value of 0.133 indicating a positive influence on PLEW with employee performance.

The WEPL shows a significance value of 0.000 < 0.05, which means that this dimension significantly influences employee performance. The regression coefficient (β) shows a value of 0.445 indicating a positive effect of WEPL on employee performance, improving employee performance.

Based on the results, it can be concluded that the dimensions of WBL that influence employee performance are PLIW and WEPL. Meanwhile, those that do not affect employee performance are Work Interference with Personal Life and PLEW.

5 DISCUSSION

Based on the results of research that has been done using multiple linear regression analysis on employees of PT. X Branch Malang obtained results stating that the hypothesis is accepted and the WBL variable significantly influences employee performance. WBL has four forming dimensions, namely WIPL, PLIW, WEPL, and PLEW where these dimensions have an influence of 26% on the performance of employees of PT. X Malang Branch. In the four dimensions of WBL, two supporting dimensions significantly influence employee performance, namely PLIW and WEPL, and two other supporting dimensions that do not significantly influence employee performance.

The analysis that has been carried out in the research partially to determine the influence of the dimensions of WBL on employee performance. The results are found in the first dimension of WIPL, which illustrates that work can interfere with activities or activities in employees' personal lives such as still doing office work at home during weekends. Thus, the employee can hardly divide time between himself and his family. This dimension shows that it does not significantly affect employee performance. This is in line with research conducted by Angin & Saragih (2021), which states that work cannot interfere with personal life because employees can distinguish between responsibilities at work and outside work, so that conflicts or stress in a job will not interfere with their personal lives.

The second dimension of PLIW shows that this dimension significantly influences employee performance. This dimension illustrates that employees have problems in their personal or family lives that can affect their work lives. However, this dimension shows that personal life that interferes with work life has a positive impact it can improve employee performance. This is due to several factors, namely the individual's responsibility, job satisfaction, and organizational commitment. This is in line with research conducted by Wolor, Kurnianti, Zahra, & Martono (2020) which states that personal problems do not negatively impact work life, but instead increase employee performance. This is because the employee is committed to his work and to the company. In addition, employees are also satisfied with the work they do every day.

The next dimension PLEW illustrates that employees' personal lives can increase productivity and individual performance in their work. Like getting attention from a partner or being able to take time to do hobbies and also feel satisfaction in personal life. Although it has a positive impact, this dimension shows that it does not significantly affect employee performance. This is in line with research conducted by Chiekezie et al. (2016) that the high and low level of employee personal life balance does not impact worker performance; therefore, there are several different methods to improve worker performance. In addition, the balance of personal life with work whether or not the employee will still carry out his work as well as possible because they have monthly and annual targets that must be achieved.

The last dimension, WEPL, shows a significant influence on employee performance. This dimension provides an illustration of the extent to which work life has a positive impact on personal life, such as after work being able to spend time with family or doing hobbies. Besides, employees can complete tasks according to deadlines, so employees feel happy because there are no more responsibilities to complete on other days. This is in line with Mendis & Weerakody's (2014) research that a good WBL can make employees and companies feel mutually beneficial to each other. In that way, employees will feel less stressed and happier at work so employees feel an increase in employee performance and the company benefits because it can realize company goals together.

It can be concluded that each dimension of WBL has a different effect on employee performance, namely the negative influence of WIPL and PLIW on employee performance and the positive influence of PLEW and WEPL with employee performance.

Research that has been done has 74% of other supporting factors WBL that can affect employee performance. One of the other factors outside of this research variable is the organizational commitment factor. This is because the personal life dimension interference with work positively influences employee performance and is contrary to the research hypothesis. This organizational commitment factor is in line with research conducted by Greenberg (1996) where the researcher explains that organizational commitment can be considered to determine employee performance. Employees who have high organizational commitment will produce good work performance and reduce absenteeism. Organizational commitment also encourages employees to maintain their jobs and show excellent performance results. Therefore, employees with a high organizational commitment will continue to update their knowledge related to their work, and also continue to strive to meet the company's targets to achieve the company's goals together.

6 CONCLUSION AND IMPLICATION

The conclusions of this study are results that show a WBL dimension that significantly affects employee performance. This is evidenced by the two supporting variables of WBL, namely PLIW and WEPL, which significantly affect employee performance. This means that the personal life of employees can negatively impact work–life, which can reduce the quality of employee performance. In addition, work life can positively influence employees' personal life, which will improve the quality of the employee's performance. Meanwhile, the two supporting variables of WBL, namely WIPL and PLEW, do not significantly affect employee performance. There is a 26% influence on WBL on employee performance and other factors influence the rest. Then the implications of this research on related companies can provide information to their employees on how much influence the dimensions of WBL have on employee performance. Especially in personal life which is included in work life and vice versa. In this case, the company can pay more attention to the balance between the personal life and work of each employee, and it is hoped that the WBL can provide benefits for both parties between employees and the company to be able to achieve goals together and get prosperity. The implementation of WBL can be done in various ways depending on the policies that apply to a company. Employees themselves are expected to be able to manage emotions, and manage time well to implement WBL properly and can provide improvements to the quality of employee performance. In addition, further research can conduct research with different settings in this study, such as research subjects, adding research variables or using research variables that have never been studied to determine employee performance. Further research is expected to add more relevant supporting data in order to strengthen the hypothesis obtained. It is hoped that this will add to the picture of the WBL with employee performance from a different point of view.

REFERENCES

Angin, I. I., & Saragih, E. H. (2021). Pengaruh Work Life Balance Terhadap Kinerja Karyawan Generasi X dan Y di Jakarta. *Journal of Emerging Business Management and Entrepreneurship Studies*, 1, 48–57. doi:10.34149/jebmes.v1i1.14.

Campbell, J. P. (1990). Modeling the Performance Prediction Problem in Industrial and Organizational Psychology in: M. D. Dunnette & L. M. Hough (eds.), *Handbook of Industrial and Organizational Psychology*. Palo Alto, Calif: Consulting Psychology Press.

Chiekezie, O. M., Nzewi, N. H., Emejulu, G., & Chukwujama, N. C. (2016). Work–life Balance and Job Performance of Selected Commercial Banks in Anambara State, Nigeria. *EPRA International Journal of Economic and Business Review*, 4(6), 115–127.

Fisher, G. G., Bulger, C. A., & Smith, C. S. (2009). Beyond Work and Family: A Measure of Work/nonwork Interference and Enhancement. *Journal of Occupational Health Psychology*, 14, 441–456. doi:doi:10.1037/a0016737.

Flippo, E. B. (2001). *Manajemen personalia*. Jakarta: Erlangga.

Greenberg, J. (1996). *Managing Behavior in Organizations: Science in Service to Practice*. New York: Prentice Hall.

Gunawan, G., Nugraha, Y., Sulastiana, M., & Harding, D. (2019). Reliabilitas dan Validitas Konstruk Work Life Balance di Indonesia. *Jurnal Penelitian dan Pengukuran Psikologi*, 8, 88–90. doi:10.21009/JPPP.082.05.

Hasbi, M. (2020). *Pengaruh Kepemimpinan dan Kompensasi Terhadap Kinerja Karyawan Pada PT*. Sumber Alfaria Trijaya, Tbk area Talang Kelapa. 1–8. Retrieved from http://repository.univ-tridinanti.ac.id/id/eprint/1741.

Kalliath, T., & Brough, P. (2008). Editorial: Achieving Work–life Balance. *Journal of Management and Organization*, 14, 224–226. doi:10.5172/jmo.837.14.3.224.

Koopmans, L., Bernaards, C. M., Hildebrandt, V. H., Vet, H. C., & Beek, A. J. (2014). Measuring Individual Work Performance Identifying and Selecting Indicators. *Work*, 48, 229–238. doi:10.3233/WOR-131659.

Mendis, M. D., & Weerakody, W. A. (2014). The Relationship Between Work Life Balance and Employee Performance: With Reference to Telecommunication Industry of Sri Lanka. *Kelaniya Journal of Human Resource Management*, 9, 95–117.

Obiageli, O. L., Uzochukwu, O. C., & Ngozi, C. D. (2015). Work–life Balance and Employee Performance in Selected Commercial Banks in Lagos State. *European Journal of Research and Reflection in Management Sciences*, 3, 63–77. Retrieved from http://www.idpublications.org/.

Widyastuti, T., & Hidayat, R. (2018). Adaptation of individual work performance questionnaire (IWPQ) into Bahasa Indonesia. *International Journal of Research Studies in Psychology*, 7, 101–112. doi:10.5861/ijrsp.2018.3020.

Wolor, C. W., Kurnianti, D., Zahra, S. F., & Martono, S. (2020). The Importance of Work Life Balance on Employee Performance Millenial Generation in Indonesia. *Journal of Critical Reviews*, 7(9), 1103–1108.

Families Mental Health and Challenges in the 21st Century – Arifin et al. (eds)
© 2024 The Author(s), ISBN 978-1-032-51442-0

Emotional intelligence and student life satisfaction: Mediated by student engagement

Silfiasari*
Faculty of Psychology, Universitas of Muhammadiyah Malang, Malang, Indonesia

ABSTRACT: Student life satisfaction is important for student success in school. According to various studies, students who achieve life satisfaction are students who have emotional intelligence and are actively involved in school. This study focuses on the relationship between emotional intelligence and student life satisfaction mediated by student engagement. A total of 295 students were studied, measured using the Schutte Emotional Intelligence Scale (SEIS), Multidimensional Student Life Satisfaction Scale (MSLSS), and The Student Engagement in Schools Questionnaire (SESQ). Analysis of the data using PROCESS from Hayes and got the results of BootLLCI 0.08 and BootULCI 0.21 which means that the value of zero is not included in this range and it can be concluded that there is an influence of emotional intelligence on life satisfaction in school mediated by student engagement.

Keywords: student life satisfaction, emotional intelligence, student engagement

1 INTRODUCTION

In the scope of education, students often experience stress personally, socially, academically, and economically (Freire *et al.* 2016). Yang & Farn (2005) explain that school burnout refers to the fatigue syndrome that comes from school and excessive educational demands on students. They are required to become self-aware and learn to act independently, be responsible individuals, improve their talents, and be creative and flexible (Doğan & Celik 2014). With so many demands, students are expected to be able to complete their assignments well. When students can complete their tasks well, students will achieve prosperity. Well-being has a close relationship with life satisfaction (Diener *et al.* 1985). Life satisfaction is an individual's perception of all aspects of his life, which includes family, community, social relations, and the environment (Diener *et al.* 1985). Individuals who have a level of satisfaction with their high life are more successful in their relationships with people at work, earn more money and overcome the disease as well (Lewis *et al.* 2011).

Life Satisfaction in school is important because having a close relationship with positive educational outcomes (Antaramian 2017). Student life satisfaction is students' perception of their relationship in the school climate (Doğan & Celik 2014). According to Huebner (1994), student life satisfaction is a general evaluation of the quality of life of students in certain domains, namely family, friends, school, self, and environment. Several studies have confirmed that student life satisfaction is related to academic competence, school experience, and attitudes toward teachers and schools (Mehdinezhad 2015). Peer relationships can also affect student life satisfaction at school (Doğan & Celik 2014). Student life satisfaction is influenced by student relationships with other students at school (Leung & Zhang 2000).

*Corresponding Author: silfiasari@umm.ac.id

DOI: 10.1201/9781003402381-19

Students who have high life satisfaction tend to have positive attitudes toward their teachers and schools (Gilman & Huebner 2006). At school, students will spend a lot of time with peers and teachers. When students spend time in a fun and meaningful way, their attitude toward school will be positive. But on the other hand, when students spend time at school unpleasantly, their attitude toward school will be negative.

Life satisfaction at school has two important components, namely an emotional component and a cognitive component (Mehdinezhad 2015). The emotional component consists of positive and negative influences on student lives at school, while the cognitive component is related to how much students are satisfied with some aspects of their lives. Ardahan (2012) explained that student life satisfaction is closely related to happiness. Happiness is defined as being healthy, educated, optimistic, carefree, religious, and smart.

One of the important roles in student life satisfaction is emotional intelligence (Foumany 2015). Since the 1990s, emotional intelligence had a great influence on the academic field (Ardahan 2012). Emotional intelligence is an individual's ability to understand emotions correctly and can solve problems well through understanding and information collected (Sarican *et al.* 2015). Life satisfaction is one of the most studied variables related to emotional intelligence. Individuals with high emotional intelligence are considered to have a great capacity to see, use, and understand emotions in themselves and in others, which supports the creation of life satisfaction (Kong *et al.* 2012). Pellitteri (2010) stated that efficient ability in emotional intelligence, understanding emotions well, and understanding feelings are some of the factors that can support the creation of life satisfaction.

Several studies explain the relationship between emotional intelligence and life satisfaction in students. Research conducted by Sarican *et al.* (2015) explained that emotional intelligence has a relationship with life satisfaction. Research conducted by Murphy (2006) explained that emotional intelligence affects life satisfaction in students in Florida. Research conducted by Palmer *et al.*, (2002) stated that emotional intelligence has a relationship with life satisfaction. Meanwhile, research conducted by Kong *et al.* (2012) on students in China stated that emotional intelligence affects life satisfaction in students mediated by self-esteem and social support.

Among the factors that affect student life satisfaction, there is one factor related to emotional intelligence, namely student engagement in school. Student engagement in school refers to a positive, satisfying state of mind, dedication to the school, having passion, and resilience in learning (Mehdinezhad 2015). Research conducted by Urquijo & Extremera (2017) explained that there is a relationship between emotional intelligence and student engagement. The emotional intelligence of students has more clarity about real academic behavior, including engagement in school. Research conducted by Drummond (2014) explained that there is a relationship between emotional intelligence and student engagement in school. Students who have emotional intelligence have perseverance in the teaching and learning process so they are active in activities at school. Another study by Maguire *et al.* (2016) explained the same result that emotional intelligence is a predictor of student engagement in school. Students who have high emotional intelligence will have engagement in good academic settings.

Student engagement also plays an important role in student life satisfaction. Research conducted by Antaramian (2017) stated that students with high life satisfaction on average show positive engagement in school. Research conducted by Proctor *et al.* (2009) stated that parent-peer relationships, school success, and engagement in school affect adolescent life satisfaction. Several studies have shown that student engagement is an important facilitator in high life satisfaction (Antaramian 2015).

Research conducted by Rastogi & Pati (2018) stated that there is a positive relationship between student engagement and student life satisfaction. Another study conducted by Lewis *et al.* (2011) stated the importance of student engagement in schools during the transition period between elementary and high school to support student life satisfaction. Research conducted by Heffner & Antaramian (2015) explained that one of the predictors of student life satisfaction is student engagement in school.

Based on the discussion above, the hypothesis in this study is:

H_1: There is a relationship between emotional intelligence and student life satisfaction
H_{2a}: There is a relationship between emotional intelligence and student engagement
H_{2b}: There is a relationship between student engagement and student life satisfaction with the presence of emotional intelligence
H_{2c}: There is a relationship between emotional intelligence and student life satisfaction with the presence of student engagement
H_{2d}: There is a relationship between emotional intelligence and student life satisfaction mediated by student engagement.

This study aims to examine the relationship between emotional intelligence and student life satisfaction mediated by student engagement. The results of this study are expected to contribute knowledge and insight regarding the concept of emotional intelligence, student life satisfaction, and student engagement. In addition, this study is also expected to add references to the role of the mediating variable, namely student engagement influencing the relationship between emotional intelligence and student life satisfaction, which has not been widely studied in previous studies.

2 METHODS

This study is a quantitative approach with a correlational design. The quantitative approach is research by collecting data directly through questionnaires and the results can be measured directly using data analysis methods (Bacon-Shone 2015). The correlational design is used to see the relationship between two or more variables (Gogtay & Thatte 2017). Characteristics of the subjects in this study were high school students in grades X, XI of all majors (science, social studies, language) with an age range of 15 to 18 years totaling 295 students. This research was conducted at SMAN 1 Sumberpucung. Characteristics of research subjects in detail can be seen in Table 1.

Table 1. Characteristics of research subjects.

Characteristics	N	Percentage
Sex		
Male	97	33
Female	198	67
Age		
15	24	8.14
16	114	38.64
17	152	51.52
18	5	1.7
Class		
X	99	33.56
XI	196	66.44
Total	295	100

Student life satisfaction was measured using *the Multidimensional student life Satisfaction Scale* (MSLSS) developed by Huebner (1994). After a tryout, this scale has 23 items and consists of 5 domain aspects, including family, friends, school, self, and the environment. High scores on this scale indicate high student life satisfaction. The MSLSS scale has a reliability value of 0.91. An example of an item on this scale is "My friends help me when I need help" (Huebner & Gilman 2002).

Emotional intelligence was measured using the emotional intelligence scale *Schutte Emotional Intelligence Scale* (SEIS) developed by Schutte (1998). After a tryout, this scale has 22 items and three aspects, namely appraisal and expression of emotions, regulation of emotions, and utilization of emotions. High scores on this scale indicate high emotional intelligence. The SEIS scale has a reliability level of 0.90. An example of an item on this scale is "I know when to talk about myself to others" (Jonker & Vosloo 2004).

Student engagement in schools was measured using *The Student Engagement in Schools Questionnaire* (SESQ) scale developed by Hart *et al.*, (2011). After a tryout, this scale has 32 items and consists of 5 aspects, namely affective engagement in interest in learning, affective engagement in interest in school, behavioral engagement in effort and persistence, behavioral engagement in extracurricular activities, and cognitive engagement. High scores on this scale indicate high student engagement. The SESQ scale has a reliability value of 0.93. An example of an item on this scale is "I play an active role in school activities such as sports week and school picnic" (Hart *et al.* 2011).

3 DISCUSSION

Based on the results of the data analysis that has been carried out, the average value and standard deviation for each variable are obtained. The emotional intelligence variable has a value mean of 3.19 with a standard deviation of 0.33. This explains that the average emotional intelligence of students is high. The student engagement variable has a value mean of 3.07 and a standard deviation of 0.30. This explains that the average student engagement is high. For the student life satisfaction variable, the value is a mean of 3.30, and the standard deviation is 0.29. This explains that the average life satisfaction of students is high.

3.1 *Hypothesis testing*

The first hypothesis (H1) in this study is that there is a relationship between emotional intelligence and student life satisfaction (total effect). From the results of data analysis that has been carried out, the total effect coefficient value is 0.314 with $p < 0.05$ which indicates that H_1 is accepted, that is, there is a positive relationship between emotional intelligence and student life satisfaction. The higher the emotional intelligence possessed by students, the higher the student's life satisfaction. Emotional intelligence lies in the ability of individuals to distinguish and react to their own emotions and can convey emotions well to others. Therefore, emotional intelligence has a positive contribution to student life satisfaction at school.

Students who have good emotional intelligence can adjust to their school environment and can socialize well with school friends so that these students will be satisfied with their life at school. Students who can realize emotions and can manage their emotions well tend to have good communication skills with other people, and this can lead to life satisfaction. When students can manage negative emotions, students can think positively and have a good mood so that their life satisfaction becomes high. This is in line with research conducted by

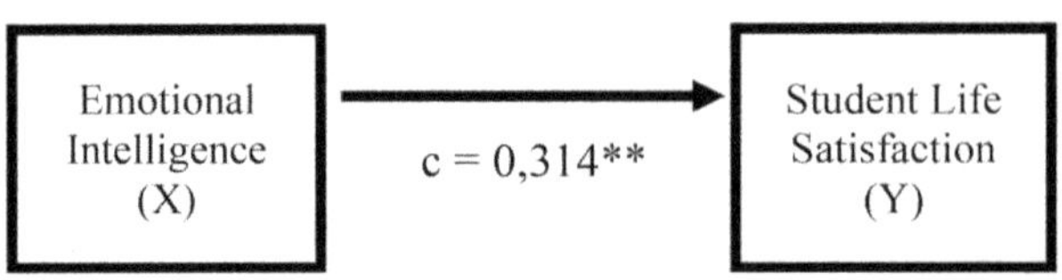

Figure 1. The results of the total effect of the relationship between emotional intelligence and student life satisfaction.

Largen (2004) which explains that emotional intelligence and life satisfaction show a positive and significant relationship in the student sample.

Students who can identify the emotions that exist in themselves and others can solve problems well have good interpersonal communication, and can interpret emotions well tend to be satisfied with their lives at school. These students can manage their emotions which causes students to have many friends and be like in the school environment and the neighborhood where they live. Emotional intelligence also focuses on using emotions well in thinking and acting.

The second hypothesis (H_{2a}) in this study is that there is a relationship between emotional intelligence and student engagement (path a). From the results of data analysis that has been carried out, the path coefficient value of a is 0.465 with $p < 0.05$ which indicates that H_{2a} is accepted, and there is a positive relationship between emotional intelligence and student engagement. Student engagement in school refers to student interest in learning at school, student participation in school activities, and student persistence in doing the tasks assigned by the school. Emotional intelligence is important as a predictor of student engagement because emotional intelligence plays an important role in human resources (Basínska & Drozdowska 2013) and helps students to function well. Students who have good emotions can regulate emotions well, tend to develop their skills at school, can adapt well to various interpersonal and intrapersonal demands at school so that they have an active engagement in school activities. On the other hand, when students are unable to regulate their emotions, they tend to develop fewer skills at school so they have passive engagement in school activities. This result is not different from the research conducted by Tekie (2014) which explains that there is a relationship between emotional intelligence and student engagement. Another study that explains the relationship between emotional intelligence and student engagement is the research conducted by Valehzaghard *et al.* (2000).

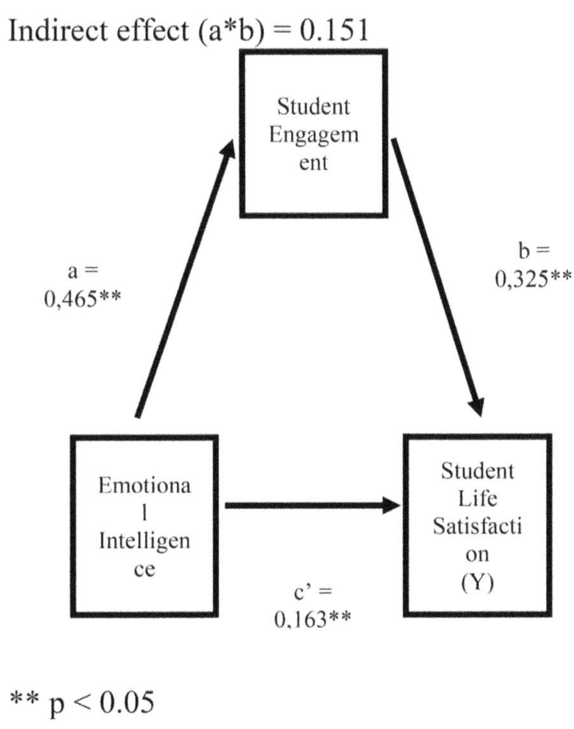

Figure 4. Result of an indirect effect.

The third hypothesis (H_{2b}) in this study is that there is a relationship between student engagement and student life satisfaction with the presence of emotional intelligence (path b). From the results of data analysis that has been carried out, the path coefficient value of b is 0.325 with $p < 0.05$ which indicates that H_{2b} is accepted, and there is a positive relationship between student engagement and student life satisfaction with the presence of emotional intelligence. Student engagement in school is one of the important factors in student life satisfaction. In the scope of the school, when students are actively involved in activities at school, students can establish good relationships with various important components in the school. These students have a role in activities at school so that their life satisfaction is high. On the other hand, when students are not actively involved in school activities, their life satisfaction is low. These results are in line with research conducted by Hakimzadeh *et al.*, (2016) on students in Iran which explains that student engagement in school is a predictor of student success and student life satisfaction. Another study conducted by Korobova & Starobin (2015) explained that there is a relationship between student engagement in school and student life satisfaction for American students and international students.

The fourth hypothesis (H_{2c}) in this study is that there is a relationship between emotional intelligence and student life satisfaction with the presence of student engagement (path c'). From the results of data analysis that has been carried out, the path coefficient value of c' is 0.163 with $p < 0.05$ which indicates that H_{2c} is accepted, and there is a positive relationship between emotional intelligence and student life satisfaction with the presence of student engagement. Students who have good emotional intelligence will have high life satisfaction compared to students who have low emotional intelligence. This is because emotional intelligence is one of the important factors that can affect student life satisfaction.

The fifth hypothesis (H_{2d}) is that there is a relationship between emotional intelligence and student life satisfaction mediated by student engagement. From the results of the data analysis that has been carried out, the mediation coefficient value in this study is 0.151 with a 95% confidence level. From the results of the Hayes test that has been carried out, the value of the indirect *effect* not standardized by bootstrap is 0.151 with a 95% confidence level and is between 0.089 to 0.218. Because zero is not included in this range, it can be concluded that there is a relationship between emotional intelligence and student life satisfaction mediated by student engagement. The mediation that occurs is partial because student engagement does not play a full role as a mediating variable. Emotional intelligence still plays a full role in life satisfaction even though there is no student engagement.

Students who can manage their emotions well can adapt to the environment, and are good at getting along with friends, tend to be actively involved in activities at school, such as diligently participating in extracurricular activities, being active in organizations, and being active in learning activities in class so that These students tend to feel satisfied with their life at school. The value of R^2 in this study amounted to 0.211, suggesting that emotional intelligence and the engagement of students can explain student life satisfaction is only 21.1%, the rest is influenced by other variables not examined. In other studies, no one has used student engagement as a mediating variable in assessing the relationship between emotional intelligence and student life satisfaction. This research can be used as a new reference in assessing student engagement in emotional intelligence and student life satisfaction.

Limitations in this study are limited subjects. The subjects studied in this study were public high school students so they did not represent the population of students in private high schools. In addition, this study did not analyze the differences in student life satisfaction between male and female students.

4 CONCLUSION

The results showed that there was a relationship between emotional intelligence and student life satisfaction mediated by student engagement. Students who can regulate their emotions well and can socialize with the surrounding environment, especially the school environment, tend to have an active engagement in activities at school so that their life satisfaction is high. On the other hand, when students are unable to regulate their emotions, students tend to be passive towards school activities, and their life satisfaction is low.

Based on the results of this study, students are expected to be able to develop their talents and interests by being actively involved in school activities, such as participating in routine extracurricular activities, participating in school organizations such as OSIS and MPK, and being actively involved in classroom learning activities so that students will be satisfied with life at school.

For teachers, it is hoped that they can help students achieve life satisfaction by using fun learning methods when in class, helping students who have difficulties in learning, and also providing direction to students regarding the potential that exists within them. In addition to aiming to achieve student life satisfaction, it also aims to increase the enthusiasm of students to have an active engagement while at school.

For schools, it is expected to facilitate students in developing their talents, such as holding various extracurricular activities. Schools can also provide comfortable classroom facilities, such as providing projectors, *sound systems*, and wall magazines in each class so that student learning activities in class can be fun. That way students will view the school positively so that students will achieve life satisfaction. In addition, schools can also add activities related to spirituality and social service activities, to increase student ability to understand themselves and understand the circumstances and emotions of others.

For further research, it is expected to examine student life satisfaction from other variables, such as social support, hope, and other independent variables. In addition, it can also examine the variables of emotional intelligence and student life satisfaction but with different mediating variables. You can also expand the research subject area, for example on the subject of high school students in other areas, or the subject of junior high school students. Further research is also expected to pay attention to aspects that exist in the variables studied, so that there is no overlapping, or there are aspects that overlap between one variable and another.

REFERENCES

Antaramian, S. (2015). Assessing Psychological Symptoms and Well-being: Application of a Dual-factor Mental Health Model to Understand College Student Performance. *Journal of Psychoeducational Assessment*, *33*(5), 419–429. https://doi.org/10.1177/0734282914557727

Antaramian, S. (2017). The Importance of Very High Life Satisfaction for Student Academic Success. *Antaramian, Cogent Education*, *4*, 1–10. https://doi.org/10.1080/2331186X.2017.1307622

Ardahan, F. (2012). Life Satisfaction and Emotional Intelligence of Participants / Nonparticipants in Outdoor Sports: Turkey Case. *Social and Behavioral Sciences*, *62*, 4–11. https://doi.org/10.1016/j.sbspro.2012.09.003

Bacon-Shone, J. (2015). *Introduction to Quantitative Research Methods*. Graduate School, The University of Hong Kong. https://doi.org/10.13140/2.1.4466.3040

Basínska, M. A., & Drozdowska, M. (2013). Emotional Intelligence as an Indicator of Satisfaction with Life of Patients with Psoriasis. *Postepy Dermatologii i Alergologii*, 365–372. https://doi.org/10.5114/pdia.2013.39435

Diener, E., Emmons, R. A., Larsen, R. J., & Griffin, S. (1985). The Satisfaction with Life Scale. *Journal of Personality Assessment*, *49*(1), 71–75.

Doğan, U., & Celik, E. (2014). Examining the Factors Contributing to Student Life Satisfaction. *Educational Sciences: Theory & Practice*, *14*(6), 2121–2128. https://doi.org/10.12738/estp.2014.6.2058

Drummond, H. A. D. (2014). *Trait Emotional Intelligence and Academic Engagement: An Investigation of Non-Academic Factors Related to Student Engagement and Persistence*. University of Toronto.

Foumany, G. H. E. (2015). The Relationship Between Emotional Intelligence and Life Satisfaction and the Mediatory Role of Resiliency and Emotional Balance among the Students of Zanjan University. *Acta Medica Mediterranea*, *31*, 1351–1357.

Freire, C., Ferradás, M. D. M., Valle, A., Núñez, J. C., & Vallejo, G. (2016). Profiles of Psychological Well-being and Coping Strategies Among University Students. *Journal Frontiers in Psychology*, *7*, 1–11. https://doi.org/10.3389/fpsyg.2016.01554

Gilman, R., & Huebner, E. S. (2006). Characteristics of Adolescents Who Report Very High Life Satisfaction. *Journal of Youth and Adolescence*, *35*(3), 311–319. https://doi.org/10.1007/s10964-006-9036-7

Gogtay, N. J., & Thatte, U. M. (2017). Principles of Correlation Analysis. *Journal of The Association of Physicians of India*, 65, 78–81.

Hakimzadeh, R., Besharat, M. A., Khaleghinezhad, S. A., & Jahromi, R. G. (2016). Peers' Perceived Support, Student Engagement in Academic Activities and Life Satisfaction: a Structural Equations Modeling Approach. *School Psychology Internatonal*, 1–15. https://doi.org/10.1177/0143034316630020

Hart, S. R., Stewart, K., & Jimerson, S. R. (2011). The Student Engagement in Schools Questionnaire (SESQ) and the Teacher Engagement Report Form-new (TERF-N): Examining the Preliminary Evidence. *Contemporary School Psychology*, *15*, 67–79. http://www.casponline.org/pdfs/pdfs/2011_journal_all_001-144-b.pdf#page=69

Heffner, A. L., & Antaramian, S. P. (2015). The Role of Life Satisfaction in Predicting Student Engagement and Achievement. *Journal of Happiness Studies*. https://doi.org/10.1007/s10902-015-9665-1

Huebner, E. S. (1994). Preliminary Development and Validation of a Multidimensional Life Satisfaction Scale for Children. *Psychological Assessment*, *6*(2), 149–158. https://doi.org/10.1037/1040-3590.6.2.149

Huebner, E. Scott, & Gilman, R. (2002). An Introduction to the Multidimensional Student Life Satisfaction Scale. *Social Indicators Research*, *60*, 115–122.

Jonker, C. S., & Vosloo, C. (2004). The Psychometric Properties of the Schutte Emotional Intelligence Scale. *Empirical Research*, 34(2), 21–30.

Kong, F., Zhao, J., & You, X. (2012). Emotional Intelligence and Life Satisfaction in Chinese University Students: The Mediating Role of Self-esteem and Social Support. *Personality and Individual Differences*, *53*(8).

Korobova, N., & Starobin, S. S. (2015). A Comparative Study of Student Engagement, Satisfaction, and Academic Success Among International and American Students. *Journal of International Students*, *5*(1), 72–85.

Largen, J. (2004). *The Relationship Between Stress, Satisfaction, and Emotional Intelligence in College Students.*

Leung, J. P., & Zhang, L. W. (2000). Modelling Life Satisfaction of Chinese Adolescents in Hong Kong. *International Journal of Behavior Development*, *24*(1), 99–104.

Lewis, A. D., Huebner, E. S., Malone, P. S., & Valois, R. F. (2011). Life Satisfaction and Student Engagement in Adolescents. *Journal of Youth Adolescence*, *40*, 249–262. https://doi.org/10.1007/s10964-010-9517-6

Maguire, R., Egan, A., Hyland, P., & Maguire, P. (2016). Engaging Students Emotionally: The Role of Emotional Intelligence in Predicting Cognitive and Affective Engagement in Higher Education. *Higher Education Research & Development*, 1–15. https://doi.org/10.1080/07294360.2016.1185396

Mehdinezhad, Vali. (2015). Student Satisfaction with Life and its Relation to School Burnout. *Bulgarian Journal of Science and Education Policy*, *9*(2), 254–265.

Murphy, K. (2006). The Relationship Between Emotional Intelligence and Life Satisfaction After Controlling for Self- esteem, Depression, and Locus of Control Among Community college students. *Dissertation and Thesis*, 1–241.

Palmer, B., Donaldson, C., & Stough, C. (2002). Emotional Intelligence and Life Satisfaction. *Personality and Individual Differences*, 33, 1091–1100.

Pellitteri, J. (2010). The Relationship Between Emotional Intelligence and Ego Defense Mechanisms. *The Journal of Psychology Interdisciplinary and Applied*, *136*(2), 182–194. https://doi.org/10.1080/00223980209604149

Proctor, C., Linley, P. A., & Maltby, J. (2009). Youth Life Satisfaction Measures: A Review. *The Journal of Positive Psychology*, *4*(2), 128–144. https://doi.org/10.1080/17439760802650816

Rastogi, A., & Pati, S. P. (2018). Student Engagement in Indian Context: UWES-S Validation and Relationship with Burnout and Life Satisfaction. *Journal Work Organisation and Emotion 89*, *9*(1), 89–106.

Sarican, H., Celik, I., & Coskun, L. (2015). The Relationship between Emotional Intelligence, Hope and Life Satisfaction in Preschool Preserves Teacher. *The International Journal of Research in Teacher Education*, *6*(1), 1–9.

Schutte, N. S., Malouff, J. M., Hall, L. E., Haggerty, D. J., Cooper, J. T., Golden, C. J., & Dornheim, L. (1998). Development and Validation of a Measure of Emotional Intelligence. *Personality and Individual Differences*, *25*, 167–177. https://doi.org/10.1016/S0191-8869(98)00001-4

Tekie, Y. T. (2014). *Adult Attachment, Emotional Intelligence, and Resilience as Correlates of Social Engagement, Academic Engagement, and Confidence of Persistence in College Students* (Issue June). https://doi.org/10.13140/RG.2.1.3921.4242

Urquijo, I., & Extremera, N. (2017). Academic Satisfaction at University: The Relationship between Emotional Intelligence and Academic Engagement. *Electronic Journal of Research in Educational Psychology*, *15*(3), 553–573.

Valehzaghard, H. K., Khodaei, R., & Chegini, N. S. (2000). Predictors of Academic Engagement Among Adolescents in Tehran. *International Journal of Humanities and Social Science Invention*, *2*(6), 45–52.

Yang, H., & Farn, C. K. (2005). An Investigation the Factors Affecting Mis Student Burnout in Technical-vocational College. *Computer in Human Behavior*, *21*, 917–932. https://doi.org/10.1016/j.chb.2004.03.001

Families Mental Health and Challenges in the 21st Century – Arifin et al. (eds)

The effect of family function on the self-acceptance of the broken home adolescent

Annisah Nurul Azizah*, Cahyaning Suryaningrum & Susanti Prasetyaningrum
Fakultas Psikologi, Universitas Muhammadiyah Malang, Malang, Indonesia

ABSTRACT: Broken-home adolescents tend to have a low level of self-acceptance, then to achieve a high self-acceptance process, individuals need reinforcement from people in their surrounding environment, one of which is the family. An optimally functioning family will help broken-home teens reach a high level of self-acceptance for their family background that is different from their friends in general. This study aimed to determine the effect of family functioning on self-acceptance in broken-home adolescents. The research design uses correlational quantitative. The data collection method used the non-probability sampling purposive sampling technique. The research subjects were 116 adolescents, with a frequency of 84 women and 32 men, female subjects. The age range of respondents is 17–21 years old. In addition, the respondent's family background is a broken home with the provision of living with a nuclear family or a large family with a frequency of 91 subjects living with their nuclear family and 25 subjects living with extended family. The measuring instrument used is the Berger scale (Self-Acceptance Questionnaire) and The General Functioning 12-item subscale (GF12) of The McMaster Family Assessment Device (FAD) scale. Test data analysis using simple linear regression which shows the influence of family functioning on self-acceptance of broken-home adolescents ($\beta = 0.613$; $p<0.001$), which means that the more the family functions optimally, the higher the level of self-acceptance of broken-home adolescents. Other results show that gender has an effect on the level of self-acceptance of broken home adolescents $p = 0.011$ ($p < 0.005$) with male ($M = 111.750$; $SD = 22.02$) and female ($M = 100.952$; $SD = 19.524$) scores. Different test based on living with a nuclear family or extended family does not affect the self-acceptance of broken-home adolescents. This means that only gender and family functioning in general influence the self-acceptance of broken-home adolescents.

Keywords: Broken home adolescents, Family functioning, Self-acceptance

1 INTRODUCTION

Research related to psychological problems in adolescents is increasing because when viewed from Erikson's theory, individuals enter the early stages of identity versus identity confusion. One of the things that are very influential on the process of discovering individual identity is when a family experiences dysfunction or a family experiences divorce, adolescents will experience a lot of pressure and confusion, and adolescents do not have ideal figures to help them become psychologically mature individuals (Santrock 2012).

One of the impacts of parental divorce is the low level of self-acceptance in adolescents. This is in line with the research conducted by Sari (2013) and Fahrudin (2012) that adolescents with broken home families tend not to accept themselves as victims of their parent's

*Corresponding Author: 1annisahnurul216@gmail.com

DOI: 10.1201/9781003402381-20

divorce. They also feel unable to face the realities of life they experience, and teenagers cannot accept criticism from those around them. Another study conducted by Andriani *et al.* (2020) that adolescents from broken-home families have poor self-acceptance marked by feelings of inferiority, withdrawing from their friends, and assessing themselves subjectively. Bastaman (2007) explained that self-acceptance is the initial stage for individuals in the process of self-development from a meaningless life stage to a meaningful life stage. This is considered very important because when an individual cannot accept himself for the bad experiences that have befallen him in the past, there is a delay in developing himself.

One of the variables that can have an impact on increasing self-acceptance of broken home adolescents is a family that works optimally. In addition, research conducted by Nisa & Sari (2019), Stanescu & Romer (2011) and Wulandri & Fauziah, (2019) explained that family functioning has a significant effect on the level of self-acceptance of broken home adolescents. Previous research has revealed that good family functioning does not only come from complete families according to research conducted by Azhari *et al.* (2019). They revealed that family functions in broken home families in Tanjung Hulu Village are quite good, this is marked by the ongoing functioning of socialization, affection functions, and the function of life in the family because families with broken home backgrounds are still able to carry out aspects of family functioning even with a single parent condition caused by divorce or death. In addition, research conducted by Yee & Sulaiman (2017) explained that family functions will continue to run optimally when existing family members can carry out aspects of family functioning well, in which good family functioning is not only determined by whether or not the parents are complete but how maximally the aspects of family functioning can be carried out.

This study aimed to determine the effect of family functioning on the self-acceptance of broken-home adolescents. This research is expected to be able to provide an overview to teenagers with broken home family backgrounds that the functioning of the family is very important to help the physical, social, and emotional development of adolescents in undergoing the development process to create individuals who can accept themselves and do not look down on themselves with bad experiences that have befallen him.

2 METHOD

This research design uses non-experimental correlational quantitative which emphasizes the analysis of quantitative data or numbers collected through the measurement stages by filling in the scale and processed by statistical analysis methods (Azwar 2018). The data collection method used the purposive non-probability sampling technique. The research subjects were 116 adolescents, with a frequency of 84 women and 32 men. The age range of respondents is 17–21 years with the majority aged 20 and 21 years. In addition, the respondent's family background is a broken home with the provision of living with a nuclear family or a large family with a frequency of 91 subjects living with their nuclear family and 25 subjects living with extended families. The measuring instrument used is the Berger scale (Self-Acceptance Questionnaire) totaling 36 items with a Cronbach's Alpha reliability index of 0.913. This scale consists of 36 items with 5 answer choices with a score of 1–5 and consists of favorable and unfavorable items where a score of 5 is for a very appropriate answer, a score of 4 for an appropriate answer, a score of 3 for a doubtful answer, a score of 2 for an inappropriate answer, and a score of 1 for a very inappropriate answer; however, the value is given if the item is favorable, and if the item is unfavorable then the scoring system be the other way around. and the 12-item General Functioning (GF12) subscale The McMaster Family Assessment Device (FAD) scale consists of 12 items with 4 answer choices namely, STS (Strongly Disagree), TS (Disagree), S (agree), and SS (Strongly agree), with Cronbach's Alpha reliability index of = 0.908. Data analysis used simple linear regression analysis.

3 RESULT

Participants in this study were 116 teenagers from broken homes. This study shows that family functioning has a significant influence on the self-acceptance of broken-home adolescents ($\beta = 0.613$; $p<0.001$).

In Table 1, it is known that the hypothesis of this study was accepted so it can be concluded that family functioning has a significant influence on the self-acceptance of broken-home adolescents. The results of other studies explain the effective contribution of the R Square (R^2) value of 0.375 which means that the variable of self-acceptance can be influenced by family functioning by 37.5% while the other 62.5% is influenced by other variables, Table 2 explained that there are differences in the results of the level of self-acceptance by gender with a value of $p = 0.011$ ($p < 0.05$) (M = 111.750; SD = 22.023 and female (M = 100.952; SD = 19.524). Male have a difference in the mean value of 10.798, which means that male (M = 111.750; SD = 22.023) adolescent self-acceptance is higher than female (M = 100.952; SD = 19.524) with significant differences ($t = -2.569$; $p<0.05$).

Table 1. Regression.

Variable	B	β	R^2	F	p
Family functioning	1.725	0.613	0.375	68.521	<0.001

Table 2. Different tests based on gender.

	Male		Female			
Variable	M	SD	M	SD	t	p
Self-acceptance	111.750	22.023	100.952	19.524	−2.569	0.011

4 DISCUSSION

The results of the analysis conducted revealed that this research can answer the proposed hypothesis, namely that there is a significant influence between family functioning on the self-acceptance of broken home adolescents. This is in line with research conducted by Walęcka-Matyja (2015) explaining that the maximum function of the family has a significant effect on the level of self-acceptance of adolescents, where families who can meet primary and secondary needs, provide support and affection will be able to support adolescents to improve positive self-image to achieve a good level of self-acceptance. In addition, research conducted by Nisa & Sari (2019), Stanescu & Romer (2011), and Wulandri & Fauziah (2019) explained that family functioning has a significant effect on the level of self-acceptance of broken home adolescents. The results of other studies explain the effective contribution of the R Square (R^2) value of 0.375 which means that the variable of self-acceptance can be influenced by family functioning by 37.5% while the other 62.5% is influenced by other variables, where one of the influential variables is support. This is in line with research conducted by Noviana and Sakti (2015) which explains that the higher the level of peer attachment, the higher the level of self-acceptance in accelerated students. According to Hurlock (Ardilla & Herdiana 2013), other factors that affect a person's self-acceptance include (1) understanding of oneself, (2) realistic expectations, (3) the absence of obstacles in the environment, (4) a pleasant society, (5) not having severe emotional disturbances, (6) the level of success experienced, (7) identification of someone who can adapt well, (8) having a broad perspective on oneself (9) parenting good, (10) good self-concept.

In addition, the results of the study reveal that there are differences in self-acceptance of broken-home adolescents based on gender, where the level of self-acceptance of male adolescents has a mean value difference of 10,798, which means that the self-acceptance of male adolescents is higher than that of women. This is in line with research conducted by Yusuf, (2020) which explains that male adolescents who are going through puberty tend to have a good level of self-acceptance because they are not worried about the changes that occur in themselves, while adolescent girls will tend to be worried about the changes they experience both physically and psychologically.

5 CONCLUSION AND IMPLICATION

The conclusion of this study is the accepted hypothesis in which there is a significant effect of family functioning on the self-acceptance of broken-home adolescents. Besides, the gender of the respondent also affects the level of self-acceptance of broken-home adolescents. Moreover, other factors can also affect the level of self-acceptance of broken-home adolescents, namely social support, peer support, and the prevailing culture in the surrounding environment. Further researchers can explore other variables that affect their self-acceptance; self-understanding, self-concept, parenting, emotion, adjustment, peer support, and gender. Further researchers can also provide interventions on the subject to increase the level of self-acceptance, the form of intervention that can be given is in the form of individual counseling or peer group counseling at school.

REFERENCES

Andriani, N., Romli, E., & Arizona. (2020). Pengaruh Kontrol Diri Terhadap Penerimaan Diri Siswa Broken Home Di Sma Negeri 10 Palembang. *Journal of the Counseling Guidance Study Program*, *7*(1), 47–52.

Ardilla, F., & Herdiana, I. (2013). Penerimaan Diri pada Narapidana Wanita. *Fakultas Psikologi, Universitas Airlangga, Surabaya*, *2*(01).

Azhari, T. D., Bahari, Y., & Ulfah, M. (2019). Pelaksanaan Fungsi Keluarga dalam Mendidik Anak Pada Keluarga Broken Home di RW 04 Kelurahan Tanjung Hulu. *Jurnal Pendidikan Dan Pembelajaran Khatulistiwa*, *8*(3), 1–8. https://jurnal.untan.ac.id/index.php/jpdpb/article/view/32310

Bastaman, H. D. (2007). *Logoterapi: Psikologi untuk Menemukan Makna Hidup dan Meraih Hidup Bermakna*. PT Raja Grafindo Persada.

Fahrudin, A. (2012). Keberfungsian Keluarga: Konsep dan indikator pengukuran dalam penelitian. *Informasi*, *17*(2), 75–81.

Nisa, H., & Sari, M. Y. (2020). Peran Keberfungsian Keluarga Terhadap Penerimaan Diri Remaja. *Psikoislamedia: Jurnal Psikologi*, *4*(1), 13–25.

Noviana, Siti & Sakti, H. (2015). Kata Kunci: Hubungan Antara Peer Attachment Dengan Penerimaan Diri Pada Siswa-Siswi Akselerasi. *Journal Empati*, *4*(April), 114–120.

Santrock, J. W. (2012). *Life-Span Development: Perkembangan Masa Hidup* (Jilid 1). Erlangga.

Sari, L. K. (2013). Penerimaan Diri Pada Remaja Korban Perceraian Orangtua. *Journal of Chemical Information and Modeling*, *53*(9), 1689–1699.

Stanescu, D. F., & Romer, G. (2011). Family Functioning and Adolescents' Psychological Well-Being in Families with a TBI Parent. *Psychology*, *2*(07), 681–686. https://doi.org/10.4236/psych.2011.27104

Walęcka-Matyja, K. (2015). Adolescent Personalities and Their Self-acceptance within Complete Families, Incomplete Families, and Reconstructed Families. *Polish Journal of Applied Psychology*, *12*(1), 59–74. https://doi.org/10.1515/pjap-2015-0004

Wulandri, D., & Fauziah, N. (2019). Pengalaman Remaja Korban Broken Home (Studi Kualitatif Fenomenologis). *Empati*, *8*(1), 1–9.

Yee, N. Y., & Sulaiman, W. (2017). Resilience as Mediator in the Relationship between Family Functioning and Depression among Adolescents from Single Parent Families. *Akademika*, *87*(01), 111–122. https://doi.org/10.17576/akad-2017-8701-08

Yusuf, F. (2020). *Perbedaan Penerimaan Diri tentang Pubertas antara Remaja Pria dan Wanita*.

Families Mental Health and Challenges in the 21st Century – Arifin et al. (eds)
© 2024 The Author(s), ISBN 978-1-032-51442-0

A brief sketch of various ideologies regarding gender relations in marriage from the perspective of feminism

Bintang Sasmita Wicaksana*
Laboratory of Psychology, Faculty of Psychology, University of Muhammadiyah Malang, Malang, Indonesia

ABSTRACT: The issue of injustice, especially gender inequality, places women in a disadvantaged condition in marriage. Knowledge of gender relations in marriage is in many ways driven by ideology. Through a literature review from the perspective of feminism theory, the author reveals that there are five kinds of ideology in marriage, namely, (1) liberal feminism, (2) radical feminism, (3) Marxist feminism, (4) socialist feminism, and (5) Muslim feminism. It is hoped that further research can be carried out empirically to inquire about the ideal description of gender relations in marriage and compare it with the various ideologies that have been mentioned.

Keywords: Marriage, ideology, feminism

1 INTRODUCTION

The study of injustice and efforts to realize social justice in social life is one of the interesting topics of discussion in every discussion regarding the formation of the concept of social change in the human social system (Fakih 2020). Mainly including Karl Marx's thoughts on analysis and his class theory which helps to understand the various forms of economic inequality and their implications for more extensive social systems (Woodfin & Zarate 2008). Then also the successors of Marx's thought such as the Frankfurt School which sued the methodology and knowledge system a la positivism as a source of injustice through criticisms of the doctrine of the objectivity and neutrality of scientific knowledge. In short, science can and has become a tool to perpetuate injustice (Bottomore 2019).

However, perhaps a study on gender inequality needs to be redeveloped in its discourse, not to mention that it is still far from the spotlight, and even gets resistance from both men and women. According to the identification made by Fakih, in 2020, there are several causes for the emergence of resistance, namely: (1) questioning the status of women is basically questioning the established systems and structures; (2) there are many misunderstandings when issues related to women are questioned; (3) discussions on gender issues actually discuss issues regarding power relations which are very personal in nature. In the end, understanding the conceptual framework of gender relations between men and women becomes complicated. In fact, discussing issues regarding gender relations is a fundamental effort in solving human relations problems (Sartre 2018).

The problem is that gender differences give birth to issues of gender inequality. Gender inequality is a system and structure that does not benefit and makes victims, both men and women. Sources that can smoothen gender inequality come from various entities such as government policies, beliefs, religious interpretations, traditional beliefs, habits, and even

*Corresponding Author: bintangsasmitawicaksana@umm.ac.id

DOI: 10.1201/9781003402381-21

scientific assumptions. Manifestations of gender inequality occur in various forms such as: (1) marginalization or the process of economic impoverishment that occurs not only in the workplace, but also at the domestic (household) level, society, culture, or even the state, (2) subordination or the notion of being unimportant in political decisions, (3) stereotypes or negative labeling, (4) violence due to the inequality of power that exists in society (5) longer and more workloads, and (6) dissemination of the ideology of gender role values, which has become so deeply rooted that it becomes the most difficult thing to change (Fakih 2020).

Cases related to gender inequality, both domestic and at work, involve women in a disadvantageous position. One of them is the fact that women with careers, especially those who are married, experience gender inequality which is manifested in the division of roles and division of labor in the family that is not balanced, causing women to be trapped in a situation of double burden that must be borne. Women who are married and working are required to play a double (often triple) role, namely women as wives and mothers who are responsible for domestic affairs (family) and also play a role as breadwinners (participation in the public sphere). At first glance, the role of the husband will increase due to changes and adjustments in the division of tasks and roles in domestic affairs. However, a lot also happens, even though the wife has a dual role, the husband is less involved in domestic affairs, and often tensions arise in the household (Rahmawati 2015). The study of this problem becomes interesting and research on this topic will explore issues related to gender inequality experienced by career women.

In the Indonesian context, marriage is generally considered a sacred social institution by placing men as leaders and women as followers. However, the incidence of gender violence in the domestic sphere often occurs with the assumption that the dominant role of men is considered to be the root cause of the oppression that occurs in the household. Aggressive behavior shown in the household can be suspected as the ideology of gender relations between men and women that can trigger gender injustice in the household context. Then, why ideology? Why is ideology important in moving one's behavior to commit gender injustice in the sphere of marriage?

1.1 *Ideology: An introduction*

A common tradition in psychology, especially concepts closely related to mental health and family, is its claim of neutrality, objectivity, and apolitical (Wijaya Mulya 2021). In short, psychology is considered distant from power and ideology. In fact, the practice is the opposite that science, especially social science (in this case psychology is also included), is not value-free, not universal, and not sterile from ideological burdens (Heryanto 2005). Regarding this, Michel Foucault provides an explanation through a framework known as *power/knowledge*.

In practice, power requires an instrument that can normalize individuals in society through discipline and norms so that uncontrolled actions such as violence can be avoided through obedience and order. This arises through a control mechanism that manifests in a form of knowledge that is represented in a structured way through a language called discourse. Discourse can be in the form of opinions, laws, morality, views of life, and ideologies that can influence patterns of thought and action in accordance with the wishes of the discourse controller (Yudhisthira 2020).

For Foucault, power relations in all human life are ideologies. Every relationship is an effort to dominate each other, to suppress each other (Yudhisthira 2020). Ideology can be referred to as 'a set of knowledge, views, awareness, tastes, or values, and general tendencies in favor of certain social interests, with or without the awareness of the parties concerned' (Heryanto 2005). Althuser (2006) emphasized that ideology does not only appear in the relationship between the state and the people but also in everyday relationships between people. For him, the practice of ideology that exists in humans is not something that is realized. The interesting thing is that ideology can be infiltrated into society through various media that are related to the structure of society such as religion, family, education, mass media, and so on.

Through mechanisms like this, ideology helps humans to understand and explain the world (Takwin). Through ideology, humans can determine actions in their daily lives, how to solve problems, and make decisions through the information processing process, one of which is influenced by ideology. In social psychology, Ajzen and Fishbein discussed their theory of Planned Behavior (*Theory of Planned Behavior*) that belief, in this case including ideology, has a close relationship with behavior (Ramdhani 2011). This theory states that three main components can shape a person's behavioral intentions, namely, (1) *attitude*, (2) *subjective norms*, and (3) *perceived behavioral control.* This shows that the thought process shown through the theory of planned behavior produces and determines what decisions and actions need to be taken so that these decisions are also influenced by ideology.

2 METHODS

This research uses a literature review design as a way of extracting data which is part of qualitative research (Creswell 2010). A literature review is research that critically reviews knowledge, ideas, or findings obtained from academically oriented literature and formulates theoretical and methodological contributions on various specific topics (Cooper 2010). A literature review is useful for providing an overview of a particular issue or research problem (Snyder 2019).

In a literature review or what is commonly referred to as a literature study, the data collected as a reading source is through various books, journal articles, and magazines that are relevant to the research objectives (Nazir 2011). The book that is the center of this paper is the work of Mansour Fakih with titled Gender Analysis and Social Transformation and Muslim Feminism in Indonesia by Alimatul Qibtiyah.

Ramdhani *et al.* (2014) provided an explanation of the stages in conducting a literature review which consists of four stages, namely, (1) selecting topics for review, (2) searching and selecting relevant articles, (3) analyzing and synthesizing literature, and (4) organization of review writing. Of the steps that must be taken in the literature review, one that must be considered is the preparation of a synthesis of conceptual or empirical articles that are relevant to the research conducted.

3 RESULT AND DISCUSSION

There is often a view that feminism is a resistance movement against men, which gets a hard stamp as an effort to rebel against social institutions that exist in society such as marriage and household which are considered against nature. Misunderstandings like this need to be corrected so that public acceptance of more egalitarian gender relations in marriage becomes easier.

It is necessary to approach feminism as an idea that consists of many compositions of thought and approaches. Qibtiyah has a closeness of thought that places women as someone who is "the same or different" from men contained in the concept, as well as the debate, which is called *sameness-difference* (Bacchi 1990). To clarify this position, the maximizers–minimizers theory was introduced by Catherine Stimpson (Chodorow 1978). What are maximizers and minimizers? Feminist schools that emphasize maximizers position call for the importance of gender differences. On the other hand, feminists with minimizers position will emphasize the importance of minimizing differences between the sexes. In this regard, many criticisms have been leveled at this theory, particularly those focused on maximizers. One of which is given by Epstein (1988) that provides empirical evidence from the biological and social sciences that this thinking is 'weak and cannot be used as a reference to the issue of gender equality equality gender' (Qibtiyah 2020).

In essence, various feminist movements have the same starting point, namely from the point of assumption and awareness about the oppression and exploitation of women which need to be ended for the sake of 'equality, dignity, and freedom to control body and life, both inside and outside the country. outside the house' (Fakih 2020). However, there is a diversity of understandings and views on what, why, and how women experience oppression and exploitation. With the various ideologies, paradigms, and theories used, this study will briefly describe the various ideologies in feminism that are applied in the context of marriage. The short sketch that will be presented below is a simplification of the results of the readings carried out by the researchers as well as a reflection on gender relations in marriage from each ideology.

3.1 *Liberal feminism*

From the perspective of liberal feminism, the basic assumption is rooted in the view that freedom and equality are rooted in rationality and the separation between the private and public worlds (Fakih 2020). The main postulate of liberal feminism focuses on fighting for 'equal opportunities and equal rights' which is important because there are no essential differences between men and women. The ability to think rationally is something that is natural, the same as gender.

This school is a critique of liberalism as a mecca of political theory whose doctrine points are autonomy, equality, moral values, and individual freedom. However, the opposite happens to women who experience discrimination (Fakih 2020). Therefore, equal opportunities and rights must be opened as widely and freely as possible so that there are no differences between men and women, and equality can be obtained. This ideology has never questioned the discrimination that occurs due to patriarchal ideology or analysis of class, politics, economy, and gender (Fakih 2020).

Therefore, the main character of this ideology in solving the problem of relations between women and men is to prepare women to be able to compete in the world of free competition. Because of their individualistic ideological character, women's backwardness is considered because women themselves are unable to compete even though they have been given equal opportunities. In addition, the strong belief in traditional values is also a factor in women's backwardness so that involving women in the development and industrialization agenda is a way to increase the status of women (Fakih 2020).

3.2 *Radical feminism*

Historically, this movement emerged as a logical consequence of the rise of *sexism* in the West in the 1960s, which was marked by fighting sexual violence and pornography. The interesting and important thing in this movement is that radical feminism considers the root of the oppression of men against women as the gender of the men themselves and their patriarchal ideology. Therefore, biologically and politically men are the biggest problem for women (Fakih 2020).

According to the perspective of radical feminism, patriarchy is an ideology of oppression by men against women. Patriarchy is an ideology that contains a sexual hierarchical system that places men in a superior position with economic privilege. Men's control over women, one of which is sexual relations, is interpreted as a form of oppression against women. Unfortunately, the analysis of gender relations is reductionistic in nature whose emphasis is only on seeing this problem as a natural biological difference (Fakih 2020).

For a revolution to take place in this movement, ideally every woman needs to take action to change their own lifestyle, experience, and relationship with men. Therefore, change through revolution against women's oppression can be carried out on a personal or subjective level, which differs diametrically from the Marxist tradition in how to achieve the goals of social change. What is unfortunate is the form of struggle based on competition to overcome men by giving the impression that being a man alone is a mistake (Fakih 2020).

Marriage, to groups with this ideology, seems to be an oppressive institution because it puts women in a position that is subordinated to men and at the same time becomes an agent of the proliferation of patriarchal ideology. This ideology offers a perspective, namely that women can rebuild cultural alternatives so that they can reshape their lives outside the concept of patriarchy. Although it has an essential contribution to the fight against gender injustice, one of which is through patriarchal theorizing, the criticism addressed to this ideology is to avoid issues related to race, social class, religion and instead to mainstream homosexual relationships while criticizing heterosexual relations (Fakih 2020). The concept of feminism is considered too radical, especially by sociologists, even among feminists themselves. Liberal feminist figures do not fully agree with the thoughts from the perspective of radical feminism (Marzuki 2007).

3.3 *Marxist feminism*

In contrast to radical and liberal feminism, Marxist feminism sees that the problem of oppression of women is part of the problem of class oppression. The framework of criticism of capitalism is also placed in the problems faced by women. In Karl Marx's view, the relationship that exists between husband and wife has something in common with the relationship between the bourgeoisie and the proletariat. Not only that, the status of women can also be a benchmark for the level of community progress (Fakih 2020).

Marxism, especially in the tradition of orthodox Marxism, has an understanding that the economic aspect of social reality determines non-economic aspects such as cultural and political aspects of social life (Fakih 1995). This tradition emphasizes the importance of class struggle in achieving social change. From a Marxist perspective, society consists of two important elements, namely the base and the superstructure. It is better known that the economy is the foundation which is an important factor of social change. Meanwhile, the superstructure consists of education, culture, and ideology which are the second priority because they are determined by the economy. In short, economic determinism is a fundamental foundation in the orthodox Marxist tradition (Fakih 1995). In Indonesia, especially since the dominance of the new order in running the government, Marxism is considered a forbidden teaching which is described as a source of 'immorality and anti-Pancasila' (Laksana 2017) so discourses on Marxism are limited, especially in school and university life.

Assumptions a la Marxism suggest that in the era of capitalism, the relationship between capitalists, husbands, and wives only benefits the capitalists. Men go through a process that makes them exploited in the workplace. Then, only the husband who has been exploited returns home and establishes a relationship with his wife (Fakih 2020).

Women are also considered profitable for a capitalistic system. Capitalism, whose main logic is to accumulate capital as much as possible, sees women as workers, which is advantageous because (1) the wages received by women can be reduced cheaper than male workers, (2) create an unlimited supply of labor reserves so that the bargaining position between capitalists with the workers becomes stronger and threatens the solidarity of the workers (Fakih 2020).

In this perspective, the capitalist system is actually the root of the problem of gender inequality that occurs so that to destroy this oppressive system, the solution that needs to be done is structural by carrying out a class structure transformation called a 'revolution' process and disconnection with the international capitalist network. From this, it can be seen that women's emancipation can be realized if and only if women are involved and stop taking care of the household, which is only realized through the industrialization process so that a change in women's status can occur through a socialist revolution while eliminating domestic work (Fakih 2020).

The role of the state is considered essential in the context of domestic affairs (Fakih 2020). Changes that are structural in nature by involving the government which should pay for housework and child care are proposals put forward by Marxist feminists. Not only that, they demand that women get the same amount of wages that men receive for the same work. However, severe criticism has been addressed to an ideology because it makes women, in the end, still dependent on the presence of the state (Qibtiyah 2019).

3.4 *Socialist feminism*

Feminism offers what the women's movement hopes for. In this perspective, oppression of women occurs in any class and revolution does not necessarily overcome this. Therefore, the classical Marxist doctrine which focuses on economic exploitation is rejected as the basis for gender oppression. However, on the contrary, if feminism does not use class analysis, it is also problematic so that criticism of capitalism which exploits class at the same time must also be accompanied by criticism of domination, subordination, and marginalization of women (Fakih 2020).

Although this ideology is considered a fusion between Marxist feminism and radical feminism, it seems that its style is still different from radical feminism. For socialist feminism, injustice does not only stem from gender differences, production, and reproduction activities of society but also from social construction. Therefore, what is being fought is the 'construction of the vision and ideology of society as well as structures and systems of justice built on gender bias' (Fakih 2020).

According to Karen Sacks, women experience a change in roles from being free and equal to being subordinated and dependent (Fakih 2020). What drives this change is an understanding of the private sphere in which the family becomes an institution that perpetuates and exploits such a position and what happens is the division of labor based on sex which is an attempt to institutionalize inter-gender dependence that benefits men (Fakih 2020). In the end, women, according to Gayle Rubin, accept this as gender identity through their interpretation of Sigmund Freud's theory (Fakih 2020).

3.5 *Muslim feminism*

Discussions about Islam and feminism always end in a tough and complicated debate. The problem revolves around whether feminism is in line with Islamic teachings or not, and whether it is possible to combine two 'thoughts' from two different civilizations between East and West. This is what Miriam Cooke is trying to illuminate that it is necessary to distinguish between so-called Islamic and Muslim feminism. Islamic feminism can be termed as using 'Islamic tradition as a consideration for feminism' (Qibtiyah 2019). Then, Muslim feminism is different from Islamic feminism which means that a feminist who is Muslim but does not necessarily combine Islamic teachings with feminism (Qibtiyah 2019). An interesting thing that deserves to be discussed is the review conducted by Saputra (2022) on the book written by Etin Anwar with the title 'Islamic Feminism: Genealogy, Challenges, and Prospects in Indonesia' which divides the Islamic feminism movement into several periods, namely, (1) Emancipation, (2) Association, (3) Development, (4) Integration, and (5) Dissemination.

By continuing to explore historical approaches, finding space for interpretation by deconstructing, and using hermeneutics to find out what messages are meant in Islamic teaching texts regarding gender equality, Muslim feminists believe that Islamic teachings have a misunderstanding of gender equality because the exclusive interpretation of Islamic teachings, very male perspective, and detrimental to women. They believe that the teachings of Islam are just the opposite. Islam highly values men and women and there are sources of reference for discussing gender equality, that in Islam men and women are equal and complementary (Qibtiyah 2019). This is in line with the opinion of Sirry (2018) which reveals an interesting thing that the main obstacle to the involvement of feminists is the 'monopoly tendency of men who control religious institutions' so that it does not provide space for the presence of women in forums involving official religious leaders. The note given by Sirry (2018) is that the feminist movement also needs to struggle with issues surrounding religious pluralism.

Presumably, it is difficult to understand Muslim feminism if you do not understand Islamic methodology in understanding Islamic teachings. In Islamic methodology, some approaches are understood by its followers in understanding religious texts, including (1) textual which is a way of understanding religious texts literally and also normatively with the claim that religious teachings never change, and (2) contextual and hermeneutic Approaching religious texts does not just refer to the Qur'an and Hadith but creatively expands the scope of its interpretation by considering space and time. From here, later by Qibtiyah, Muslim thought will be grouped into three different groups, namely conservative, moderate, and progressive (Qibtiyah 2019).

For example, in the context of an equal division of roles between the domestic sphere and work, conservative groups will think that this is related to the concept of nature which God has given and cannot change that men must work and women take care of domestic affairs so that the gender view is What applies is that women cannot work with their main responsibilities being to serve their husband's needs, take care of their children, and take care of their husband's property. Moderate groups have a similar tendency but are more flexible in assuming that women can contribute to economic affairs provided they are considered as additional income for the family. It is different with more progressive groups who argue that although men and women are naturally different, in domestic and public affairs they have the same role because in essence, the roles played by men and women are social constructions (Qibtiyah 2019).

4 CONCLUSION AND IMPLICATION

Marriage has a basic ideology and if drawn from the perspective of feminism, there are 5 main themes, namely: (1) liberal feminism, which focuses on fighting for 'equal opportunities and equal rights' which is important because there are no essential differences between men and women. (2) radical feminism, which considers the root of men's oppression of women to be the gender of men themselves and their patriarchal ideology. fighting for 'equal opportunities and equal rights' which is important because there is no essential difference between men and women; (3) Marxist feminism, which sees that the root cause of injustice in the household is economics; (4) socialist feminism, which sees and tries to synthesize between radical feminism and Marxism; (5) Islamic feminism, which has a variety of groups consisting of conservative, moderate, and contextual.

Further research that needs to be developed is the need to provide a critical description of the dynamics between ideologies in each school of feminism thought by conducting empirical research in order to try to describe in depth the ideal picture of marriage in married couples and then compare it with various thoughts that have been proposed and discussed above.

REFERENCES

Althuser, L. (2006). Ideology and Ideological State Apparatuses (Notes towards an Investigations). *The anthropology of the state: A reader*, 9(1).

Bacchi, C. L. (1990). *Same Difference: Feminism and Sexual Difference*. Sydney, Australia: Allen & Unwin.

Bottomore, T. (2019). *Mazhab Frankfurt: Gagasan dan kritik*. Yogyakarta: Penerbit Independen.

Chodorow, N. (1978). *The Reproduction of Mothering: Psychoanalysis and the Sociology of Gender*. University of California Press.

Cooper, H. (2010). *Research Synthesis and Meta-analysis: A Step-by-step Approach*. 5th edn. Thousand Oaks, CA: SAGE Publications, Inc.

Creswell, J. W. (2010). *Research Design: Pendekatan Kualitatif, Kuantitatif, Dan Mixed*. Yogjakarta: PT Pustaka Pelajar.

Epstein, C. F. (1988). *Deceptive Distinctions: Sex, Gender, and the Social Order*. Yale University Press; Russell Sage Foundation.

Fakih, M. (1995). The Role of Nongovernmental Organizations in Social Transformation: a Participatory Inquiry in Indonesia. *Doctoral dissertation*. Retrieved from https://scholarworks.umass.edu/dissertations_1/2259

Fakih, M. (2020). *Analisis Gender Dan Transformasi Sosial*. Yogyakarta: INSISTPress.

Heryanto, A. (2005). Ideological Baggage and Orientations of the Social Sciences in Indonesia. In V. R. Hadiz, & D. Dhakidae (Eds.), *Social Science and Power in Indonesia* (pp. 57–89). Singapore: Institute of Southeast Asian Studies.

Laksana, B. (2017). Pendidikan, Pembangunan Dan Kesadaran Kritis. Retrieved from https://indoprogress.com/2017/07/pendidikan-pembangunan-dan-kesadaran-kritis/

Marzuki, M. (2007). Kajian Awal Tentang Teori-Teori Gender. *Jurnal Civics*, 4(2), 67–77. doi:https://doi.org/https://doi.org/10.21831/civics.v4i2.6032

Nazir, M. (2011). *Metode Penelitian*. Bogor: Ghalia Indonesia.

Qibtiyah, A. (2019). *Feminisme Muslim di Indonesia*. Yogyakarta: Suara Muhammadiyah.

Rahmawati, A. (2015). Harmoni Dalam Keluarga Perempuan Karir: Upaya Mewujudkan Kesetaraan Dan Keadilan Gender Dalam Keluarga. *Palastren*, 8, (1) 1–34.

Ramdhani, A. *et al.* (2014). Writing a Literature Review Research Paper: A Step-by-step Approach. *International Journal of Basic and Applied Science*, 3(1), 47–56. Retrieved from https://digilib.uinsgd.ac.id/5129/1/08IJBAS%283%29%281%29.pdf

Ramdhani, N. (2011). Penyusunan Alat Pengukur Berbasis Theory of Planned Behavior. *Buletin Psikologi*, 19 (2), 55–69. doi:https://doi.org/10.22146/bpsi.11557

Saputra, M. A. (2022). Review of Feminisme Islam: Genealogi, Tantangan, dan Prospek di Indonesia. *Mimikri: Jurnal agama dan kebudayaan*, 8(1), 243–257.

Sartre, J. (2018). *Eksistensialisme dan Humanisme*. (Y. Murtanto, Trans.) Yogyakarta: Pustaka Pelajar.

Sirry, M. (2018). *Islam Revisionis: Kontestasi agama zaman radikal*. Yogyakarta: Suka Press.

Snyder, H. (2019). Literature Review as a Research Methodology: An Overview and Guidelines. *Journal of Business Research*, 104, 333–339. doi:doi:10.1016/j.jbusres.2019.07.039

Takwin, B. (n.d.). *Bahasa, Ideologi Dan Pembuatan Keputusan: Sebuah Usulan Kajian Ideologi Dalam Psikologi Sosial*. Retrieved from staff.ui.ac.id: https://staff.ui.ac.id/system/files/users/bagus-t/publication/tentangideologidalampsikologisosial.doc

Wijaya Mulya, T. (2021). Kontestasi Diskursif di Balik Konsep dan Praktik Kesehatan Mental: Kekuasaan, Kolonialisme, dan Kapitalisme. In HIMPSI, T. Susana, A. Supratiknya, J. Abraham, & R. Ardi (Eds.), *Kesehatan Jiwa Dan Resolusi Pascapandemi di Indonesia (Seri Sumbangan Pemikiran Psikologi Untuk Bangsa 5)* (pp. 701–723). Jakarta: Himpunan Psikologi Indonesia.

Woodfin, R., & Zarate, O. (2008). *Marxisme Untuk Pemula*. Yogyakarta: Resist Book.

Yudhistira, A. (2020). *Dilarang gondrong: Praktik Kekuasaan Orde Baru Terhadap Anak Muda Awal 1970an*. Tangerang Selatan: Marjin Kiri.

Families Mental Health and Challenges in the 21st Century – Arifin et al. (eds)
© 2024 The Author(s), ISBN 978-1-032-51442-0

The development of academic dishonesty scale for online learning

Muhamad Salis Yuniardi* & Miqdad Daly Ahmad
Psychology Faculty, University of Muhammadiyah Malang, Malang, Indonesia

ABSTRACT: Education, which previously was conducted face-to-face, is forced to be done online in times of the Covid-19 pandemic. However, online learning has weaknesses, for example, the lack of monitoring by the teachers and the increase of academic dishonesty attempts by the students. The current existing measuring instrument for academic dishonesty is not yet covering the field of online learning. Therefore, this research is aimed to develop an academic dishonesty scale for online learning. The scale developed consists of four dimensions: intentional cheating, fabrication, plagiarism, and the facilitation of academic dishonesty or helping someone to do such things. It has 22 items and was in the Likert model. This research was conducted by involving 62 students from two different universities. The content validity was discussed by a series of meetings involving expertise both in psychometric and educational psychology. Based on the statistical analysis, the scale of online academic dishonesty has Cronbach's reliability of 0.861 and has a good discriminatory power. These all provide evidences that this scale has good validity and reliability and thus, could be used to investigate academic dishonesty specifically during online learning.

1 INTRODUCTION

Behaviors of academic dishonesty have a negative impact on the individuals who perform it, other people, and educational institutions. Students who perform academic dishonesty can obstruct their character development, the more often they do academic dishonesty, the more difficult for them to develop moral personality and ethics (Whitley & Keith-Spiegel 2002). Moreover, if more students are doing academic dishonesty, the quality of an institution can decrease (Heriyati & Ekasari 2020). It also affects people's trust in an educational institution (Rujoiu & Rujoiu 2014). If a habit of academic dishonesty is not handled quickly, it can spread to the next educational stage or even at the workplace. Those who are used to doing academic dishonesty tend to be dishonest to people who rely on their profession (Rujoiu & Rujoiu 2014).

When defined, academic dishonesty is an act of violating the submission of an assignment in general which is done by students to acquire grades, and which the assignment is intended to show the student's knowledge and understanding about the context or process of that assignment (Lambert *et al.* 2003).

Various factors can cause students to do academic dishonesty, whether from individual or institutional factors. On the individual level, academic dishonesty could happen because students have low self-efficacy, which makes them not want to put effort into academic activity (Finn & Frone 2004; Murdock *et al.* 2001), a presence of opportunity to do an act of academic dishonesty (Heriyati & Ekasari 2020; Lewellyn & Rodriguez 2015), poor time management (Lambert *et al.* 2003; Park 2010), students rationalizing assumptions or making excuses, such

*Corresponding Author: salis@umm.ac.id

DOI: 10.1201/9781003402381-22

as assuming that cheating has no effect to their friends, or thinking that cheating does not do any harm to their friends (Ashworth *et al.* 1997), and students feeling pressured to get a better score (Ameen, Guffey & McMillan 1996; Heriyati & Ekasari 2020; Lewellyn & Rodriguez 2015). Meanwhile, on institutional level, academic dishonesty could happen because the institution gives too many assignments (Ameen *et al.* 1996), teachers badly monitor their students (Heriyati & Ekasari 2020), the lack of teacher's discipline toward students who perform academic dishonesty (Guo 2011; Smith *et al.* 2002), and thinking of teachers that academic dishonesty is normal behavior (Whitley & Keith-Spiegel 2002).

In addition, there are also strong factors that present in academic dishonesty, which are technological factors. The rate of academic dishonesty increases because of the presence of technology that helps students to cheat with ease (Born 2003; Park 2010). The use of internet-based technology can benefit the institution and the teachers, however, the abuse of internet-based technology by the students could make them able to perform plagiarism effortlessly in academic activity (Scanlon 2004). Since the Covid-19 outbreak, the use of technology for education has been increasing, where education which was previously conducted face-to-face are becoming online, on the other hand, online learning actually causes an increase in academic dishonesty among students such as cheating in online examination with various form of cheating they can do with the internet (Jankea *et al.* 2021). The increase in academic dishonesty during online learning can be caused by the decrease in teachers' monitoring ability and the increase in students' opportunity to perform the act, where a lack of monitoring can increase academic dishonesty (Heriyati & Ekasari 2020)so with the opportunities that are caused by the nature of online learning (Heriyati & Ekasari 2020; Lewellyn & Rodriguez 2015).

There are various forms of academic dishonesty. The first one is cheating such as seeing each others' answers during an examination (Akbulut *et al.* 2008; Faradiena 2019; Iyer & Eastman 2008; Jurdi *et al.* 2011; Pavela 1978). Another one is plagiarism such as copying essays from the internet (Akbulut *et al.* 2008; Faradiena 2019; Iyer & Eastman 2008; Jurdi *et al.* 2011; Pavela 1978). The third one is outside help such as asking for help from another person to finish an assignment (Faradiena 2019; Iyer & Eastman 2008). The fourth one is falsification such as data manipulation (Akbulut *et al.* 2008; Faradiena 2019). The fifth one is fraudulence, which is an act of giving fake excuses to a teacher to obtain certain advantages (Akbulut *et al.* 2008; Jurdi et al 2011).

In order to measure academic dishonesty, a form of scale is needed, however, the development of an academic dishonesty scale that focuses on online learning has not received much progress. Therefore, this research aims to develop a scale for academic dishonesty in online learning using the typology of academic dishonesty by Pavela (1978).

1.1 *Academic dishonesty*

Lambert, Ellen, and Taylor (2003) define academic dishonesty as an act of violating the assignment submission in general which is done by students to acquire grades, and the assignment is intended to show the student's knowledge and understanding of the context or process of that assignment. According to Pavela (1978), there are four types of academic dishonesty, the first one is intentional cheating, such as using sources, information, or assistance that is prohibited in an academic activity. The second one is a fabrication, such as faking citations in an academic activity. The third one is plagiarism or copying an idea without citing the source. And the fourth one is the facilitation of academic dishonesty or helping someone to do such a thing.

1.2 *Dimensions of academic dishonesty*

There are several dimensions of academic dishonesty. The first dimension is cheating, for example, seeing each other's answers in an examination (Akbulut *et al.* 2008; Faradiena 2019; Iyer & Eastman 2008; Jurdi *et al.* 2011; Pavela 1978). Another dimension is plagiarism, for example, copying essays from the internet (Akbulut *et al.* 2008; Faradiena 2019; Iyer & Eastman 2008; Jurdi *et al.* 2011; Pavela 1978). The third dimension is outside help, for example, asking for help from another person to finish an assignment (Faradiena 2019; Iyer & Eastman 2008). The fourth dimension is falsification, for example, data manipulation (Akbulut *et al.* 2008; Faradiena 2019). The fifth dimension is fraudulence, which is an act of giving fake excuses to teachers in order to obtain certain advantages (Akbulut *et al.* 2008; Jurdi *et al.* 2011).

2 METHOD

2.1 *The stage of scale development*

The research began with a literature study of academic dishonesty-related papers, then the researcher formulated a guide of an interview that is intended to find out the behaviors of academic dishonesty in online learning. The interview was conducted by involving undergraduate and postgraduate students who are doing online learning that is caused by the Covid-19 outbreak.

The results of the literature review and the interview were extracted into five dimensions: cheating, plagiarism, outside help, falsification, and fraudulence. The cheating dimension consists of six items, the plagiarism dimension consists of three items, the outside help dimension consists of four items, the falsification dimension consists of three items, and the fraudulence dimension consists of six items. In total, there are 22 items. The response of the scale has six options: "never", "rarely", "sometimes", "often", and "always".

2.2 *Research subjects*

The subjects of the current research were undergraduate students from two different universities. The sampling technique used in this research is accidental sampling. Accidental sampling is a technique of choosing a sample by chance (Sugiyono 2017). From the sampling result, there are 68 students and 16 of them are male and 48 are female.

3 RESULTS

3.1 *Descriptive analysis*

The researcher has done a descriptive analysis of the sample of this research. The results of the analysis are in Table 1.

Table 1. Demography.

Demography	Mean	F	%
Sex			
Male		16	25
Female		48	75
Age	20.2		
18		3	4.7
19		13	20.3
20		26	40.6
21		18	28.1
22		2	3.1
23		1	1.6
30		1	1.6

Based on the descriptive analysis, the total sample is 64 people. Of 68 people, 16 of them, or approximately 25 percent, are males. Meanwhile, the rest 48 people, or approximately 75% are females. The average age of the sample in this research is 20.2 years, with 18 years as the youngest age that consists of 3 people. The oldest age is 30 years which consists of only 1 person. The largest sample with a total of 26 people or approximately 40.6% is 20 years old.

3.2 *Reliability and discriminatory power analysis*

In order to determine the consistency of this academic dishonesty scale, the researcher conducted a reliability test. The result is given in Table 2.

Table 2.

	Cronbach's α Efficiency
Academic dishonesty	0.891

Based on the reliability test, the coefficient of Cronbach's α is 0.891. In other words, this academic dishonesty scale for the online learning field has high reliability.

To examine discriminatory power, the researcher conducted a discriminatory power assessment by looking at the item-rest correlation. And the result of the discriminatory power assessment given in the following table.

Table 3.

Item	Item-rest correlation
C1	0.749
C2	0.660
C3	0.620
C4	0.573
C5	0.567
C6	0.470
P1	0.500
P2	0.554
P3	0.454
OH1	0.178
OH2	0.536
OH3	0.708
OH4	0.675
FA1	0.225
FA2	0.440
FA3	0.547
FR1	0.481
FR2	0.014
FR3	0.430
FR4	0.476
FR5	0.451
FR6	0.219

Based on the discriminatory assessment, the following results were found. The discriminatory assessment was used to examine the discriminatory power of each item, whether they are good or bad. According to Azwar (2015), items that have coefficiency > 0.3 are the items that have good discriminatory power. Almost all items have a discriminatory power score of more than 0.3. Of the 22 items, 18 items have good discriminatory power, meanwhile, there are four items that have bad discriminatory power. Items that have bad discriminatory power are OH1, FA1, FR2, and FR6.

4 DISCUSSION

The academic dishonesty scale that exists since before the pandemic era is the one that specialized in offline education, it can be acknowledged by the manual cheating behavior that occurred in an offline environment (Iyer & Eastman 2008; Pavela 1978). The academic dishonesty scale that is specialized for offline education is considered not suitable for online classes or lectures. In this research, the researcher developed an academic dishonesty scale that specialized for online learning with bachelor students as the research object. The development was done by interviewing some students from various universities to identify dishonesty behaviors in online learning. Then, the researcher analyzed the data and started the development of an academic dishonesty scale for online learning. After that, the researcher conducted the reliability scale and discriminatory power analysis on every item in the academic dishonesty scale.

The online dishonesty scale consists of 22 items that are divided into five dimensions. Those five dimensions are cheating, plagiarism, outside help, falsification, and fraudulence. The cheating dimension consists of six items, the plagiarism dimension consists of three items, the outside help dimension consists of four items, the falsification dimension consists of three items and the fraudulence dimension consists of six items.

Based on the reliability test, the researcher acquired Cronbach's α of 0.891 or greater than 0.60. Therefore, it may be inferred that this academic dishonesty scale has a good reliability. Reliability means how far a scale can be consistent or trusted if being used again to test other populations outside the research (Hardani *et al.* 2020; Sugiyono 2017). Therefore, it can be said that this academic dishonesty scale for online learning is consistent and can be trusted.

From the discriminatory assessment involving 22 items, 18 items have good discriminatory power. On the other hand, four items have bad discriminatory power, which are OH1, FA1, FR2, and FR6. The discriminatory power shows the correlation between item function and scale function in revealing individual differences (Azwar 2015). In other words, all items, except OH1, FA1, FR2, and FR6, are the items that can show the correlation between the item function and scale function.

The following is the explanation of items that have low discriminatory power. Item OH1 is about how often students ask lecturers for personal advantages. Item FA1 is about how often they are faking citations. Item FR2 is about how often students use cheating-related services from other people. FR6 is about the use of cheating tricks when performing online presentations. Those four items have low discriminatory power which could be caused by the rare occurrence of those behaviors in Indonesia. This is also supported by research which shows that cultural differences can affect the behaviors of academic dishonesty (Hendy *et al.* 2021), and these items were arranged using literature from different cultures and from the results of interviews with several subjects. In other words, those four items may not suitable to be applied in Indonesia to the behavior of online academic dishonesty.

5 CONCLUSION

In this research, the researcher developed an academic dishonesty scale that is specialized for online learning. This was based on the fact that there was yet no academic dishonesty scale that specialized for online learning because the previous researches still use academic dishonesty scales that specialized for offline education. From the reliability analysis and the discriminatory power assessment, results that are reliable for the online academic dishonesty scale were acquired. Moreover, most of the items in the online academic dishonesty scale also have good discriminatory power. However, some items have low discriminatory power. The weakness of this study is that the validity of the items of the scale has not been tested yet, therefore the next research is suggested to do an in-depth validity test to this online academic dishonesty scale.

REFERENCES

Akbulut, Y., Şendağ, S., Birinci, G., Kılıçer, K., Şahin, M. C., & Odabaşı, H. F. (2008). Exploring the Types and Reasons of Internet-triggered Academic Dishonesty Among Turkish Undergraduate Students: Development of Internet-Triggered Academic Dishonesty Scale (ITADS). *Computers & Education*, *51*(1), 463–473. doi:10.1016/j.compedu.2007.06.003

Ameen, E. C., Guffey, D. M., & McMillan, J. J. (1996). Accounting Student's Perceptions of Questionable Academic Practices and Factors Affecting Their Propensity to Cheat. *Accounting Education*, *5*(3), 191–205. doi:10.1080/09639289600000020

Ashworth, P., Bannister, P., & Thorne, P. (1997). Guilty in Whose Eyes? University Students' Perceptions of Cheating and Plagiarism in Academic Work and Assessment. *Studies in Higher Education*, *22*(2), 187–203. doi:10.1080/03075079712331381034

Azwar, S. (2015). *Penyusunan Skala Psikologi* (Vol. 2). Yogyakarta: Pustaka Pelajar. doi:9799289084

Bashir, H., & Bala, R. (2018). Development and Validation of Academic Dishonesty Scale (ADS): Presenting a Multidimensional Scale. *International Journal of Instruction*, *11*(2), 57–74. doi:10.12973/iji.2018.1125a

Born, A. D. (2003). How to Reduce Plagiarism. *Journal of Information Systems Education*, *14*, 223.

Eastman, J. K. (2008). Academic Dishonesty: An Exploratory Study Examining Whether Insurance Students Are Different From Other College Students. *Risk Management and Insurance Review*, *11*(1), 209–226. doi:10.1111/j.1540-6296.2008.00138.x

Faradiena, F. (2019). Uji Validitas Alat Ukur Ketidakjujuran Akademik. *Jurnal Pengukuran Psikologi dan Pendidikan Indonesia*, *8*(2), 88–104. doi:DOI: http://dx.doi.org/10.15408/jp3i.v8i2.13316

Finn, K. V., & Frone, M. R. (2004). Academic Performance and Cheating: Moderating Role of School Identification and Self-Efficacy. *The Journal of Educational Research*, *97*(3), 115–121. doi:10.3200/JOER.97.3.115-121

Guo, X. (2011). Understanding Student Plagiarism: An Empirical Study in Accounting Education. *Accounting Education*, *20*(1), 17–37. doi:10.1080/09639284.2010.534577

Hardani, Auliya, N. H., Andriani, H., Fardani, R. A., Ustiawaty, J., Utami, E. F., ... Istiqomah, R. R. (2020). *Metode Penelitian Kualitatif & Kuantitatif* (1 ed.). Yogyakarta: Pustaka Ilmu. doi:978-623-7066-33-0

Hendy, N. T., Montargot, N., & Papadimitriou, A. (2021). Cultural Differences in Academic Dishonesty: A Social Learning Perspective. *Journal of Academic Ethics*, *19*, 49–70. doi:https://doi.org/10.1007/s10805-021-09391-8

Heriyati, D., & Ekasari, W. F. (2020). A Study on Academic Dishonesty and Moral Reasoning. *International Journal of Education*, *12*(2), 56–62. doi:10.17509/ije.v12i2.18653

Iyer, R., & Eastman, J. K. (2008). The Impact of Unethical Reasoning on Academic Dishonesty: Exploring the Moderating Effect of Social Desirability. (2, Ed.) *Marketing Education Review*, *18*. doi:DOI: 10.1080/10528008.2008.11489034

Jankea, S., Rudertb, S. C., Petersena, Ä., Fritzc, T. M., & Daumillerc, M. (2021). Cheating in the Wake of COVID-19 How Dangerous is Ad-hoc Online Testing for Academic Integrity? doi:10.31234/osf.io/6xmzh

Jurdi, R., Hage, S. H., & Chow, H. P. (2011). Academic Dishonesty in the Canadian Classroom: Behaviours of a Sample of University Students. *Canadian Journal of Higher Education*, *41*(3). doi:DOI: https://doi.org/10.47678/cjhe.v41i3.2488

Lambert, K. D., Ellen, N., & Taylor, L. (2003). Cheating—what is it and why do it: A study in New Zealand Tertiary Institutions of the Perceptions and Justifications for Academic Dishonesty. *Journal of American Academy of Business*, *3*.

Lewellyn, P. G., & Rodriguez, L. (2015). Does Academic Dishonesty Relate to Fraud Theory? A Comparative Analysis. *American International Journal of Contemporary Research*, *5*(3), 1–6.

Murdock, T. B., Hale, N. M., & Weber, M. J. (2001). Predictors of Cheating among Early Adolescents: Academic and Social Motivations. *Contemp Educ Psychol*, *26*(1), 96–115. doi:10.1006/ceps.2000.1046

Park, C. (2010). In Other (People's) Words: Plagiarism by University Students-literature and Lessons. *Assessment & Evaluation in Higher Education*, *28*(5), 471–488. doi:10.1080/02602930301677

Pavela, G. (1978). Judicial Review of Academic Decision-making After Horowitz. *School Law Journal*, *55*(8), 55–75.

Pavela, G. (1978). Judicial Review of Academic Decisionmaking After Horowitz. *NOLPE School Law Journal*, *1*, 55–75.

Rujoiu, O., & Rujoiu, V. (2014). Academic Dishonesty and Workplace Dishonesty. An overview. *Management Challenges for Sustainable Development*, 928–938.

Scanlon, P. M. (2004). Student Online Plagiarism: How Do We Respond? *College Teaching*, *51*(4), 161–165.

Smith, K. J., Davy, J. A., Rosenberg, D. L., & Haight, G. (2002). A Structural Modeling Investigation of the Influence of Demographic and Attitudinal Factors and In-class Deterrents on Cheating Behavior Among Accounting Majors. *Journal of Accounting Education*, *20*(1), 45–65. doi:10.1016/S0748-5751(01)00026-4

Sugiyono. (2017). *Metode Penelitian Kuantitatif, Kualitatif dan R&D* (1 ed.). Alfabeta. doi:9798433645

Whitley, B. E., & Keith-Spiegel, P. (2002). *Academic Dishonesty An Educator's Guide*. London: Lawrence Erlbaum Associate.

Families Mental Health and Challenges in the 21st Century – Arifin et al. (eds)
© 2024 The Author(s), ISBN 978-1-032-51442-0

The relationship between mental health status and quality of life among COVID-19 survivors in Malaysia

Siti Nursyahirah Md Yusof & Nor Firdous Mohamed*
Department of Psychology and Counselling, Faculty of Human Development, Sultan Idris Education University, Malaysia

ABSTRACT: The purpose of this research is to examine the correlation between mental health status and quality of life, as well as the gender differences in mental health status and quality of life among COVID-19 survivors in Malaysia. The design of this research is a quantitative cross-sectional survey design using an online questionnaire. The questionnaire includes demographic information, the Patient Health Questionnaire (PHQ-9), and the World Health Organization Quality of Life—Brief Version (WHOQOL-BREF) questionnaire. This research recruited 243 COVID-19 survivors (n = 67 men and n = 176 women) by using the purposive sampling technique. The findings indicated that there is a significant correlation between the PHQ-9 score (mental health status) and the rate of quality of life ($r(175) = -368$, $p< .001$), physical health ($r(175) = -.372$, $p< .001$), psychological ($r(175) = -.597$, $p< .001$), social relationship ($r(175) = -.291$, $p< .001$), and environment ($r(175) = -.303$, $p< .001$). In conclusion, mental health status may affect one's quality of life, while there are no gender differences in mental health status and quality of life.

Keywords: mental health status, quality of life, COVID-19 survivors

1 INTRODUCTION

The COVID-19 pandemic has impacted and led to significant changes in various aspects of people's life. People who are infected with the coronavirus disease will not only experience a decline in their physical health but will also have an impact on their mental health. Depression, anxiety, post-traumatic stress disorder, bipolar disorder, or schizophrenia are predicted to affect one out of every five people (22%) who live in a conflict-affected area as almost everyone who is affected by a disaster will endure psychological distress (WHO 2018). The pandemic caused by COVID-19 has had a significant influence on people's quality of life as well as their physical and mental health (Algahtani *et al.* 2021).

Changes in a person's mental health status may have a significant association with alterations in that person's quality of life. This may be shown by the study that was carried out by Suryavanshi *et al.* (2020) as during the COVID-19 pandemic, the researchers identified a significant prevalence of symptoms of depression and anxiety, as well as low quality of life among Indian healthcare professionals. In addition, the study that was carried out by Pieh *et al.* (2020) found that women have a lower mental quality of life and lower overall mental health than males do. Besides, much research has been conducted on both mental health and quality of life; however, there have been relatively few studies conducted on the quality of life in COVID-19 (Méndez *et al.* 2021). As a result, this study is being carried out

*Corresponding Author: norfirdous@fpm.upsi.edu.my

DOI: 10.1201/9781003402381-23

to investigate the following issue: The relationship between mental health status and quality of life among COVID-19 survivors in Malaysia.

2 METHODS

A total of 243 respondents were recruited among COVID-19 survivors in Malaysia through the purposive sampling technique. The research is a quantitative cross-sectional survey design. A structured survey questionnaire was given to the respondents to collect the data. The instruments being used in this research were the Patient Health Questionnaire-9 (PHQ-9) and the World Health Organization Quality of Life-Brief Version (WHOQOL-BREF).

2.1 *Procedure*

The survey consisting of the informed consent form, demographic information, 9-item Patient Health Questionnaire, and 26-item WHOQOL-BREF assessment were prepared using Google Forms. The respondents self-administered the questionnaire in the Google Form. The respondents answered the questionnaire after being granted informed consent. The process of respondents' recruitment was repeated until the required sample of respondents had been achieved. After the required sample data was fulfilled, the researcher eliminated the response from the respondent who had not fulfilled the requirement of the sample. Then, the collected data underwent a series of statistical analyses and was further analyzed. The respondents were recruited by using a purposive sampling technique.

2.2 *Research ethics approval*

This research was granted research ethics from Human Research Ethics Committee at Sultan Idris Education University. The reference number for the research ethics approval is 2021-0414-01.

3 RESULTS

3.1 *Descriptive analysis*

3.1.1 *Demographic information*

Table 1 presents the number of samples (n) and percentages (%) on age, gender, ethnicity, marital status, and severity of illness of COVID-19.

Table 1. Participants' demographic information (n = 243).

		N	Percentages (%)
Age			Mean = 27.62
Gender	Male	67	27.6
	Female	176	72.4
Race	Malay	223	91.8
	Chinese	7	2.9
	Indian	2	.8
	Bumiputera Sabah	8	3.3
	Bumiputera Sarawak	2	.8
	Others	1	.4
Marital Status	Single	172	70.8

(*continued*)

Table 1. Continued

		N	Percentages (%)
	Married	68	28.0
	Widowed	1	.4
	Divorced	2	.8
The severity of Illness of COVID-19	Category 1 (No symptoms)	74	30.5
	Category 2 (Mild symptoms, no pneumonia)	136	56.0
	Category 3 (Symptomatic, with pneumonia)	26	10.7
	Category 4 (Symptomatic, Pneumonia, Requiring supplemental oxygen)	6	2.5
	Category 5 (Critically ill with multi-organ involvement)	1	.4

Notes: N = 243

Most of the samples recruited in this study were females with 72.4% (n = 176) and only 27.6% of the samples were males (n = 67). The age of the participants' ranges from 18 to 63 years old, and the mean age of the participants was M = 27.62 years old. Considering ethnicity, most of the participants were dominated by Malay at 91.8% (n = 223) and the majority of the samples are still single at 70.8% (n = 172). Among the severe cases of COVID-19 infections, almost half of the participants were infected in Category 2 (with mild symptoms and no pneumonia) with 56.0% (n = 136).

3.2 *Mental health status*

Figure 1 illustrates the number of samples (n) and percentages (%) for each level of severity of depression by using PHQ-9. The participants with none to minimal depression were placed under no depression, whereas participants with mild, moderate, moderately severe, and severe depression were placed under the category of has depression. Based on Figure 1, most of the participants are regarded as having depression and only 27.2% (n = 66) were regarded as having no depression. There was a total of 83 (34.2%) of the COVID-19 survivors who had mild depression, followed by 66 (27.2%) who have none to minimal depression. About 60 participants (24.7%) of the COVID-19 survivors have moderate depression, whereas 23 of the participants (9.5%) have moderately severe depression and only 11 (4.5%) have severe depression. However, in terms of mental health status, the lower the score of PHQ-9 represents the higher the mental health status.

3.3 *Quality of life*

Based on Figure 2, 46.1% (n = 112) of the total sample of COVID-19 survivors rated that they are having a good quality of life and 35.8% (n = 87) of them were having a neutral quality of life as they rated their quality of life as "neither poor nor good". Finally, about 13.6% (n = 33) of the respondents were having a very good quality of life and only 4.5% (n = 11) reported having poor quality of life. There were no participants who rated their quality of life as poor.

On the other hand, Table 2 shows the mean, standard deviation, and minimum and maximum values for the domains of WHOQOL-BREF. It is reported that the highest mean score for the WHOQOL-BREF domain is the environment domain (M = 67.31, SD =

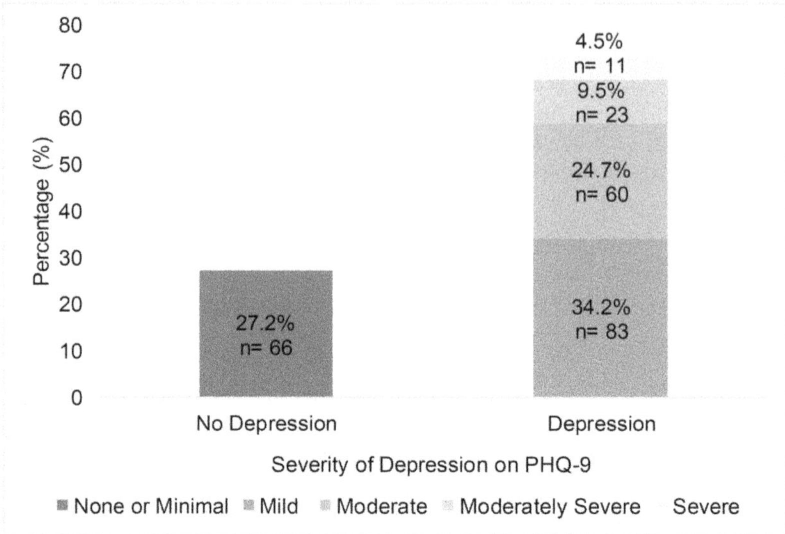

Figure 1. PHQ-9 Scores of Participants (n = 243).

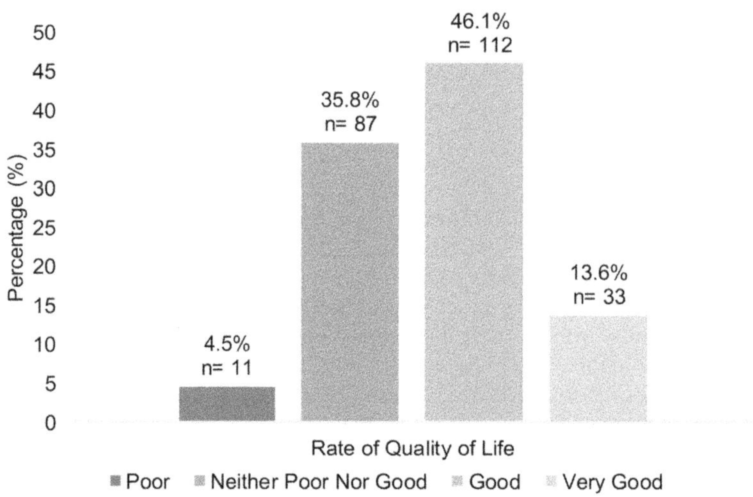

Figure 2. Rate of quality of life in percentage, % (n = 243).

Table 2. Domains of World Health Organization quality of life—Brief version (n = 243).

Quality of Life	Mean	SD
Physical health	65.99	15.438
Psychological	62.81	18.374
Social relationships	63.55	20.298
Environment	67.31	16.432

16.432). The highest mean indicates that the COVID-19 survivors have the highest environmental quality of life compared to other domains. It is followed by the physical health domain (M = 65.99, SD = 15.438). There is a slight difference in the mean between social relationships and psychological domains. The mean for the social relationships domain is 63.55 (M = 63.55, SD = 20.298) while for the psychological domain is 62.81 (M = 62.81, SD = 18.374).

3.4 *Inferential analysis*

Table 3 illustrates the correlation relationship between mental health status by using PHQ-9, rate of quality of life, and WHOQOL-BREF domains among COVID-19 survivors with depression. There were 177 COVID-19 survivors who reported having depression. Based on Table 3, the results show that the PHQ-9 score is significantly related to the rate of quality of life and all dimensions of WHOQOL-BREF ($p < .001$). The PHQ-9 score reported having a negatively moderate correlation with psychological ($r(175) = -.597$, $p<.001$), and a negatively weak correlation with the rate of quality of life ($r(175) = -368$, $p<.001$), physical health ($r(175) = -.372$, $p<.001$), and environment ($r(175) = -.303$, $p<.001$). However, the PHQ-9 score has a negatively weak correlation with the social relationship domain ($r(175) = -.291$, $p<.001$). Therefore, the null hypothesis is rejected. There is a significant negative correlation between the PHQ-9 score and quality of life among COVID-19 survivors in Malaysia.

Table 3. The relationship between mental health status by using PHQ-9, rate of quality of life, and WHOQOL-BREF domains among COVID-19 survivors with depression (N = 177).

	PHQ-9	1	2	3	4	p-value
1. Rate of quality of life	$-.368^{**}$	–	–	–	–	$.000^{***}$
2. Physical health	$-.372^{**}$	$.406^{**}$	–	–	–	$.000^{***}$
3. Psychological	$-.597^{**}$	$.640^{**}$	$.615^{**}$	–	–	$.000^{***}$
4. Social relationships	$-.291^{**}$	$.502^{**}$	$.439^{**}$	$.592^{**}$	–	$.000^{***}$
5. Environment	$-.303^{**}$	$.453^{**}$	$.598^{**}$	$.599^{**}$	$.582^{**}$	$.000^{***}$

**Correlation is significant at the 0.01 level (2-tailed)
***p<.001

3.5 *Comparison of rate of quality of life and WHOQOL-BREF domains between COVID-19 survivors with no depression and has depression*

Table 4 illustrates the comparison of the rate of quality of life and WHOQOL-BREF domains between COVID-19 survivors with no depression and has depression.

Based on Table 4, it is reported that there was a significant difference in the rate of quality of life ($t(241) = 6.010$, $p<.001$), physical health domain ($t(241) = 6.769$, $p<.001$), psychological domain ($t(241) = 6.080$, $p<.001$), social relationships ($t(241) = 5.540$, $p<.001$), and environment domain ($t(241) = 5.176$, $p<.001$) between COVID-19 survivors who have depression and no depression. There is a significant difference in all domains of quality of life between the COVID-19 survivors who have depression and no depression. The COVID-19 survivors who did not suffer from depression reported rating their quality of life higher than those who suffered from depression. Those who have no depression are also reported to have a higher quality of life in all four domains of WHOQOL-BREF compared to those who have depression.

Table 4. Comparison of quality of life between COVID-19 survivors with no depression and has depression (n = 243).

	No depression N = 66		Has depression N = 177				
	M	SD	*M*	SD	df	t	Sig.
Rate of Quality of Life	4.14	.762	3.52	.692	241	6.010	.000***
Physical health	76.08	14.572	62.23	14.033	241	6.769	.000***
Psychological	73.76	17.105	58.72	17.158	241	6.080	.000***
Social relationships	74.70	20.558	59.39	18.612	241	5.540	.000***
Environment	75.80	16.777	64.14	15.171	241	5.176	.000***

***p<.001

The psychological domain and social relationships domains had a large mean difference. Table 4 illustrates that the COVID-19 survivors who have no depression (M = 73.76, SD = 17.105) have a higher quality of life in the psychological domain compared to those who have depression (M = 58.72, SD = 17.158). For the social relationship domain, COVID-19 survivors who suffered from depression (M = 59.39, SD = 18.612) have a lower quality of life compared to those who do not suffer from depression (M = 74.70, SD = 20.558).

4 DISCUSSION

4.1 *The relationship between mental health status and quality of life among COVID-19 survivors in Malaysia*

The null hypothesis (H_{01}) is there is no significant relationship between the mental health status and quality of life among COVID-19 survivors in Malaysia. From this study, the findings suggest that the PHQ-9 score has a negatively weak correlation with the rate of quality of life (r(175) = –.368, p<.001). The negative relationship denotes that a decrease in PHQ-9 score is associated with an increase in the rate of quality of life and vice-versa. In terms of mental health status, the person with a low PHQ-9 score is having a high mental health status, and thus, they will rate high in their quality of life. Therefore, COVID-19 survivors who have low PHQ-score (high mental health status) have a higher rate of quality of life.

The PHQ-9 score has a negatively moderate correlation with psychological (r(175) = –.597, p<.001) and negatively weak correlation with the physical health (r(175) = –.372, p<.001), and environment (r(175) = –.303, p<.001). However, the PHQ-9 score has a negative very weak correlation with social relationships (r(175) = –.291, p<.001). The negative correlation denotes that a decrease in the PHQ-9 score is associated with an increase in the quality of life for each domain and vice-versa. In terms of mental health status, the person with a low PHQ-9 score will have a high mental health status and thus, they will have a high quality of life in the domain of physical health, psychological, social relationship, and environment. Therefore, the findings rejected the null hypothesis and accepted the alternate hypothesis. This finding confirms the alternate hypothesis that there is a significant relationship between mental health status and quality of life among COVID-19 survivors.

The result of this study is in accordance with research by Ma *et al.* (2020), as the research found depression was related to a decrease in quality of life and the COVID-19 patients who had depression reported a poorer quality of life compared to those who did not have depression. Furthermore, the research by Raman *et al.* (2021) also discovered that the patient who had a greater burden of self-reported symptoms of depression also suffered a substantial decrease in all dimensions of quality of life in comparison to the controls. Consistent with the current study, the person with depression (low mental health status) reported to have a lower rate in their quality of life and also have a lower quality of life in all four domains of WHOQOL-BREF.

5 CONCLUSION AND IMPLICATION

In conclusion, the present study was able to determine the relationship between mental health status and quality of life among COVID-19 survivors in Malaysia. This finding is in line with the past research that found that mental health status was related to quality of life. Several limitations in this study needed to be addressed. The present research also relied on the respondent's self-reported responses in the form of an online survey; therefore, there would be a chance for the individual to recollect bias in their responses.

REFERENCES

Algahtani, F. D., Hassan, S. U. N., Alsaif, B., & Zrieq, R. (2021). Assessment of the Quality of Life during COVID-19 Pandemic: A Cross-Sectional Survey from the Kingdom of Saudi Arabia. *International Journal of Environmental Research and Public Health, 18*(3), 847. https://doi.org/10.3390/ijerph18030847

Ma, Y. F., Li, W., Deng, H. B., Wang, L., Wang, Y., Wang, P. H., Bo, H. X., Cao, J., Wang, Y., Zhu, L. Y., Yang, Y., Cheung, T., Ng, C. H., Wu, X., & Xiang, Y. T. (2020). Prevalence of Depression and its Association with Quality of Life in Clinically Stable Patients with COVID-19. *Journal of Affective Disorders, 275*, 145–148. https://doi.org/10.1016/j.jad.2020.06.033

Méndez, R., Balanzá-Martínez, V., Luperdi, S., Estrada, I., Latorre, A., González-Jiménez, P., Feced, L., Bouzas, L., Yépez, K., Ferrando, A., Hervás, D., Zaldívar, E., Reyes, S., Berk, M., & Menéndez, R. (2021). Short-term Neuropsychiatric Outcomes and Quality of Life in COVID-19 Survivors. *Journal of Internal Medicine, 290*(3), 621–631. https://doi.org/10.1111/joim.13262

Pieh, C., Budimir, S., & Probst, T. (2020). The Effect of Age, Gender, Income, Work, and Physical Activity on Mental Health During Coronavirus Disease (COVID-19) lockdown in Austria. *Journal of Psychosomatic Research, 136*, 110186. https://doi.org/10.1016/j.jpsychores.2020.110186

Raman, B., Cassar, M. P., Tunnicliffe, E. M., Filippini, N., Griffanti, L., Alfaro-Almagro, F., Okell, T., Sheerin, F., Xie, C., Mahmod, M., Mózes, F. E., Lewandowski, A. J., Ohuma, E. O., Holdsworth, D., Lamlum, H., Woodman, M. J., Krasopoulos, C., Mills, R., McConnell, F. A. K., ... Neubauer, S. (2021). Medium-term Effects of SARS-CoV-2 Infection on Multiple Vital Organs, Exercise Capacity, Cognition, Quality of Life and Mental Health, Post-hospital Discharge. *EClinicalMedicine, 31*, 100683. https://doi.org/10.1016/j.eclinm.2020.100683

Suryavanshi, N., Kadam, A., Dhumal, G., Nimkar, S., Mave, V., Gupta, A., Cox, S. R., & Gupte, N. (2020). Mental Health and Quality of Life Among Healthcare Professionals During the COVID-19 Pandemic in India. *Brain and Behavior, 10*(11). https://doi.org/10.1002/brb3.1837

World Health Organization. (2018, March 30). *Mental Health: Strengthening Our Response*. https://www.who.int/news-room/fact-sheets/detail/mental-health-strengthening-our-response

The relationship between extraversion and loneliness among upsi undergraduate students during COVID-19 pandemic

Syafiqah Najwa Binti Ahmad Zabidi*
Department of Psychology and Counseling, Faculty of Human Development, Sultan Idris Education University, Tanjung Malim, Malaysia

ABSTRACT: The current study aimed to investigate the relationship between extraversion and loneliness among undergraduate students during the COVID-19 pandemic and also to investigate the difference between gender in level of extraversion and level of loneliness among undergraduate students. The quantitative study was conducted through online questionnaires in Google Form and was participated by 152 students (20 males and 132 females) of Bachelor of Psychology at Sultan Idris Education University (UPSI) by using a convenience sampling technique. The instruments used were Introversion Scale and UCLA Loneliness Scale Version 3. The result revealed that there was a weak significant positive relationship between extraversion and loneliness among undergraduate students. Apart from that, there was a significant difference between gender in level of extraversion but no significant difference between gender in level of loneliness among undergraduate students.

Keywords: Extraversion, Loneliness, COVID-19 Pandemic

1 INTRODUCTION

The COVID-19 pandemic has changed the way people live around the world. Public life restrictions measures were executed globally including in Malaysia in order to stop the spread of the deadly virus. A new set of norms replaced the old ones including sanitizing hands, wearing masks as well as physical distancing. This also includes measures that aimed to reduce physical contact such as reducing family and friends' visitation, maintaining physical distance, and restriction on the social gathering. While adapting to the new norms, the COVID-19 pandemic has caused various psychological impacts on general population (Xiong *et al.* 2020). University students are also one of the groups that are affected (Browning *et al.* 2021). Loneliness is one of the increasing mental health concerns since the outbreak of the virus (Killgore *et al.* 2020). In the United States of America (USA), 61% of young people aged 18 to 25 years old reported experiencing loneliness during the COVID-19 pandemic (Weissbourd *et al.* 2020). Meanwhile, in Malaysia, it is reported that 31% of Malaysians aged 16 to 64 years old feel lonely since the COVID-19 pandemic started which is 10% higher than before the pandemic started (Snapchat's Friendship Time Capsule 2020). In Malaysia, some universities have allowed students to return to the campus where students and whose programs need practical approaches are prioritized. The remaining students are continuing their classes online. Both of these situations have altered students' social lifestyle that they practiced before the COVID-19 pandemic such as social gatherings and joining physical university activities. All the students are obligated to follow Standard Operational Procedure (SOP) at campus that was implemented by the government including wearing

*Corresponding Author: syafiqahazabidi@gmail.com

DOI: 10.1201/9781003402381-24

mask, physical distancing, and avoiding crowded places. As for students who undergo online learning, they are not able to meet their friends to hang out, study together, and do recreational activities. To conclude, university students' life is not the same as before the COVID-19 pandemic. University students which are mostly young adults are the most vulnerable group to experience mental health deterioration including loneliness during the COVID-19 pandemic (Browning *et al.* 2021; Bratz & Olsson 2021; Groarke *et al.* 2020; Weissourd *et al.* 2020). Loneliness is a serious mental health concern as it can lead to depression and suicidal ideation (Killgore *et al.* 2020; Lee *et al.* 2020). According to Holt-Lunstad *et al.* (2015), it is reported that loneliness serves a 26% chance of mortality (Sîrbu & Dumbravă 2019). Personality specifically extraversion can predict loneliness (Sîrbu & Dumbravă 2019). A meta-analysis between loneliness and extraversion showed that extraversion was negatively correlated to loneliness (Buecker *et al.* 2020). Extraversion is synonymous with engaging in and enjoying social interaction as well as participating in social activities (as cited in Gubler *et al.* 2020; John *et al.* 2008; Lee *et al.* 2008; Srivastava *et al.* 2008). Hence, extraversion will be a unique variable to investigate its association with loneliness in the context of the COVID-19 pandemic. The majority of studies that exploring the association between extraversion and loneliness in the context of the COVID-19 pandemic were done in Europe and more to the general population as samples (Bratz & Olsson 2021; Gubler *et al.* 2020; Landmann & Rohmann 2021; Wei 2020). Therefore, the current study used undergraduate university students as sample to fulfill the research gaps in the relationship between extraversion and loneliness.

The outcome of the study will add to the existing body of literature in the field of research regarding personality and loneliness during the COVID-19 pandemic. This knowledge can be referenced in the future. It can also give an insight to fellow researches on studies regarding personality and loneliness in Southeast Asia specifically in Malaysia as the existing studies are focused on European countries (Bratz & Olsson 2021; Gubler *et al.* 2020; Landmann & Rohmann 2021; Wei 2020). Apart from that, many researchers used the general population as samples instead of university students (Bratz & Olsson 2021; Gubler *et al.* 2020; Landmann & Rohmann 2021; Wei 2020). In addition, the finding will serve as information to universities' psychologists and counselors on the COVID-19 pandemic's impact on students. It will also give them an idea of what happened to the students during the COVID-19 pandemic as well as helping them to better understand the issue of loneliness among university students. Thus, personality-based intervention can be created to help undergraduates that have a higher risk of experiencing loneliness during the COVID-19 pandemic. This can overcome the worrying consequences of loneliness as well as reduce other mental health concerns that are rising from loneliness such as depression and suicidal ideation.

2 METHODS

A total of 152 students (20 males and 132 females) of Bachelor of Psychology at Sultan Idris Education University (UPSI) participated in the quantitative research by using a convenience sampling technique. The study also used a cross-sectional design. The data were collected by distributing online questionnaires to the participants. The instrument used to measure extraversion was Introversion Scale and to measure loneliness was the UCLA Loneliness Scale Version 3.

3 PROCEDURE

The informed consent and the questionnaires including demographic information, Introversion Scale and UCLA Loneliness Scale Version 3 were constructed in Google Form in four sections respectively. The Google Form link was spread through WhatsApp. Each batch leader from the UPSI Bachelor of Psychology with Honour program was contacted through WhatsApp private messaging to aid in distributing the link. They were added to a WhatsApp group created by the researcher. A text was sent in the WhatsApp group introducing the researcher's name, current semester, and program as well as explaining the objective of the study. Then, they were given the Google Form link and asked to forward it to their respective batch WhatsApp group. A reminder was sent in the same WhatsApp group every three days which remind each batch leader to forward the Google Form link to their respective batch WhatsApp group again. After the required sample of participants was achieved, the link was closed and the data were analyzed by using Statistical Package for the Social Sciences (SPSS) version 23.

4 RESULTS

4.1 *Demographic information*

Table 1. Summary of demographic information (N = 152).

Variables	Frequency (N)	Percent (%)
Gender		
Female	132	86.8
Male	20	13.2
Age group		
19–20	19	12.5
21–22	94	61.8
23–24	29	19.1
25–26	10	6.6
Semester		
1	13	8.6
2	9	5.9
3	27	17.8
4	2	1.3
5	34	22.4
6	6	3.9
7	55	36.2
8	6	3.9

Based on Table 1, there were two genders in the current study which are male and female. The result showed the participants consist of 20 male participants with 13.2% and 132 female participants with 86.8%. The majority of the participants age is 22 years old which consists of 69 participants with 45.4% and the least number of participants was 26 years old which consists of 3 participants with 2.0%. The most number of participants after 22 years old was 21 years old which consists of 25 participants with 16.4%. Participants who were 24 years old consist of 17 participants with 11.2% while 20 years old consist of 13 participants with 8.6%. Participants of age 23 years old consist of 12 participants with 7.9% and 25 years old consist of 7 participants with 4.6%. Lastly, participants of age 19 years old consist of 6 participants with 3.9%. The majority of the participants were in their seventh semester which consists of 55 participants with 36.2% and the least number of participants were in their fourth semester which consists of 2 participants with 1.3%. The most number of participants after semester seven was in their fifth semester which consists of 34 participants with 22.4% and semester 3 which consist of 27 with 17.8%. Participants who were in the first semester consisted of 13 participants with 8.6% and 9 participants were in their second semester with 5.9%. Lastly, participants who were in their sixth semester and eighth semester consisted of 6 participants with 3.9%.

4.2 *Prevalence*

4.2.1 *The level of extraversion among undergraduate students*

Table 2. Prevalence of level of extraversion among undergraduate students.

Level of Extraversion	Frequency (N)	Percent (%)
Low Extraversion	7	4.6
Moderate Extraversion	132	86.8
High Extraversion	13	8.6

4.2.2 *The level of loneliness among undergraduate students*

Table 3. Prevalence of level of loneliness among undergraduate students.

Level of Loneliness	Frequency (N)	Percent (%)
Low Loneliness	3	2.0
Moderate Loneliness	35	23.0
High Loneliness	114	75.0

4.3 *Independent T-test*

4.3.1 *The difference in level of extraversion between male and female undergraduate students*

Table 4. Group statistics of level of extraversion and gender.

	Gender	N	Mean	Std. Deviation	Std. Error Mean
Level of Extraversion	Male	20	37.55	9.501	2.124
	Female	132	33.53	8.108	.706

4.3.2 *The difference in level of loneliness between male and female undergraduate students*

Table 5. Group statistics of level of loneliness and gender.

	Gender	N	Mean	Std. Deviation	Std. Error Mean
Level of Loneliness	Male	20	49.40	12.800	2.862
	Female	132	51.08	11.523	1.003

Table 6. Independent samples test for level of extraversion and gender.

		Levene's Test for Equality of Variances		t-test for Equality of Means						
									95% Confidence Interval of the Difference	
		F	Sig.	T	df	Sig. (2-tailed)	Mean Difference	Std. Error Difference	Lower	Upper
Level of Extraversion	Equal variances assumed	.524	.470	2.019	150	.045	4.020	1.991	.086	7.954
	Equal variances not assumed			1.796	23.384	.085	4.020	2.239	-.607	8.646

Table 7. Independent samples test for level of loneliness and gender.

		Levene's Test for Equality of Variances		t-test for Equality of Means						
									95% Confidence Interval of the Difference	
		F	Sig.	T	df	Sig. (2-tailed)	Mean Difference	Std. Error Difference	Lower	Upper
Level of Loneliness	Equal variances assumed	.108	.742	-.597	150	.551	−1.676	2.806	−7.219	3.868
	Equal variances not assumed			-.553	23.900	.586	−1.676	3.033	−7.936	4.585

Based on Table 6, an independent sample t-test was conducted to compare level of extraversion between male and female undergraduates. There was a significant difference in level of extraversion between male (M = 37.55, SD = 9.501) and female (M = 33.53, SD = .8.108) conditions; t (150) = 2.019, p = .045. The null hypothesis was rejected and the alternate hypothesis was accepted.

Based on Table 7, an independent sample t-test was computed to compare the difference in the level of loneliness between male and female undergraduates. There was no significant difference in the level of loneliness between male (M = 49.4, SD = 12.8) and female (M = 51.08, SD = 11.52) conditions; t (150) = .597, p = .551. The null hypothesis was accepted and the alternate hypothesis was rejected.

4.4 *Correlation*

Based on Table 8, the mean for the introversion score was M = 34.06 and for the loneliness score was M = 50.86. The standard deviation for the introversion score was SD = 8.382 and for loneliness score was SD = 11.667. Both introversion score and loneliness score had an equal sample size, n = 152. A Pearson Correlation was computed to assess the relationship between extraversion and loneliness among undergraduate students. Based on Table 9, there was a weak significant positive correlation between extraversion and loneliness, $r = .323$, $n = 152$, $p = .000$.

Table 8. Descriptive statistics for correlation between extraversion and loneliness.

	Mean	Std. Deviation	N
Introversion Score	34.06	8.382	152
Loneliness Score	50.86	11.667	152

Table 9. Correlation between extraversion and loneliness.

		Introversion Score	Loneliness Score
Introversion Score	Pearson Correlation	1	.323**
	Sig. (2-tailed)		.000
	N	152	152
Loneliness Score	Pearson Correlation	.323**	1
	Sig. (2-tailed)	.000	
	N	152	152

**. Correlation is significant at the 0.01 level (2-tailed).

5 DISCUSSION

5.1 *Level of extraversion among undergraduate students*

Based on the findings, 86.8% of undergraduate students were moderate extraversion. 8.6% were high extraversion while 4.6% were low extraversion. Thus, majority of the undergraduate students in UPSI are moderate extraversion. This finding was parallel with previous study by Nayak (2016) where 72.8% of medical students in India were moderate extraversion. However, this finding contradicted previous studies done by Simonenko (2017). The majority of medical students in Ukraine Simonenko (2017) were high extraversion which was at 39.6%. Contradiction with previous studies' findings may occur due to COVID-19 life restriction as students. Most undergraduate students at UPSI were still undergoing online classes. Thus, their social life as students was not similar before the pandemic when they were not allowed to attend physical classes and curricular activities. Krautter (2021) explained that changes to low extraversion during the COVID-19 pandemic among students were because of restrictions on extroverted behavior during the lockdown. However, during the duration of the current study, lockdown has been lifted but citizens need to follow new set of norms aimed to reduce physical contact. For example, students may have a small meet up with neighborhood friends at the restaurant but they have to wear a mask, avoid shaking hands, and apply hand sanitizer adequately as well as practice social distancing. The leniency of the rules compared to lockdown may cause extraversion levels among undergraduate students to be moderate.

5.2 *Level of loneliness among undergraduate students*

Based on the findings, 75.0% of undergraduate students were high loneliness, 23.0% were moderate loneliness and 2.0% were low loneliness. Thus, majority of undergraduate students at UPSI were feeling very lonely. This finding was in line with previous findings that suggested young adults (mostly university students) were the most vulnerable group to mental health deterioration including loneliness during the COVID-19 pandemic (Browning *et al.* 2021; Bratz & Olsson 2021; Groarke *et al.* 2020; Weissbourd *et al.* 2020). However, this finding contradicted previous studies by Labrague *et al.* (2020) and Hussin *et al.* (2021). Both Labrague *et al.* (2020) and Hussin *et al.* (2021) found that the majority of college students in Central Philippines and public university's students in Malaysia were moderately lonely respectively. The contradiction with both previous studies may be because of the time of the study conducted. Labrague *et al.* (2020) collected the data during the second month of the mandatory lockdown in the Philippines whereas Hussin *et al.* (2021) collected the data during the first month of the Movement Control Order (MCO) in Malaysia. This showed that both previous studies were done during early couple of months of the COVID-19 pandemic. Gubler *et al.* (2020) suggested that loneliness did not decrease for over six weeks of public life restrictions during the COVID-19 pandemic. Weber *et al.* (2022) did a longitudinal study and found that loneliness among university students were significantly increased from July 2020 to November 2020. High loneliness among undergraduate students may be because of the factor of the COVID-19 pandemic duration. According to Shankar *et al.* (2011, as cited in Pietrabissa & Simpson 2020), prolonged isolation can lead to loneliness and vice versa. The current study was done after almost two years after MCO started in Malaysia. Although strict lockdown (MCO) was over months ago, citizens may still recover from being isolated for a long time. However, some university students were still not able to practice their usual routine and lifestyle such as attending physical class, participating in university curricular activities as well as social activities with other students which may lead them to feel very lonely.

5.3 *Differences between male and female in level of extraversion among undergraduate students*

Based on the findings, there was a significant difference in level of extraversion between male and female undergraduate students. Thus, the null hypothesis was rejected and the alternate hypothesis was accepted. This finding was supported by previous studies done by Chauhan and Sharma (2015) and Olowookere *et al.* (2020) where there was a significant difference in extraversion between male and female student. However, it contradicted in terms of which gender had higher extraversion. Both previous studies found that male students had high extraversion compared to female students. This contradiction can be explained by the number of male and female students in the sample. Chauhan and Sharma's (2015) sample of both genders were equally balanced where both males and females were 50 while Olowookere *et al.* (2020) collected data from 98 male students and 103 female students. The current study's sample consisted of an imbalanced number of male and female students where male students were 20 and female students were 132.

5.4 *Differences between males and females in level of loneliness among undergraduate students*

Based on the findings, there was no significant difference in the level of loneliness between male and female undergraduate students. Thus, the null hypothesis was accepted and the alternate hypothesis was rejected. This finding was supported by a previous study conducted by Chauhan and Sharma (2015) where female students experienced higher loneliness than male students. However, it contradicted in terms of the significant difference. Chauhan and Sharma (2015) found that there was a significant difference between male and female students in loneliness. This contradiction can also be explained by the contradiction of the finding in 5.4 which was the difference of sample in number of male and female students.

5.5 *Discussion on the relationship between extraversion and loneliness among undergraduate students*

Based on the findings, there was a weak significant positive correlation between extraversion and loneliness among undergraduate students. Thus, the null hypothesis was rejected and the alternate hypothesis was accepted. The current finding was contradicted by previous studies before and during the COVID-19 pandemic where extraversion was found to be negatively correlated with extraversion (Bratz & Olsson 2021; Buecker *et al.* 2020; Chauhan & Sharma 2015; Dalal *et al.* 2018; Keldal & Atli 2016; Wei 2020). One of the possible explanations for the reason behind the contradiction was the context of the study. Buecker *et al.* (2020), Dalal *et al.* (2018), Chauhan and Sharma (2015), and Keldal and Atli (2016) conducted the previous studies before the COVID-19 pandemic. The current study was conducted during the COVID-19 pandemic in Malaysia. Another explanation for the contradiction with previous studies which was conducted during the COVID-19 pandemic is the time of the studies and also the sample population (Bratz & Olsson 2021; Wei 2020). Bratz and Olsson (2021) conducted the study in 2021, while in Sweden Wei (2020) conducted the study between late April to early May 2020 and the sample came from the United Kingdom, Canada, Australia, Germany and United States of America. Bratz and Olsson explained their study's limitation was less generalizable to other countries with different contextual conditions. The rules given by the government to handle the spread of COVID-19 were different at that time than in Malaysia. Furthermore, although a mandatory lockdown or MCO was lifted, there was still an SOP that aimed to restrict physical contact during the time of the current study conducted as discussed in 5.2. Based on Eyesenck's Three-Factor PEN Model, individuals with high extraversion prefer an external stimulating environment in order to achieve certain brain activation level. Thus, individuals with high extraversion enjoy seeking social interaction and are described as liking parties, energetic, adventurous, and enthusiastic (Eyesenck & Eyesenck 1964; John & Srivastava 1999). Due to public life restrictions in Malaysia, some universities had undergone online learning to avoid the spread of the virus. Students were unable to undergo usual physical activities with classmates and friends just like before the COVID-19 pandemic such as attending physical classes, participating in university curricular activities as well as social activities with other students. Students with high extraversion may struggle to seek physical social interaction and activities in a highly stimulating environment which then may lead to experiencing loneliness. Aschwanden *et al.* (2020) mentioned that people with high extraversion reported that disruption of daily activities due to the COVID-19 pandemic which can lead to loneliness was a big concern for them. However, the current finding was partially supported by previous study done by Gubler *et al.* (2021). Gubler *et al.* (2021) found a weak negative correlation between extraversion and loneliness but emphasized that extraversion may lose its protective value for loneliness in a limited engagement of social activities during the COVID-19 pandemic which served as a low stimulating environment to individuals with high extraversion. Gubler *et al.* (2021) explained that activities like meeting with friends and family as well as talking to colleagues were temporarily unavailable due to the COVID-19 pandemic which makes individuals with high extraversion feel burdened and lonely. Due to the public life restrictions in Malaysia, undergraduate students with high extraversion were unable to physically interact with each other and seek a physical company which may lead to feeling of loneliness. Apart from that, the finding was also partially supported by Landmann & Rohmann (2021). Previous studies found that extraversion was positively correlated to physical loneliness but negatively correlated with both emotional and social loneliness. Young individuals who had high extraversion missed the company of others which may be because of limited opportunities to maintain social contact during the COVID-19 pandemic (Landmann & Rohmann 2021). Individuals with high extraversion may have no problem in confiding to others through various types of online communication by having large social networks but were unable to engage in physical social interaction. However, the current study did not include types of loneliness. Thus, it was unable to define what kind of loneliness undergraduate students at UPSI experienced.

6 CONCLUSION

In conclusion, the study showed that the highest frequency of level of extraversion among undergraduate students was moderate. This showed a moderate level of extraversion may cause by the leniency of rules during the COVID-19 pandemic compared to the mandatory lockdown for Malaysian except for essential departments. The majority of undergraduate students showed a high level of loneliness caused by the COVID-19 pandemic public life restriction. The frequency of high level of loneliness among undergraduate students was at an alarming rate. Hence, universities psychologists, and counselors are encouraged to assess loneliness among students and formulate a program and intervention to help students who experience high loneliness in order to avoid detrimental physical and mental health effects. The significant difference found in the level of extraversion between gender added to the existing body of literature in the field of research regarding personality differences between gender among students in Malaysia. No significant difference between gender in loneliness shed light that loneliness can be experienced at the same level regardless of gender. A positive correlation between extraversion and loneliness disputed most previous findings which found extraversion to be negatively correlated with loneliness. It indicated that individuals with high extraversion are often described as social and optimistic experience loneliness in the context of COVID-19. This can shed light that under tough conditions, individuals especially undergraduate students with high extraversion can have a negative experience. Furthermore, it confirmed Eyesenck's Three-Factor PEN Model where individuals with high extraversion seek external stimulation to achieve certain brain-level activity. Individuals with high extraversion were unable to be in a high stimulating environment as frequently as they do before COVID-19 pandemic which can lead to feeling of loneliness. Several limitations in the study include the total sample of male and female students were not fairly balanced. Out of 152, 20 were males and 132 were females. This may affect the result of the independent t-test which compared gender differences in extraversion and loneliness levels. Second, the sample was homogenous. The participants were from a specific program in UPSI which was Bachelor of Psychology students. Hence, the result cannot be generalized to all UPSI undergraduate students. Future researchers can consider a heterogenous sample, not just undergraduate students in a specific program but also various programs in various universities in Malaysia, in order to ensure that the finding truly represent Malaysian undergraduate students. In terms of sample size, it is recommended for future research to collect a larger sample size which has a balanced number of male and female participants in order to have more accurate results in analyzing the differences between gender by utilizing an independent t-test. Apart from that, it is encouraged for future research to include dimension of loneliness including physical, social, and emotional loneliness in order to find out what kind of loneliness undergraduate students experience during the COVID-19 pandemic in Malaysia due to public life restrictions.

REFERENCES

Aschwanden, D., Strickhouser, J. E., Sesker, A. A., Lee, J. H., Luchetti, M., Stephan, Y., Sutin, A. R., & Terracciano, A. (2020). Psychological and Behavioural Responses to Coronavirus Disease 2019: The Role of Personality. *European Journal of Personality*, 10.1002/per.2281. Advance online publication. https://doi.org/10.1002/per.2281

Bratz, B., & Olsson, O. (2021). *Mental Health During the covid19 Pandemic in Sweden, as an Interaction Between Personality and Contextual Factors* [Bachelor's Thesis, Lund University]. *LUP Student Papers.* https://lup.lub.lu.se/student-papers/search/publication/9049444

Browning, M. H., Larson, L. R., Sharaievska, I., Rigolon, A., McAnirlin, O., Mullenbach, L., Cloutier, S., Vu, T. M., Thomsen, J., Reigner, N., Metcalf, E. C., D'Antonio, A., Helbich, M., Bratman, G. N., & Alvarez, H. O. (2021). Psychological Impacts From COVID-19 Among University Students: Risk Factors Across Seven States in the United States. *PLOS ONE, 16*(1), e0245327. https://doi.org/10.1371/journal.pone.0245327

Buecker, S., Maes, M., Denissen, J. J. A., & Luhmann, M. (2020). Loneliness and the Big Five Personality Traits: A Meta-Analysis. *European Journal of Personality*, *34*(1), 8–28. https://doi.org/10.1002/per.2229

Chauhan, B., & Sharma, S. (2015). To Explore the Loneliness as it related to Personality Traits among Emerging Adults Studying Medicine. *The International Journal of Indian Psychology*, *3*(1), 28–37. https://doi.org/10.25215/0301.149

Chen, X., Qiu, N., Chen, C., & Zhai, L. (2021). Personality Traits, Loneliness, and Affect Among Boxers. *Frontiers in Psychology*, *12*, 609153. https://doi.org/10.3389/fpsyg.2021.609153

Citra, M., & Zainil, Y. (2021). Language Learning Strategies in Speaking Classroom Activity: Extrovert and Introvert Learners. *Advances in Social Science, Education and Humanities Research*, *539*, 161–166. https://www.atlantis-press.com/article/125954682.pdf

Coghlan, D., & Brydon-Miller, M. (2014). *The SAGE Encyclopedia of Action Research* (Vols. *1–2*). SAGE Publications Ltd. 10.4135/9781446294406

Dalal, N., Kathad, A., Masurkar, S., Pavithran, S., & Soumyanarayan, T. (2018). Correlation Studies Between Big Five Personality Traits, Age and Loneliness. *Indian Journal of Mental Health*, *5*(1), 32–37. http://dx.doi.org/10.30877/IJMH.5.1.2018.32-37

Ergüneş, H. (2018). The Review of Five Emperical Studies: To What Extent Contemporary Findings Provide Biological Evidence for Eysenck's PEN Model? *KOÇ University Undergraduate Psychology Journal*, (4), 18–24. https://kuupj.ku.edu.tr/downloads/KUUPJ_4.pdf

Eysenck, H. J. (1983). Psychophysiology and Personality: Extraversion, Neuroticism and Psychoticism. *Individual Differences and Psychopathology* (pp. 13–30). https://hanseysenck.com/wp-content/uploads/2019/12/1983_eysenck_-_psychophysiology_and_personality_extraversion_neuroticism_and_psychoticism_physiological.pdf

Eysenck, H. J., & Eysenck, S. B. G. (1964). *Manual of the Eysenck Personality Inventory*. Hodder and Stoughton. https://genepi.qimr.edu.au/contents/p/staff/1964_Eysenck_Manual.pdf

George, J. L. (1984). A Comparison of Chronic and Transient Loneliness on the Variables of Anxiety, *Depression and Self-Esteem. LSU Historical Dissertation and Theses. 4015*. https://digitalcommons.lsu.edu/gradschool_disstheses/4015/

Gosztyla, T., & Prokopiak, A. (2019). Mediating Role of Lack of Support for the Relationship Between Extraversion and Sense of Loneliness in Parents of Children with Autism Spectrum Disorder and Parents of Children with Intellectual Disability. *Italian Journal of Psychopathology*, *25*(2), 51–57. https://www.researchgate.net/publication/334163469_Mediating_role_of_lack_of_support_for_the_relationship_between_extraversion_and_sense_of_loneliness_in_parents_of_children_with_autism_spectrum_disorder_and_parents_of_children_with_intellectual_disab

Groarke, J. M., Berry, E., Graham-Wisener, L., McKenna-Plumley, P. E., McGlinchey, E., & Armour, C. (2020). Loneliness in the UK During the COVID-19 Pandemic: Cross-sectional Results from the COVID-19 Psychological Wellbeing Study. *PloS one*, *15*(9), e0239698. https://doi.org/10.1371/journal.pone.0239698

Gubler, D. A., Makowski, L. M., Troche, S. J., & Schlegel, K. (2020). Loneliness and Well-Being During the Covid-19 Pandemic: Associations with Personality and Emotion Regulation. *Journal of Happiness Studies*, 1–20. https://dx.doi.org/10.1007%2Fs10902-020-00326-5

Harlendea, C. Z., & Kartasasmita, S. (2021). The Relationship between Loneliness and Problematic Internet Use Among Young Adults who are Social Media Users. *Advances in Social Science, Eduucation and Humanities Research*, *570*, 2352–5398. https://dx.doi.org/10.2991/assehr.k.210805.059

Heinrich Heine Universität Düsseldorf. (2017). *G*Power 3.1 Manual*. Retrieved from https://www.psychologie.hhu.de/fileadmin/redaktion/Fakultaeten/Mathematisch-Naturwissenschaftliche_Fakultaet/Psychologie/AAP/gpower/GPowerManual.pdf

Hussin, S. H., Daud, A. I. A., Taibi, M., & Hussin, S. R. (2021). Loneliness, Coping Strategies and Perceived Social Support among Students of Public Universities in Malaysia during the COVID-19 MCO. *International Journal of Business and Society*, *22*(3), 1402–1419. https://doi.org/10.33736/ijbs.4311.2021

Itzick, M., Kagan, M., & Zychlinski, E. (2020). The Big Five Personality Traits as Predictors of Loneliness among Older Men in Israel. *The Journal of Psychology*, *154*(1), 60–74. https://doi.org/10.1080/00223980.2019.1653250

Jie, C. H., Yi, S. H., & Yu, W. X. (2016). *The Relationship between Loneliness, Life Satisfaction and Internet Addiction among Young Adults* (Degree's thesis, Tunku Abdul Rahman University). http://eprints.utar.edu.my/2322/1/PY-2016-1105308.pdf

John, O. P., Robins, R. W., & Pervin, L. A. (2008). *Handbook of Personality: Theory and Research* (3rd ed.). Guilford Press. https://www.ocf.berkeley.edu/~johnlab/bigfive.htm

John, O. P., & Srivastava, S. (1999). The Big-Five trait taxonomy: History, measurement, and theoretical perspectives. In L. A. Pervin & O. P. John (Eds.), *Handbook of personality: Theory and research, 2*, 102–138. New York: Guilford Press. https://pages.uoregon.edu/sanjay/pubs/bigfive.pdf

Keldal, G., & Atli, A. (2016). University Students' Personality Traits as Predictors of their Loneliness Levels. *Çukurova University Faculty of Education Journal, 45*(2), 131–145. https://www.researchgate.net/publication/312041122_University_Students'_Personality_Traits_as_Predictors_of_their_Loneliness_Levels

Killgore, W., Cloonan, S. A., Taylor, E. C., & Dailey, N. S. (2020). Loneliness: A signature mental health concern in the era of COVID-19. *Psychiatry Research, 290*, 113117. https://doi.org/10.1016/j.psychres.2020.113117

Krautter, K., Friese, M., Hart, A., & Reis, D. (2022). No party no joy?-Changes in University Students' Extraversion, Neuroticism, and Subjective Well-being During Two COVID-19 Lockdowns. *Applied psychology. Health and well-being*, 10.1111/aphw.12336. Advance online publication. https://doi.org/10.1111/aphw.12336

Labrague, L. J., De Los Santos, J., & Falguera, C. C. (2021). Social and Emotional Loneliness Among College Students During the COVID-19 Pandemic: The Predictive Role of Coping Behaviors, Social Support, and Personal Resilience. *Perspectives in Psychiatric Care, 57*(4), 1578–1584. https://doi.org/10.1111/ppc.12721

Landmann, H., & Rohmann, A. (2021). When Loneliness Dimensions Drift Apart: Emotional, Social and Physical Loneliness During the COVID-19 Lockdown and its Associations with Age, Personality, Stress and Well-being. *International Journal of Psychology*. https://doi.org/10.1002/ijop.12772

Lee, C. M., Cadigan, J. M., & Rhew, I. C. (2020). Increases in Loneliness Among Young Adults During the COVID-19 Pandemic and Association With Increases in Mental Health Problems. *The Journal of adolescent health: official publication of the Society for Adolescent Medicine, 67*(5), 714–717. https://doi.org/10.1016/j.jadohealth.2020.08.009

Lee, E. E., Depp, C., Palmer, B. W., Glorioso, D., Daly, R., Liu, J., Tu, X. M., Kim, H. C., Tarr, P., Yamada, Y., & Jeste, D. V. (2019). High Prevalence and Adverse Health Effects of Loneliness in Community-dwelling Adults Across the Lifespan: Role of Wisdom as a Protective Factor. *International Psychogeriatrics, 31*(10), 1447–1462. https://doi.org/10.1017/S1041610218002120

McCroskey, J. C. (n.d.). *Introversion Scale*. James C. McCroskey. Retrieved October 10, 2021, from https://www.jamescmccroskey.com/measures/introversion.htm

Najiya, F., & Thomas, S. (2021). Self-Esteem, Empathy and Introversion among Adolescent Readers. *The International Journal of Indian Psychology, 9*(1), 592–602. https://doi.org/10.25215/0901.059

Nayak, R. D. (2016). Relationship of Extroversion Dimension with Academic Performance of Medical Students. *The International Journal of Indian Psychology, 3*(2). 10.25215/0302.004

Neto, F., & Mullet, E. (2020). Subjective Well-being, Personality, and Loneliness as Correlates of Saudade: A Preliminary Study. *Journal of Relationships Research, 11*(4), 1–6. https://doi.org/10.1017/jrr.2020.1

Noprianto, E. (2017). Extrovert versus Introvert Students: What EFL Learning Strategies do They Use? *ASIAN TEFL Journal of Language Teaching and Applied Linguistics, 2*(2), 119–135. http://dx.doi.org/10.21462/asiantefl.v2i2.34

Olowookere, E. I., Omonijo, D. O., Odukoya, J. A., & Anyaegbunam, M. C. (2020). Exploring the Effect of Gender and Personality Characteristics on Educational Performance. *Journal of Educational and Social Research, 10*(5), 99. https://doi.org/10.36941/jesr-2020-0091

Perlman, D., & Peplau, L. A. (1981). Toward a Social Psychology of Loneliness. In S, Duck., & R., Gilmour. (Eds.), *Personal Relationships in Disorder* (pp. 31–56). Academic Press. https://peplau.psych.ucla.edu/wp-content/uploads/sites/141/2017/07/Perlman-Peplau-81.pdf

Pietrabissa, G., & Simpson, S. G. (2020). Psychological Consequences of Social Isolation during COVID-19 Outbreak. *Front. Psychol.* https://doi.org/10.3389/fpsyg.2020.02201

Reddy, M. K. D. S. M. (2020). Social Anxiety among Introvers and Extroverts across Gender among Undergraduate Students during the Pandemic COVID-19 in Bangalore. *Psychology and Education Journal, 57*(8), 1184–1191. https://doi.org/10.17762/pae.v57i8.14214

Rubaiyat, N., Islam, A., Israt, L., & Kabir, L. (2018). *An Experiential Investigation on Internet Addiction and Specific Disorders like Depression, Self-esteem and Introversion among University Students* [Degree's thesis, BRAC University]. http://dspace.bracu.ac.bd/xmlui/handle/10361/11441

Russell, D. W. (1996). UCLA Loneliness Scale (Version 3): Reliability, Validity, and Factor Structure. *Journal of Personality Assessment*, *66*(1), 20–40. https://doi.org/10.1207/s15327752jpa6601_2

Sadriyeva, G. S. K. (2021). Language Learning Strategy Use: Do Extroversion and Introversion Impact Learners' LLS Preferences? *Academic Research in Educational Sciences*, *2*(3), 1116–1123. https://doi.org/10.24411/2181-1385-2021-00515

Schutter, N., Koorevaar, L., Holwerda, T. J., Stek, M. L., Dekker, J., & Comijs, H. C. (2020). 'Big Five' personality characteristics are associated with loneliness but not with social network size in older adults, irrespective of depression. *International Psychogeriatrics*, *32*(1), 53–63. https://doi.org/10.1017/S1041610219000231

Shaughnessy, J. J., Zechmeister, E. B., & Zechmeister, J. S. (2015). *Research Methods in Psychology* (10th ed.). McGraw-Hill Education

Shettar, M., Karkal, R., Kakunje, A., Mendonsa, R. D., & Chandran, V. M. (2017). Facebook Addiction and Loneliness in the Post-graduate Students of a University in Southern India. *The International Journal of Social Psychiatry*, *63*(4), 325–329. https://doi.org/10.1177/0020764017705895

Simonenko, G. (2017). Detection of Autonomic Dysfunctions in Complex Neuropsychological Examination of NMU Students. *EUREKA: Health Sciences*, (4), 12–17. https://doi.org/10.21303/2504-5679.2017.00384

Sîrbu, A. A., & Dumbravă, A.C. (2019). Loneliness at Work and Job Performance: The Role of Burnout and Extraversion. *Human Resource Psychology*, *17*, 7–17. https://dx.doi.org/10.24837/pru.2019.1.491

Snapchat's Friendship Time Capsule Helps You Reconnect with Friends. (2020, October 28). The Sun Daily. https://www.thesundaily.my/spotlight/snapchat-s-friendship-time-capsule-helps-you-reconnect-with-friends-DJ4847544

Sun, J. J., & Chang, Y. J. (2021). Associations of Problematic Binge-Watching with Depression, Social Interaction Anxiety, and Loneliness. *International Journal of Environmental Research and Public Health*, *18* (3), 1168. https://doi.org/10.3390/ijerph18031168

Weber, M., Schulze, L., Bolzenkötter, T., Niemeyer, H., & Renneberg, B. (2022). Mental Health and Loneliness in University Students During the COVID-19 Pandemic in Germany: A Longitudinal Study. *Frontiers in Psychiatry*, *13*, 848645. https://doi.org/10.3389/fpsyt.2022.848645

Wei M. (2020). Social Distancing and Lockdown: An Introvert's Paradise? An Empirical Investigation on the Association Between Introversion and the Psychological Impact of COVID19-Related Circumstantial Changes. *Frontiers in Psychology*, *11*, 561609. https://doi.org/10.3389/fpsyg.2020.561609

Weissbourd, R., Batanova, M., Lovison, V., & Torres, E. (2021). *Loneliness in America: How the Pandemic has Deepened An Epidemic of Loneliness and What we can do about it.* Making Caring Common. https://mcc.gse.harvard.edu/reports/loneliness-in-america

World Health Organization. (2020, March 11). *WHO Director-General's opening remarks at the media briefing on COVID-19.* https://www.who.int/director-general/speeches/detail/who-director-general-s-opening-remarks-at-the-media-briefing-on-covid-19—11-march-2020

Xiong, J., Lipsitz, O., Nasri, F., Lui, L., Gill, H., Phan, L., Chen-Li, D., Iacobucci, M., Ho, R., Majeed, A., & McIntyre, R. S. (2020). Impact of COVID-19 Pandemic on Mental Health in the General Population: A Systematic Review. *Journal of Affective Disorders*, *277*, 55–64. https://doi.org/10.1016/j.jad.2020.08.001

Families Mental Health and Challenges in the 21st Century – Arifin et al. (eds)
© 2024 The Author(s), ISBN 978-1-032-51442-0

Social support and quarter-life crisis: Study on students with divorced parents

Anindita Pramesti Zahara & Sofa Amalia
Psychology, University of Muhammadiyah Malang, Malang, Indonesia

ABSTRACT: Students with divorced parents in addition to having academic demands are also less able to develop secure attachments as early adults. Meanwhile, in early adulthood, individuals are faced with new responsibilities and developmental tasks so they are vulnerable to experiencing a quarter-life crisis where one of the underlying factors is related to demands from the environment. Social support can be a barrier to stress symptoms during stressful events. The purpose of this study was to determine the relationship between social support and quarter-life crisis in college students with divorced parents. The study used a correlational quantitative approach with 192 participants. The research subjects were taken using a purposive sampling technique with the criteria that the students were male or female, aged 18–25 years, having parents divorced when they were in their teens (13–18 years). The research instrument consisted of the multidimensional scale of perceived social support (MSPSS) and the Quarter Life Crisis scale. Data analysis was done using Pearson Product Moment correlation. The results showed that there was a negative relationship between social support and quarter-life crisis in students with divorced parents ($r = -0.342$; p 0.000 <0.05). Therefore, the higher the social support, the lower the quarter-life crisis for students with divorced parents or vice versa.

Keywords: College Students, Early Adulthood, Social Support, Quarter-Life Crisis, Parent's Divorce

1 INTRODUCTION

Early adulthood will be faced with greater roles and responsibilities than in the past so this period is considered a period of worry with various choices regarding the continuation of life in the future. The task of early adult development includes the emergence of community expectations such as starting to take responsibility for oneself, choosing a partner, learning to raise a family, setting a career, to raising children (Santrock 2012). Meanwhile, the worry of carrying out one of the developmental tasks related to building romantic relationships will be felt more severe if the individual has an unpleasant past experience such as parental divorce. When parents experience divorce, there will be a vulnerability to negative judgments related to opposite-sex relationships. This can continue to increase anxiety when faced with early adult developmental tasks about starting to learn to commit more seriously to a relationship (Kurniati & Rozali 2020).

According to Wallerstein (in Berns 2016) when parents experience a divorce when the individual is in his teens, the individual will continue to feel burdened with feelings of sadness about the divorce even up to 10 to 20 years later and feel more of the impact of divorce for the future, especially in decision-making decisions in a relationship. Sarbini (2014) revealed that parental divorce will have an impact on children such as the emergence of insecurity, a

DOI: 10.1201/9781003402381-25

sense of isolation, a sense of losing one family member, a sense of neglect, a sense of not getting support, and not having satisfaction in life so that they are more prone to feeling anxiety. This is stated by Atwood & Scholtz (2008) where adults with divorced parents have a tendency to experience a quarter-century crisis that is higher because they are often worried about the task of adult development in relationships and decide to postpone marriage or live alone. On the other hand, students with divorced parents need to adjust to the conditions of a divorced family, also have responsibilities as a learner, and are also unstable and are faced with many choices regarding the future. According to Murithi (2019), students are individuals who are vulnerable to crises in their 20s which result in increased comfort in lectures in the world of work, hopes of getting a better job thanks to the education they receive, new developmental tasks, and challenges to and expectations of those around them.

The crisis in the 20s is called the quarter-life crisis in psychology. Robbins & Wilner (2001) describe the quarter-life crisis as instability, and continuous change resulting from a new developmental task so that a person feels afraid and worried about his life. Individuals who are in a quarter-life crisis and are unable to respond and get through it can experience indecision, think that their life is too stagnant, feel hopeless, emotionally unstable, feel isolated, doubt themselves, and feel afraid of failure. Meanwhile, individuals who experience a quarter-life crisis and are able to deal with it without protracted will be able to commit to a new role in their life by controlling their feelings, realizing that change is sometimes needed, being more motivated, and having a sense of satisfaction with themselves when compared to life before the crisis. According to Arnett (2004), one of the external factors of the quarter-life crisis is environmental pressure from friendship, love, and family relationships. Individuals will begin to lose and find it difficult to find true friends, begin to think about when and with whom to marry and begin to gain hope of building their own household, feel mature to still depend on their parents and the pressures and demands of life from parents regarding what steps to take in the future. Meanwhile, the thing that triggers stress during the transition period is that individuals do not feel any help from those around them in dealing with and solving problems in early adulthood which are full of various possibilities. This is in line with the theory of Zimet *et al.* (1988) which states that social support can be a barrier to stress symptoms where individuals feel cared for and are not alone in dealing with times of crisis.

If viewed based on Erikson's developmental age, early adulthood is an individual who enters the intimacy vs isolation stage. When individuals are not able to form close relationships and do not get support from the environment, it will cause a sense of isolation. Conversely, when individuals accept and develop good relationships with others, healthy relationships will be created (Santrock 2012). According to Sibua & Silaen (2020), social support can provide encouragement and encouragement when facing difficulties, so that individuals will be challenged and try to solve problems because they feel they have an environment that always supports them. Perceptions of the availability of social support, and the existence of social networks that are built between individuals will help them adapt to dealing with stressful situations during the transition period in students with divorced parents. Social support can act as self-protection and a source of coping so that individuals feel cared for, meaningful, and cared for by those around them. Meanwhile, according to Gonçalves et al (2014), low levels of social support are often assumed to be one of the factors that increase stress and high suicidal ideation in students when faced with stressful situations.

On the other hand, social support is not only obtained from family, but friends and significant others also have a role whose presence can complement and become a source of assistance. When individuals get social support from friends in the form of feelings of the same fate, sympathy, and comfort that is not obtained from their parents or support from significant others, including being willing to listen to complaints and giving confidence, it will have a positive effect and be able to reduce anxiety (Hurlock 2011). Felton & Berry (in Astuti & Hartati 2013) reveal that instrumental social support is often given by closest relations such as family, while emotional support tends to be given by relations outside the family such as friends or significant other. Therefore, social support is complementary and mutually reinforcing so its presence is needed side by side for students with divorced parents.

The purpose of this study was to determine the relationship between social support and quarter-life crisis in college students with divorced parents. The hypothesis proposed in this study is that there is a negative relationship between social support and the quarter-life crisis in students with divorced parents, where the higher the social support perceived by the individual, the lower the quarter-life crisis. On the other hand, the lower the social support perceived by the individual, the higher the quarter-life crisis.

2 METHODS

The research method used is a quantitative approach with a correlational design. A quantitative approach is a research that uses numerical data and is processed by statistical analysis to examine a particular population or sample using research instruments with the aim of testing the established hypothesis. Meanwhile, the research design used is correlational research which aims to examine the relationship between independent variables and variables. The sampling technique used purposive sampling with the criteria of university students being 18–25 years old and having parents divorced when they were in their teens (13–18 years).

In this study, two instruments were used: (a) the quarter-life crisis scale compiled by Hassler (2009) and adapted by Agustin (2012) based on the theory of Robbins & Wilner (2001). The aspects revealed in the quarter-life crisis scale are indecision in decision-making, hopelessness, negative self-assessment, being stuck in difficult situations, anxiety, feeling depression, and worried about interpersonal relationships. The results of the scale analysis consisting of 25 items show the Cronbach coefficient = 0.924; (b) multidimensional scale of perceived social support (MSPSS) compiled by Zimet et al (1988) and adapted by Sarirusadi (2021) based on Zimet's (1988) social support theory. The dimensions revealed in the MSPSS scale are family support (family support), friend support (friend support), and significant other (closest people). The results of the scale analysis consisting of 12 items show the Cronbach coefficient = 0.789.

3 RESULT

Based on the research that has been done, 192 respondents were then grouped into two categorizations, namely high and low.

Based on the categorization results from Table 1, it is known that the social support of students with divorced parents is mostly high with a percentage of 54.2%. Meanwhile, in the quarter-life crisis variable, 102 (53.1%) subjects were in the high category.

Table 1. Variable categorization.

Research variable	Categorization	Interval	Frequency	Percentage
Social Support	Low	<46	88	45,8
	High	≥46	104	54,2
Quarter Life Crisis	Low	<65	90	46,9
	High	≥65	102	53,1

Before testing the hypothesis, the first assumption test was carried out, namely the normality test of the data using the One-Sample Kolmogorov–Smirnov Unstandardized residual test using a significance level of 0.05. The data is declared normally distributed when the significant value (Sig.) > 0.05. Based on the results of the normality test, it shows a significant value (Sig.) of 0.116 (Sig. > 0.05) so this shows that the data from both variables have a normal distribution.

The second assumption test is the linearity test using the basis for making decisions on the deviation value of linearity and sig. linearity. Two variables are said to have a linear relationship if the deviation value from linearity > 0.05 and sig. Linearity < 0.05. Based on the calculation, it can be seen that the variables of social support and quarter-life crisis have a linear relationship. This is proven by the deviation value from linearity obtained by 0.139 (deviation from linearity > 0.05). The linearity obtained is 0.000 (sig. linearity <0.05) so it can be said as the second variable with relationship.

Hypothesis testing was carried out by using the Pearson product-moment correlation test through the SPSS for Windows version 21 application with the basis for making the decision, the value of sig < 0.05. If the significance value is < 0.05, the second variable is declared to be correlated, otherwise, if the significance value is > 0.05, the second variable is declared uncorrelated. Based on Table 2, the significance value of the social support variable and the quarter-life crisis variable is 0.000 (sig < 0.05) with a correlation coefficient (r) of –0.342, meaning that there is a negative relationship between social support and the quarter-life crisis. This shows that the hypothesis is accepted when the social support felt by students with divorced parents is high, the quarter-life crisis experienced will be low, on the contrary, if the social support felt by students with divorced parents will be low, the quarter-life crisis experienced an increase.

Table 2. Product moment correlation test of social support and quarter-life crisis.

		Social Support	Quarter Life Crisis
Social Support	*Pearson Correlation*	1	–0,342
	Sig. (1-tailed)		0,000
	N	192	192
Quarter Life Crisis	*Pearson Correlation*	–0,342	1
	Sig. (1-tailed)	0,000	
	N	192	192

4 DISCUSSION

Based on the research conducted, it was found that there was a significant negative relationship between social support and the quarter-life crisis ($r = -0.342; p<0.05$). This shows that the higher the social support, the lower the quarter-life crisis of students with divorced parents. On the other hand, the lower the social support, the higher the quarter-life crisis for students with divorced parents. Another study that strengthened the results of this study was conducted by Putri (2020) with the results that there was a negative relationship between social support and the quarter-life crisis in final-year students. During the quarter-life crisis, individuals will experience anxiety, feelings of instability, stress, and a sense of helplessness. Meanwhile, interpersonal relationships between individuals that continue to be well established are related to the formation of social support where the presence of social support can reduce anxiety and relieve what is feared. This study shows that one of the successes in dealing with the quarter-life crisis in final-year students is related to the availability of real action in the form of support from the environment.

Divorce is often considered a tragedy where the incident will give emotional reactions such as fear, anger, and confusion to all family members. Meanwhile, students with divorced parents are also faced with all the possibilities related to the future as well as problems that are vulnerable to experiencing a quarter-life crisis. According to Nurasmi *et al.* (2018), individuals with divorced parents who received social support will be able to overcome all adversity and be able to live a better life. This is supported by the research of Connel *et al.* (2015) where there are no significant differences in adjustment in adulthood in students from intact families and from divorced families because divorced family students receive support from lecturers, peers, family, and have developed the ability to survive in the face of life's difficulties. Thus, the presence of social support can help reduce the sense of loss and adjustment experienced by students with divorced parents during the quarter life crisis.

The quarter-life crisis faced by students in the long term causes negative consequences on life such as stress and even depression. Research conducted by Alsubaie *et al.* (2019) showed a negative correlation between social support and symptoms of depression in college students where the higher the social support, the lower the symptoms of depression. This is expressed by the opinion of Sarafino & Smith (2015) that social support provides health and well-being benefits regardless of the amount of frustration experienced. Therefore, social support can act as an external protection mechanism in the face of changes during the transition period as well as indecision that is more related to one of the developmental tasks related to building romantic relationships where at this time individuals have a tendency to feel psychological pressure which causes a quarter of life crisis.

The perceived social support from the environment in the form of motivation, information, real help, appreciation, love, and sharing joy and sadness can make individuals think that they are cared for and that other people can be relied on so that individuals tend to grow positive attitudes towards themselves and are able to face quarter life crisis by avoiding negative effects. Sarafino *et al.* (2015) explain that social support can change an individual's perspective on life problems and difficulties, feel more valuable, and be able to reduce the potential for new or prolonged stress. The presence of social support will affect individual thinking and behavior such as being more motivated to get up and evaluate previous failures so that individuals can suppress negative responses that arise during the quarter-life crisis.

5 CONCLUTION AND IMPLICATION

Based on the research that has been done, it can be said that the accepted hypothesis is that there is a negative relationship between social support and the quarter-life crisis in students with divorced parents. The higher the social support, the lower the quarter-life crisis for students with divorced parents. On the other hand, when perceived social support is low, the quarter-life crisis is high.

The existence of a relationship between social support and the quarter-life crisis in this study has implications for students with divorced parent's backgrounds to not only increase relationships with other people, but also need to start improving and developing a relationship that is being undertaken such as strengthening friendship and family relationships. One of them is by spending time together, being open and trusting with each other, and establishing regular communication considering that communication is an inseparable part of providing social support. In addition, the importance of always creating a good relationship with the environment so that it triggers people around them to be happy to provide support and are willing to share stories about the difficulties experienced by utilizing possible social support, thus students with divorced parents are able to minimize the occurrence of a quarter-life crisis. Meanwhile, suggestions for development that researchers can give to further researchers who are interested in conducting research with similar phenomena can develop research related to social support and the quarter-life crisis in terms of the length of parental divorce and can make context adjustments to the measuring tools used.

REFERENCES

Agustin, I. (2012). *Terapi Dengan Pendekatan Solution-focused Pada Individu Yang Mengalami Quarter-life Crisis*. Thesis, Universitas Indonesia.

Alsubaie, M. M., Stain, H. J., Webster, L. A. D., & Wadman, R. (2019). The Role of Sources of Social Support on Depression and Quality of Life for University Students. *International Journal of Adolescence and Youth*, *24*(4), 484–496. https://doi.org/10.1080/02673843.2019.1568887

Arnett, J. J. (2004). *Emerging adulthood: The Winding Road From the Late Teens through the Twenties*. New York: Oxford University Press.

Astuti, T. P., & Hartati, S. (2013). Dukungan Sosial Pada Mahasiswa Yang Sedang Menyusun Skripsi. *PSikologi*, *13*, 69–81.

Atwood, J. D., & Scholtz, C. (2008). The Quarter-life Time Period: An Age of Indulgence, Crisis or Both? *Contemporary Family Therapy*, *30*(4), 233–250. https://doi.org/10.1007/s10591-008-9066-2

Berns, R. M. (2016). *Child, Family, School, Community: Socialization and Support*. Stanford: Wadsworth Publisher.

Connel, B., Hayes, D. M., & Carlson, M. (2015). Relation between Parental Divorce and Adjustment in College Students. *Journal of Divorce and Remarriage*, *56*(4), 336–345. https://doi.org/10.1080/10502556.2015.1025756

Gonçalves, A., Sequeira, C., Duarte, J., & Freitas, P. (2014). Suicide Ideation in Higher Education Students: Influence of Social Support. *Atencion Primaria*, *46*(S5), 88–91. https://doi.org/10.1016/S0212-6567(14)70072-1

Hassler, C. (2009). *Are you Having a Quarter Life Crisis?* https://www.huffpost.com/entry/are-you-having-a-quarterl_b_326612

Kurniati, A., & Rozali, Y. A. (2020). Pengaruh dukungan Sosial Terhadap Kecemasan Untuk Menikah Pada Wanita Dewasa Awal Dengan Latar Belakang Orang Tua Bercerai. *JCA Psikologi*, *1*, 85–92.

Murithi, G. (2019). Psychological Factors Contributing To Quarterlife Crisis among University Students from A Kenyan University. *International Journal for Advanced Research*, *5*(3).

Nurasmi, R., Maulana, I., Farida Inli, D., Tendikat Fitri, Z., Karunia Sari, L., Kurnia Sari, N., & Putra Azis, A. (2018). Dukungan Sosial Komunitas Hamur Pada Remaja Broken Home. *Jurnal Fakultas Psikologi*, 1–9.

Putri, A. R. (2020). *Hubungan Antara Dukungan Sosial Dengan Quarter Life Crisis Pada Mahasiswa Tingkat Akhir*. Skripsi, Universitas Islam Negeri Sunan Kalijaga. http://repository.radenintan.ac.id/12581/1/SKRIPSI_PERPUS.pdf

Robbins, A., & Wilner, A. (2001). *Quarterlife Crisis: The Unique Challenges of Life in Your Twenties*. Penguin Putnam.

Santrock, J. W. (2012). *Life-span Development Perkembangan Masa Hidup Edisi Ketigabelas Jilid 2*. Jakarta: Erlangga.

Sarafino, E., Smith, T. W., King, D., & Delongis, A. (2015). *Health Psychology: Biopsychosocial Interactions – Canadian Edition*. Amerika Serikat: John Wiley & Sons, Inc.

Sarbini, W., Wulandari, K., Sos, S., Si, M., Ilmu, J., Sosial, K., Ilmu, F., Politik, I., & Unej, U. J. (2014). Kondisi Psikologi Anak Dari Keluarga Yang Bercerai (The conditions of child psychology toward family divorced). *Artikel Ilmiah Penelitian Mahasiswa*, 1–5.

Sarirusadi, R. E. (2021). *Pengaruh Dukungan Sosial Terhadap Tingkat Kecemasan Tenaga Kerja di Masa Pandemi*. Skripsi, Universitas Muhammadiyah Malang.

Sibua, R. U. R., & Silaen, S. M. J. (2020). Dukungan Sosial dan Kecerdasan Emosional (Emotional Quotient) Dengan Stres di Tengah Pandemi Covid-19 Pada Masyarakat Cempaka Putih Barat, Jakarta Pusat. *IKRA-ITH Humaniora*, *4*(3), 187–193. https://journals.upi-yai.ac.id/index.php/ikraith-humaniora/issue/view/61

Zimet, G. D., Dahlem, N. W., Zimet, S. G., & Farley, G. K. (1988). The Multidimensional Scale of Perceived Social Support. *Journal of Personality Assessment*, *52*(1), 30–41. https://doi.org/10.1207/s15327752jpa5201_2

Secure attachment of parents to the resilience of adolescents who experiencing their parent's death due to COVID-19

Uun Zulfiana & Shafa Alvita*
Faculty of Psychology, University of Muhammadiyah Malang, Indonesia

ABSTRACT: The pandemic situation has resulted in several adolescents losing their parents due to exposure to the COVID-19 virus. The ability of resilience is very important for adolescents so that they can get through challenges and difficult life circumstances. Secure attachment is thought to be important for parents and can help adolescents to form their psychological resilience so that adolescents can recover and be stronger in facing difficult conditions for them. The purpose of this study was to determine the effect of parental secure attachment on the resilience of adolescents who experienced parental death due to COVID-19. This study uses a quantitative approach with a purposive sampling technique. The subjects of this study were 65 adolescents aged 13–21 years who lost one parent due to COVID-19. The scale in this study used the Inventory of Parent and Peer Attachment Scale (IPPA) and The Connor-Davidson Resilience Scale (CD-RISC). The results of this study indicate that there is a significant positive effect between secure attachment and resilience (R^2 = 0.159; $\beta = 0.399$, p $0.001 < 0.05$). These results indicate that the higher the secure attachment in adolescents, the higher the resilience they have. The magnitude of the influence of the secure attachment variable on resilience is 15.9%.

Keywords: Adolescents, COVID-19, Resilience, Secure Attachment

1 INTRODUCTION

The pandemic situation has resulted in several adolescents losing their parents due to exposure to the COVID-19 virus. The adolescent phase is a crisis stage of an individual development because adolescence is a transition period for an individual to develop into a mature adult both physically and psychologically, many significant changes occur during this period. In reality, adolescence is usually filled with conflict, be it conflicts that come from within him, his peers, or his family, especially if the problem faced is about the loss of a parent. Ability of resilience is very important for adolescents so that they can get through the challenges and difficult life circumstances in their development period. Resilience is defined as an individual's ability to adapt to face adversity and difficult conditions (Connor & Davidson 2003). Adolescents who have lost their parents will experience various kinds of pressure due to reduced or even no more support and affection given as before. Besides, in a pandemic situation, based on research on adolescent anxiety in the pandemic era conducted by Fitria & Ifdil (2020), it was found that as many as 54% of adolescents experienced anxiety which was in the severe category during the COVID-19 pandemic.

According to Hurlock (1990), the absence of parents either due to death, unknown whereabouts, or unwanted children is one of the supporting factors that cause the child's personality development to be less than optimal. When they are in a new situation of losing

*Corresponding Author: alvitashafa@gmail.com

DOI: 10.1201/9781003402381-26

their parents, some adolescents have not been able to fully independently make adjustments to take care of their own lives, because previously adolescents have been used to being served or assisted by their parents, so in this case adolescents often try to find out and do a lot of things. things that had never been or were rarely done (Zahrina 2017).

According to Barankin & Khanlou (2007), the factors that influence the level of resilience are divided into internal factors (within oneself), family factors, and social environmental factors. The internal factors that can affect the level of resilience include self-concept, emotion regulation, way of thinking, adaptive skills, and social skills. The family factors include affection, communication, parental relationships, parenting, and support outside the family. While social environmental factors include relationships with peers.

Parents play a role in nurturing, directing, and guiding adolescents to help them overcome the conflicts they face, find identity, and how to interact and adapt to their social environment. One of the roles of parents that proven to be effective in parenting is to provide full support to adolescents. Full support from parents can provide a sense of emotional security for adolescents so that they can explore things in a positive way and can face various challenges (Peterson *et al.* 1986). This full support can be done by encouraging children to communicate their feelings, being involved in children's lives, and providing acceptance and motivation to train children to have independent thinking (Darling & Steinberg 1993; Openshaw *et al.* 1984). This is supported by Hurlock (2002) that how the role of parents in providing care, support, and how to interact with children is very influential in the formation of the child's personality.

The death of a parent will cause various kinds of impacts, especially psychological impacts for spouses and children who have been abandoned. Based on research conducted by Andriessen *et al.* (2018), the psychological impact that arises in dealing with the death of a parent in adolescents is making the adolescent's condition stressful even to depression, often blaming himself, as well as attempting suicide. This can happen because adolescents naturally feel lonely, hopeless, and afraid of facing life (Fitria 2013). The psychological impact caused can last in the long term if adolescents are not able to manage their emotions, especially their negative emotions, namely sadness. Seeing the impact, especially on the emotional aspect, adolescents who were abandoned by one of their parents of course really need attention from one of the parents who are still alive or also from foster parents, one of which is through secure attachment.

Secure attachment in this case is defined as a strong emotionally secure attachment that is formed through the interaction of parents with children as the basis for the child's psychological development. The pattern of relationships formed makes adolescents feel confident in their parents as attached figures who accompany them, are responsive, and are full of love and affection when adolescents seek protection and comfort (Bowlby 1988). In this case, the emotionally secure attachment between parents and children is not limited to infancy and childhood but continues to develop from adolescence to adulthood. Secure attachment is thought to be important for parents so that they can help adolescents to form their psychological resilience so that adolescents can recover and be stronger in facing difficult conditions for them. So secure attachment is expected to have a significant impact on increasing adolescent resilience.

Several previous studies have been conducted regarding attachment to resilience, for example, research conducted by Alvina *et al.* (2020) which in his research found that there was a significant positive relationship between attachment to fathers and adolescent resilience at SMP Negeri 5 Banjarbaru. In addition, research conducted by Febe (2014) found that there was a significant positive relationship between parental attachment and resilience in first-year students.

This research needs to be done so that single parents or foster parents have an understanding of the developmental tasks of adolescents and are able to assist in carrying out their roles and functions as good caregivers and educators because attachment also plays an important role in helping adolescents to fulfil their developmental tasks (Fadhillah & Faradina 2016). If adolescents do not get secure attachments, they will be more inclined to withdraw from the environment, are uncomfortable in a close relationship, have excessive emotions, and are also less able to control their emotions. Parents need to understand and assist their child's growth and development process when entering adolescence because secure attachment from parents will be an important foundation for the psychological development of adolescents in the future.

Secure attachment is important to always be studied and trained so that it can be one of the important factors to improve adolescents resilience. The specific purpose of this study is to determine and analyze the effect of secure attachment on the resilience of adolescents who experience parental death due to COVID-19. The practical benefits of this research are expected to make a real contribution to increasing awareness that single parent's secure attachment to adolescents has a major impact on resilience. psychology of adolescents, the real contribution can be in the form of psychological interventions to increase secure attachment so that adolescents can adapt and continue to live prosperously after having to face the traumatic event they experienced.

2 METHODS

The type of research used is non-experimental quantitative research. The measurement method in this study uses the regression analysis method. The population in this study is adolescents aged 13–21 years based on the age category of Santrock (2007). Then the sampling technique of this study uses a purposive sampling technique where sampling is carried out on certain considerations (Sugiyono 2017). The sample criteria used in this study, namely: (1) Male/ female, (2) 13–21 years old, (3) Loss of one parent due to COVID-19.

Based on the theoretical basis and the formulation of the existing research hypotheses, there are two variables studied, namely: secure attachment as the independent variable (X), and resilience as the dependent variable (Y).

This research procedure consists of three stages, starting from the research preparation stage, research implementation, and data analysis. At the research preparation stage, the researcher makes a research proposal that includes the background of the research, theoretical studies as a basis for deepening the material and develops research methods by determining the research instrument in the form of a scale by adapting the secure attachment and resilience scale.

First, the secure attachment scale in this study was measured using the Inventory of Parent and Peer Attachment (IPPA) scale instrument developed by Armsden & Greenberg (1987) and adapted by Anisyatul (2021). On this scale, there are three subscales, namely secure attachment to mother, father, and peers which are measured separately with each containing 25 items. However, in this study, researchers only measured secure attachment to parents, namely father and mother only. This IPPA scale measures the secure attachment section which consists of 3 aspects, namely trust, communication, and alienation. This study uses a Likert scale with favorable and unfavorable concepts that have a value of 1 for the answer "Never", a value 2 for the answer "Never", a value of 3 for the answer "Rarely", value of 4 for the answer "Often", and value 5 for the answer "Very Often". An example of item is "I feel that my mother does her job as a mother well" and an example of a mother's secure attachment item is "I feel that my mother does her job as a mother well". mother well" and an example of a father's secure attachment item is "I tell my father about my problems and difficulties about the problems and difficulties I experience".

Second, the resilience scale in this study used The Connor-Davidson Resilience Scale (CD-RISC) developed by Connor & Davidson (2003) and adapted by Hidayati (2018). The scale includes 5 dimensions, including personal competence, strong facing pressure, positive acceptance, self-control, and spirituality. The level of resilience can be seen from the results obtained when the subject fills out the scale, the higher the score obtained, the more resilience possessed by adolescents with parents who died due to the death of their parents. the higher the score obtained, the resilience possessed by adolescents with parents who died due to COVID-19 is high. The number of items contained in this scale totals 25 items, with which each item aims to measure resilience. each item aims to measure adolescent resilience. This study uses a Likert scale with a favorable concept that has a value of 1 for the answer "Very Unsuitable", a value of 2 for the answer "Not Suitable", a value 3 for the answer "Neutral", a value 4 for the answer "Suitable", and value 5 for the answer "Very Suitable". The example item of this scale is "I am able to adapt to sudden changes that occur in my life".

The next stage is the implementation of the research, researchers distributed research scales to adolescent subjects aged 13–21 years who lost one of their parents due to COVID-19. The distribution of the research scale was carried out for 3 months starting from December 2021 to February 2022. During the first 1 month, researchers distributed the scale only through the Google Form platform through online media which included Instagram and Twitter with the aim that this research could reach a wider range of subjects. This research is able to reach a wider range of subjects. However, because the number of subjects obtained was thought to be unable to describe the population of adolescents who lost their parents due to COVID-19 in Indonesia, in January the researcher took care of a research permit to request related data assistance from the Social Service P3AP2KB Malang City, DP3AP2KB Pasuruan Regency, Bakesbangpol East Java, and DP5A Surabaya City.

During approximately 3 months of research implementation, the next stage the researchers conducted data analysis. First, the researcher conducted a validity and reliability test. Second, the researcher conducted a normality test as a condition for the next analysis stage, namely the regression analysis test. Based on the normality test of the data using the Kolmogronov–Smirnov test, it was found that the research data consisting of 65 subjects had a significance value of $0.181 > 0.05$ which means that the data obtained were normally distributed. After knowing that the data obtained were normally distributed, the researcher conducted a regression analysis test using SPSS version 25 to analyze the significant effect between one independent variable that has dimensions on the dependent variable and describes the data that has been collected as it is without intending to make conclusions that apply to the public.

3 RESULT

Research conducted on adolescents aged 13–21 years who lost their parents due to COVID-19 obtained the results shown in Table 1.

Based on Table 1, which is a breakdown of the research subjects, it can be seen that the adolescents who were the subjects of this study consisted of 19 male adolescents (29.3%) and 46 female adolescents (70.7%). Based on the age range, there are 13 early adolescents (20%),

Table 1. Research subject.

Characteristics	Frequency	Percentage
Sex		
Male	19	29.3
Female	46	70.7
Age		
Early Adolescence (13–15 years)	13	20
Middle Ages (16–18 years)	36	55
Late Adolescence (19–21 years)	16	25
Parents Who Died Due to COVID-19		
Father	39	60
Mother	26	40

36 middle adolescents (55%), and 16 late adolescents (25%). Based on the parents who died from COVID-19, 39 adolescents lost their father (60%) and 26 adolescents lost their mother (40%).

Based on Table 2, the researchers divided the research subjects into 2 categories, namely, high and low on the secure attachment variable and the resilience variable. Of 65 adolescents, 60 adolescents (92.3%) had a high-secure attachment and 5 other adolescents (7.7%) had a low-secure attachment. Furthermore, of 65 adolescents, 62 adolescents (95.4%) had high resilience and 3 adolescents (4.6%) had low resilience.

The hypothesis test in this study is to see whether or not there is a positive influence between secure attachment on the resilience of adolescents who have lost their parents due to COVID-19. After testing the normality of the data and the results obtained that the data obtained are normally distributed, then the researchers conducted a regression analysis using SPSS version 25, the results are shown in Table 3.

Table 2. Categorization of secure attachment and research subject resilience.

	Category	Interval	Freq.	%	Mean	SD
Secure	High	123–205	60	92.3	160.7	26.25
Attachment	Low	41–122	5	7.7		
Resilience	High	66–110	62	95.4	85.38	11.67
	Low	22–65	3	4.6		

Table 3. Results of regression test analysis.

Variable	R^2	β	*F*	*(p) Sig.*
Secure Attachment	0.15	0.3	11.94	.00
Resilience	9	99	2	1

Based on Table 3, it is known that the regression analysis test results in the percentage of the influence of parental secure attachment on adolescents, the coefficient of determination (R^2) is 0.159, which means that the influence of parental secure attachment on adolescent resilience is 15.9%, while 84.1% is influenced by factors that other. The value of the regression coefficient (β) was obtained at 0.399. The calculated F value is 11.942 with a significance level of $0.001 < 0.05$, which means that there is an effect of parental secure attachment on adolescent resilience.

4 DISCUSSION

Based on the results of this study, it was obtained that the research hypothesis was acceptable, in which there was a significant positive effect of parental secure attachment on resilience in adolescents ($R^2 = 0.159$; = 0.399, p $0.001 < 0.05$). These results indicate that the higher the secure attachment of parents in adolescents, the higher the resilience they have and vice versa, and the lower the secure attachment of parents in adolescents, the lower the resilience they have.

Based on data recorded as of March 6, 2022, the death rate due to COVID-19 in Indonesia reached 150,172 people with the majority of the age group dying, ages 31–45 as many as 12.8%, ages 46–59 as many as 36%, and age 60 as many as 47.1% (Satuan Tugas Penanganan COVID-19 2022). The data shows that the majority who died from COVID-19 were in the adult to elderly age range, whereas at that age, the majority were married and living with families. This means that they left their spouse and children. This study, which was conducted on 65 adolescents who lost their parents due to COVID-19, showed that 39 adolescents (60%) lost their father and 6 (40%) lost their mother, which means that the parental status of the adolescents has changed to a single person who has to carry out two roles at once, namely the role of a father and a role of a mother, accordingly.

The death of one parent can have a major impact on the development of adolescents as well as the psychological condition of a single parent who is still alive. Adolescents will feel the love and support given is reduced or even no more love and support is given because of the loss of one parent who is more emotionally attached to them than the parent who is still alive. Meanwhile, parents will face new problems as single parents, which begin with feelings of anxiety, fear, worry, and confusion because they feel the heavy responsibility of living life alone to meet material needs, namely being the backbone of supporting children and the costs of daily life, as well as psychological needs of children such as providing a feeling of security, protection, love, and affection.

The results of a study conducted by 65 adolescents who had lost a parent due to COVID-19 found that 60 adolescents (92.3%) had a high level of secure attachment. This shows that most of the subjects of this study, between parents and adolescents, have an attitude of mutual trust with one another marked by acceptance and understanding of one another, have good communication so that they are able to express the difficulties they face, as well as a low attitude of isolation from parents who make adolescents angry, embarrassed, or annoyed to a minimum so that the relationship between parents and adolescents remains harmonious. Meanwhile, 5 other adolescents (7.7%) had a low level of secure attachment.

This is in line with previous research conducted by Anisyatul (2021) which found that most adolescents have a highly secure attachment to their parents, which indicates that most of these adolescents have a sense of security and have a good relationship with their parents. But in reality, some adolescents in Indonesia have lost their parents due to COVID-19. Losing someone who has left first and has to experience the suffering of losing someone who has been so emotionally attached since childhood is a very difficult condition and makes a teenager's life worse. If it can't be handled properly because it continues in a slump, it will lead to adolescent emotional turmoil at a later stage of development.

However, Balson (1993) says that the impact of losing a father or mother will not drown the child in protracted depression or emotional turmoil, when the abandoned family relationship is based on respect for equality, mutual support, and trust in each other. This shows the importance of awareness and the role of single parents in providing secure attachment to adolescents because, with good parent-adolescent relationships that can be established, it can be a source of fulfilling adolescent psychological needs and can also be a supporting factor that can increase adolescent resilience to rise from adversity.

Death events for adolescents will have a worse impact if the death occurs suddenly because adolescents do not have the psychological readiness to face the loss of people close to them (Suprihatin & Adina 2013). Based on research conducted by Wisnu *et al.* (2021), it was found that the majority of patients with a critical degree of COVID-19 were aged 50 and had a shorter duration of hospitalization of around 3–4 days because they had been declared dead while undergoing treatment at the hospital. This shows that some patients who died from COVID-19 can be said to have died suddenly. A meta-analysis has shown that the presence of comorbid disease in COVID-19 patients increases disease severity by about threefold (Radwan *et al.* 2020). Sudden or unexpected death will be really surprising for the people left behind because they are required to be able to adapt to sudden changes that occur in their lives in order to struggle to continue living in prosperity.

The results of the research that has been done also found that 62 adolescents (95.4%) were in the high category of resilience level. Adolescents who have a high level of resilience mean that they will be stronger in dealing with the problems they are facing, have the confidence to deal with their problems, can accept the changes that occur, and can control themselves, especially in controlling their emotions in a slumped state. Meanwhile, it was found that 3 adolescents (4.6%) were in the low category of resilience level, which means that the teenager has not been able to face the problems he is facing, lacks the confidence to deal with the problem, is less able to accept the changes that occur, and is less able to control himself in a slumped situation.

Research conducted by Lin, Sandler, Ayers, Wolchik, & Luecken (2004) found that some children have low resilience when facing the death of their parents. However, in this study, it was found that the majority of adolescents can be said to have high resilience. The high level of resilience possessed by adolescents will help them in interpreting difficult events (parental death) in a positive way, helping them to turn difficult events into their impetus to develop in terms of the ability to face problems and control themselves in a slump. Bowlby (1988) also says that humans are basically born with an innate motivation system so they tend to be able to adapt to special situations such as when responding to threats or facing difficult events because humans will naturally seek protection to increase feelings of security when adapting to these special situations.

This research has advantages and disadvantages. The advantage of this study is that it is able to prove and can be used as an illustration that the secure attachment provided by parents as the main attachment figure affects the resilience of adolescents who have lost their parents due to exposure to COVID-19. The advantages of this study are also supported by several previous studies that describe secure attachment and resilience. However, some of the shortcomings of this study are that the sample size is not large enough to describe the population of adolescents who have lost their parents due to COVID-19 in Indonesia. In addition, researchers have not compared the secure attachment of father and mother, which has a higher impact on the resilience of adolescents who have lost their parents due to COVID-19 because the data obtained is that the number of fathers and mothers who died is not equal.

5 CONCLUSION AND IMPLICATION

Based on the results of research conducted on 65 adolescents aged 13–21 years, the results showed that the research hypothesis was accepted. The effect of parental secure attachment on the resilience of adolescents who have lost their parents due to COVID-19 is 15.9%, and 84.1% is influenced by other factors. Therefore, it can be concluded that there is a significant positive effect between parental secure attachment and the resilience of adolescents who have lost their parents due to COVID-19.

This study implies that parents are expected to be able to increase awareness that a sense of secure attachment is important, especially for the resilience of adolescents who have lost one of their parents, especially due to COVID-19 because when adolescents have good resilience abilities, adolescents will be stronger in dealing with the problems faced by them. he is facing, has the confidence to face his problems, is able to accept the changes that occur, is able to control himself in a slumped state and is able to achieve the real meaning of life. The results of this study can be used as a reference for further research to carry out psychological interventions in order to increase the resilience of adolescents who have lost their parents due to COVID-19 through parental secure attachment.

Based on the description that has been presented, suggestions for development that can be given for further research are to deepen research studies related to the effect of parental secure attachment on adolescent resilience, especially in the comparison of a better father or mother to adolescent resilience, and can also use related additional variables in order to complement and deepen other factors that are thought to have a stronger influence for use in further research.

REFERENCES

Alvina, Rahmi, Firdha. (2020). Hubungan Kelekatan pada Ayah dengan Resiliensi Remaja SMPN 5 Banjarbaru. *Jurnal Kognisia*, 3(2).

Andriessen, K., Mowll, J., Lobb, E., Draper, B., Dudley, M., & Mitchell, P. B. (2018). "Don't bother about me". The Grief and Mental Health of Bereaved Adolescents. *Death Studies*, 42(10). doi: 10.1080/07481187.2017.1415393.

Anisyatul, F. (2021). Pengaruh Secure Attachment Remaja Pada Orang Tua Terhadap Kompetensi Emosi. Skripsi. Universitas Muhammadiyah Malang.

Armsden, G. C., & Greenberg, M. T. (1987). The Inventory of Parent and Peer Attachment: Relationships to Well-being in Adolescence. *Journal of Youth and Adolescence*, 16(5). doi: 10.1007/BF02202939.

Balson. (1993). *Psychology of Family*. New York: McGraw-Hill, Co.

Barankin, T & Khanlou, N. (2007). *Growing Up Resilience*. North America: Camh Publication.

Bowlby, J. (1988). *A Secure Base: Parent–child Attachment and Healthy Human Development*. New York: Basic Books.

Connor, K. M., & Davidson, M. D. (2003). Development of a New Resilience Scale: The Connor Davidson Resilience Scale (CD-RISC). *Depression and Anxiety*, 18(2), 76–82. doi: 10.1002/da.10113.

Darling, N., & Steinberg, L. (1993). Parenting Style as Context: An Integrative Model. *Psychological Bulletin*, 113, 487–496.

Fadhillah, N., & Faradina, S. (2016). Hubungan Kelekatan Orang Tua dengan Kemandirian Remaja SMA di Banda Aceh. *Jurnal Ilmiah Mahasiswa Psikologi*, 1(4), 42–51.

Febe, Tirza. (2014). Hubungan antara Attachment terhadap Orang Tua dengan Resiliensi pada Mahasiswa Tahun Pertama Angkatan 2013 Fakultas Psikologi UKSW yang Berasal dari Luar Pulau Jawa. Skripsi. Universitas Kristen Satya Wacana Salatiga.

Fitria, Adina. (2013). Grief pada Remaja Akibat Kematian Orang Tua Secara Mendadak. Skripsi. Universitas Negeri Semarang.

Fitri, L., & Ifdil, I. (2020). Kecemasan remaja pada masa pandemi Covid-19. *Jurnal EDUCATIO: Jurnal Pendidikan Indonesia*, 6(1), 1–4.

Hidayati, L. (2018). Pengaruh Iklim Organisasi dan Resiliensi terhadap Keterikatan Kerja pada Konselor Adiksi di Lembaga Medan Plus. Thesis. Universitas Sumatera Utara.

Hurlock, E. B. (1990). *Psikologi Perkembangan: Suatu Pendekatan Sepanjang Rentang Kehidupan*. Jakarta: Erlangga.

Hurlock, Elizabeth B. (2002). *Psikologi Perkembangan: Suatu Pendekatan Sepanjang Rentang Kehidupan*. Jakarta: Erlangga.

Lin, K. K., Sandler, I, N., Ayers, T. S., Wolchik, S. A., & Luecken, L. J. (2004). Resilience in Parentally Bereaved Children and Adolescents Seeking Preventive Services. *Journal of Clinical Child and Adolescent Psychology*, 33(4), 673–683. doi: 10.1207/s15374424jccp3304_3.

Openshaw, K. D., Thomas, D. L., & Rollins, B. C. (1984). Parental Influences of Adolescent Self-esteem. *Journal of Early Adolescence*, 4, 259–274.

Peterson, G. W., Stivers, M. E., & Peters, D. F. (1986). Family Versus Nonfamily Significant Others for the Career Decisions of Low-income Youth. *Family Relations: Journal of Applied Family & Child Studies*, 35, 417–424.

Radwan, NM, Mahmoud, NE., Alfaifi, AH., & Alabdulkareem, KI. (2020). Comordibities and Severity of Coronavirus Disease 2019 Patients. *Saudi Med J*, 41(11).

Reivich & Shatte. (2002). Psychosocial Resilience. *American Journal of Orthopsychiatry*, 57, 316.

Santrock, J. W. (2007). Remaja Jilid 2. *Terjemahan oleh Benedictine Widyasinta*. Jakarta: Erlangga.

Satuan Tugas Penanganan COVID-19. (2022). Peta Sebaran COVID-19. Komite Penanganan Covid-19 dan Pemulihan Ekonomi Nasional. Diakses pada tanggal 7 Maret 2022 dari https://covid-19.go.id/peta-sebaran.

Sugiyono. (2017). *Metode Penelitian Kuantitatif, Kualitatif dan R&D*. Bandung: Alfabeta.

Suprihatin, Adina. (2013). Grief pada Remaja Akibat Kematian Orangtua secara Mendadak. Skripsi. Universitas Negeri Semarang.

Wisnu, K., Made, I, & Surya, B. (2021). Hubungan Penyakit Komorbiditas Terhadap Derajat Klinis COVID-19. *Jurnal Intisari Sains Medis*, 12(2), 708–717.

Zahrina, Ashri. (2017). Pengaruh Pengalaman Traumatik Terhadap Kepedulian Sosial Ditinjau dari Jenis Kelamin. Skripsi. Universitas Muhammadiyah Malang.

Families Mental Health and Challenges in the 21st Century – Arifin et al. (eds)
© 2024 The Author(s), ISBN 978-1-032-51442-0

Psychometric property of subjective well-being in high school students: A multidimensional approach

Atika Permata Sari*
Faculty of Psychology, University Muhmmadiyah Malang, Malang, Indonesia

ABSTRACT: Subjective well-being (SWB) is a multidimensional construct that should be assessed using at least two separate measurements. Meanwhile, despite of many research related to SWB in Indonesia, many of them still used a single measurement of SWB. The purpose of this study is to assess the validity and reliability of the SWB measurement in high school students. This study included 238 participants ranging in age from 15–18 years. SWB measurements included the students' life satisfaction scale (SLSS), brief multidimensional students' life satisfaction scale (BMSLSS), positive affect negative affect scale (PANAS), and the Satisfaction with Life Scale (SWLS). The results showed that all scales had Cronbach's Alpha greater than 0.70. The results of exploratory factor analysis revealed that the model is a good fit and most of the items have adequate loading factors. As a result, the scale used in this study has adequate validity and reliability.

Keywords: BMSLSS, PANAS, SLSS

1 INTRODUCTION

Previous research has shown that late adolescence has a higher intensity of negative emotions compared to early adolescence. The incidence of depression in adolescents was found to increase drastically after puberty. In late adolescence, the prevalence of depression in one year reached 4% (Thapar *et al.* 2012). The screening results using Indonesia's Children's Depression Inventory (CDI) in 2016 support these findings. In 2016, it was found that 30% of students in grades 1 and 2 of high school in Indonesia can experience depression (Sukmasari 2016).

In addition to being associated with negative emotions and moods, other studies have found a decrease in life satisfaction from late childhood to adolescence. Girls tend to experience a more significant reduction in general life satisfaction than boys. However, a significant decrease was found in general life satisfaction, health satisfaction, and family relationships (Goldbeck *et al.* 2007). Several other studies have found that adolescents' life satisfaction decline is consistently found globally in different countries (Opshaung 2013). Older adolescents had lower life satisfaction than younger adolescents (Nee *et al.* 2016). Individuals who are at the age of 16 years get lower scores than the previous age both on measurements of general life satisfaction and life satisfaction in specific domains (Goldbeck *et al.* 2007).

The higher intensity and frequency of negative emotions and the finding of decreased life satisfaction in late adolescence indicate that late adolescence has the potential to have low SWB. SWB is a general term that refers to the overall assessment of an individual's life or emotional experience, namely satisfaction, positive affect, and low negative affect (Diener *et al.* 2018). The components in the SWB are proven to be different factors based on factor

*Corresponding Author: atikapermatasari@umm.ac.id

 DOI: 10.1201/9781003402381-27

analysis and have different relationships with other variables. Therefore each component must be assessed independently. A comprehensive SWB assessment will not be obtained by measuring only one component of this construct; several components must be measured to provide comprehensive information about SWB (Diener *et al.* 2016). Life satisfaction is the cognitive component of SWB and plays an important role in positive development as an indicator, a predictor, a mediator/moderator, and an outcome. Whereas low life satisfaction is associated with psychological, social, and behavioral problems, high life satisfaction is related to good adaptation and optimal mental health among youth. Life satisfaction and positive affect mitigate the negative effects of stressful life events and work against the development of psychological and behavioral problems among youth (Park 2004).

In Indonesia, several studies found that several high school students have low SWB levels (Hamdan & Alhamdu 2015; Prasetya 2015; Nayana 2013; Nisa 2019). Meanwhile, SWB was found to be positively related to academic achievement in school, grades, attendance, and good math skills (Bucker *et al.* 2018; Manzoor *et al.* 2014; Suldo *et al.* 2011). SWB can also improve cognitive performance because, according to the build and broaden theory, positive emotions affect creativity and imagination by increasing more flexible thinking styles and a broader mental context (Diener & Ryan 2009). Students with higher SWB were also found to show lower levels of anxiety, depression, negative self-concept, somatization, and emotional problems (Kaplan 2017). In addition, SWB was also found to predict problematic behavior in adolescents, such as antisocial behavior, alcohol consumption, smoking behavior, suicidal tendencies, and dropout rates from school (Arslan & Renshaw 2018). Apart from positively impacting the individual level, SWB also benefits society in general. A high level of SWB can contribute to a stable, productive, and effective society (Diener & Ryan 2009).

Given the positive impact of high SWB, it's important to have reliable and valid measurements of SWB in adolescents, especially in high school students. Measurement is the foundation of the reproducibility and replicability of research findings. It also serves as the basis for deriving theoretical conclusions (Gregmini 2020). Despite many findings related to students' SWB in Indonesia, many of them still used a single measurement of SWB (Hamdan & Alhamdu 2015; Ismail & Yudiana 2020; Nayana 2013; Prasetyawati *et al.* 2021). On the other side, the use of three SWB indicators, life satisfaction, positive affect, and negative affect has become well-established in the scientific literature (Proctor 2014). Previous research measured three facets of SWB but focused on the general population, not students (Akhtar 2019). The main purpose of this research is to test the reliability and validity of SWB measurement in high school students.

2 METHODS

2.1 *Research subject*

The subjects of this study are 238 high school students consisting of 152 (64%) women and 86 (36%) men aged 15–18 years old. The sampling method used is the nonprobability sampling method.

2.2 *Variables and research instrument*

The variable in this research is SWB that consisted of three different facets that are life satisfaction, positive affect, and negative affect. Data collection in this study used three different scales. The first scale is the student's life satisfaction scale (SLSS) which measures the general life satisfaction of students (Huebner 1991). SLSS consisted of 7 items such as "Hidup saya berjalan baik" and "Hidup saya lebih baik dari kebanyakan anak." The participants were asked to rate their level of agreement with the given statement by choosing between range 1 (strongly disagree) – 6 (strongly agree). The second scale is the brief multidimensional student's

Table 1. Demographic data of the research subject.

Category	Frequency	Percentage
Gender		
Man	86	64%
Woman	152	34%
Age		
15 y.o	13	5%
16 y.o	115	49%
17 y.o	101	42%
18 y.o	9	4%
Pocket money per month		
<33,5 USD	35	15%
33,5–67 USD	156	66%
67–134 USD	44	18%
>134 USD	3	1%
Parent marriage status		
Married	205	86%
Divorce	27	11%
Widowed	6	3%

life satisfaction scale (BMSLSS) which helped measure satisfaction in the life domain (Huebner *et al.* 2006). BMSLSS has 5 items that measure five different domains namely family, friends, school experience, self, and living place. BMSLSS asked the participant how satisfied they are with each of the important domains in their lives by choosing between range 1 (terrible) – 7 (delighted). The last scale is positive affect negative affect scale (PANAS) which can measure both positive affect (PA) such as "Tertarik", "Bersemangat", etc. and negative affect (NA) such as "Malu", "Gugup", etc. (Watson *et al.* 1988).

2.3 *Procedure and data analysis*

The first step in doing validation is forward translation and backward translation. Each translation process is carried out by psychology master students. Both have moderate levels of English proficiency. The translated scale was then assessed by the expert, namely three psychology lecturers. The expert assessed the representativeness of each item related to the construct domain. After getting suggestions and revising items by the input received, the process continues with testing the measuring instrument. The result of try out is then analyzed to determine the reliability and validity of the measuring instrument. The reliability was estimated using Cronbach's alpha while validity was proven using evidence based on internal structures with confirmatory factor analysis (CFA).

3 RESULT

3.1 *Reliability*

Reliability alpha for general life satisfaction is 0.758; domain life satisfaction is 0.805; positive affective is 0.78, and the negative affective is 0.846. All scales have alpha reliability that can be categorized as "acceptable" based on Geoge & Mallery (Gliem & Gliem 2003). Thus it can be said that measurement results with this tool can be trusted.

A summary of the results of the reliability estimation using Cronbach's Alpha and corrected total item correlation is presented in Table 2. Several items have a corrected total item

Table 2. Reliability and corrected total item correlation.

Facets	Scale	N	Corrected total item correlation	Cronbach's Alpha
General life satisfaction	SLSS	7	0.190 – 0.680	0.758
Domain life satisfaction	BMSLSS	5	0.528 – 0.654	0.805
Positive affect	PANAS	10	0.301 – 0.612	0.786
Negative affect	NANAS	10	0.407 – 0.705	0.846

correlation of less than 0.3. However, other literature states that the acceptable limit for the corrected total item correlation is at least 0.15 (Briggs & Cheek 1986). Items that showed corrected total item correlation less than 0.3 is a reversed item in SLSS that state "Saya ingin mengubah banyak hal dalam hidup saya" and "Saya berharap saya menjalani hidup yang berbeda." Other than those two, the rest of the items already showed a corrected total item correlation of more than 0.3.

3.2 *Validity*

3.2.1 *Validity of SLSS*

The author tested whether the 7 items were measuring one factor, general life satisfaction. The results of the CFA showed that the one-factor model was not fit, with Chi-Square = 71.09, df = 14, P-value = 0.00000, and RMSEA = 0.131. Therefore, a modification was made, where the measurement errors on several items are free to correlate with each other, as shown in Figure 1.

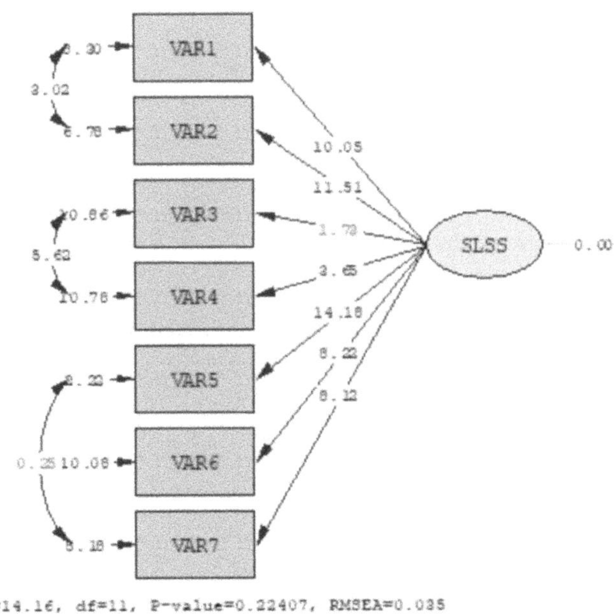

Figure 1. The results of the modified model met the goodness of fit criteria that have been set. The model showed RMSEA = 0.035 (< 0.08) with Chi-Square = 14.16. The results of other model fit tests such as CFI, GFI, AGFI, and RFI also show a value > 0.90 so the model is declared fit. Even though the model showed a good fit, there is one item that had a low loading factor, which is item number 3 which state "Saya ingin mengubah banyak hal dalam hidup saya".

Table 3. Values for the goodness of fit SLSS.

Goodness of fit indicator	Values
CFI	1.00
GFI	0.98
AGFI	0.96
RFI	0.96
NFI	0.98

3.2.2 *Validity of BMSLSS*

The author tested whether the 5 items were measuring one factor, domain life satisfaction. The results of the CFA showed a good fit with Chi-Square = 8.59, df = 14, P-value = 1.2663, and RMSEA = 0.055, as shown in Figure 2.

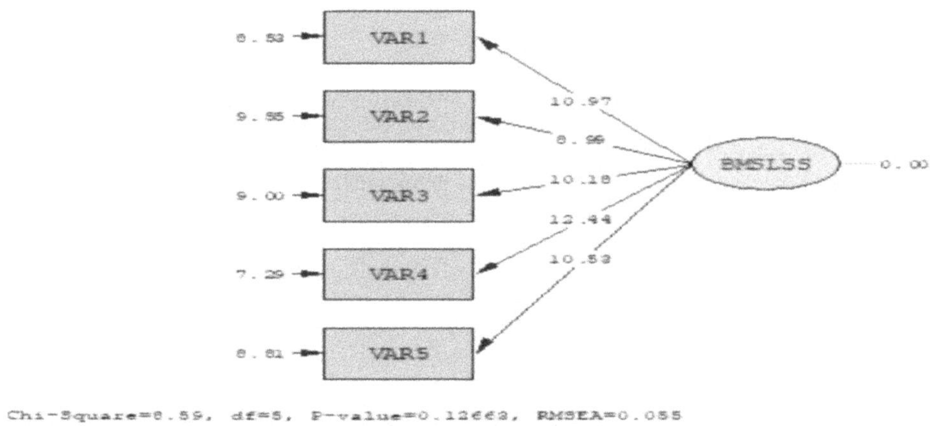

Figure 2. The results of other model fit tests such as CFI, GFI, AGFI, and RFI also show a value > 0.90 so the model is declared fit. Each item also has a significant loading factor that showed that each item had a contribution toward the latent variable.

Table 4. Values for the goodness of fit BMSLSS.

Goodness of fit indicator	Values
CFI	0.99
GFI	0.99
AGFI	0.96
RFI	0.96
NFI	0.98

3.2.3 *Validity of positive affect*

The author tested whether the 10 items were measuring one factor, positive affect. The results of the CFA showed a good fit with Chi-Square = 64.42, df = 35, P-value = 0.00178, and RMSEA = 0.060, as shown in Figure 3.

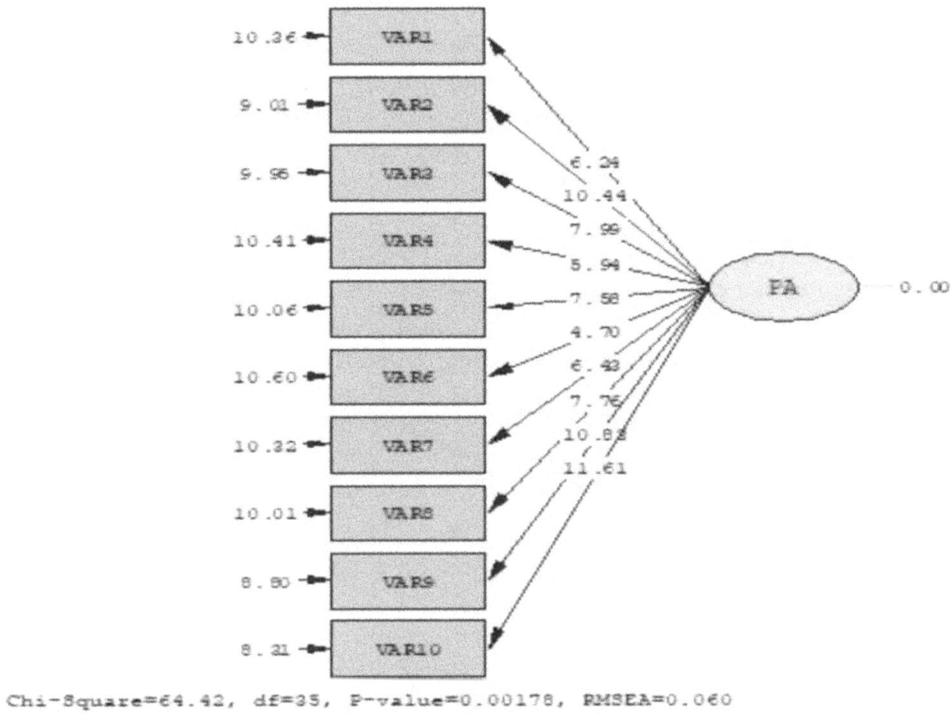

Figure 3. The results of other model fit tests such as Comparative Fit Index (CFI), Goodness of Fit Index (GFI), Adjusted Goodnes of Fit Index (AGFI), and Relative Fit Index (RFI) showed a value > 0.90 so the model is declared fit. Each item also has a significant loading factor that showed that each item had a contribution toward the latent variable.

Table 5. Values for the goodness of fit positive affect.

Goodness of fit indicator	Values
CFI	0.96
GFI	0.95
AGFI	0.92
RFI	0.91
NFI	0.93

3.2.4 *Validity of negative effect*

The author tested whether the 10 items were measuring one factor, positive affect. The results of the CFA showed that the one-factor model was not fit, with Chi-Square = 135.99, df = 35, P-value = 0.00000, and RMSEA = 0.110. Therefore, a modification was made, where the measurement errors on several items are free to correlate with each other, as shown in Figure 4.

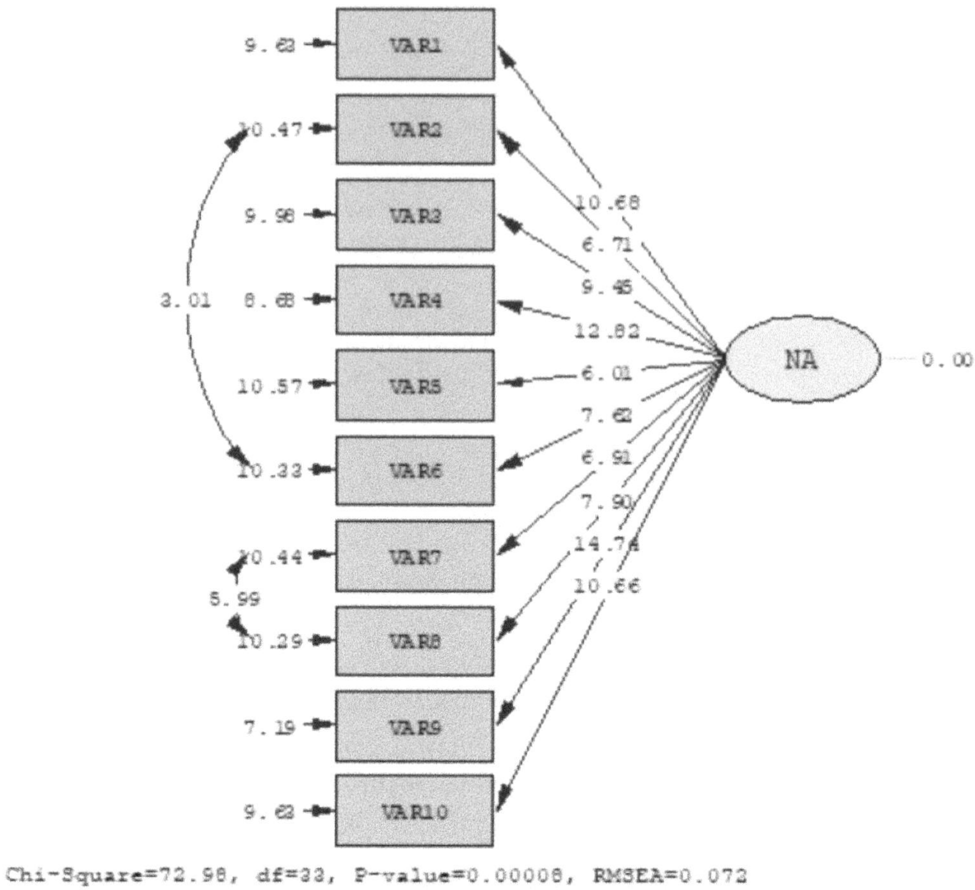

Figure 4. The results of the modified model met the goodness of fit criteria that have been set. The model showed RMSEA = 0.072 (< 0.08) with Chi-Square = 72.98 The results of other model fit tests such as CFI, GFI, AGFI, and RFI also show a value > 0.90 so the model is declared fit. The results of other model fit tests such as CFI, GFI, AGFI, and RFI also show a value > 0.90 so the model is declared fit. Each item also has a significant loading factor that showed that each item had a contribution toward the latent variable.

Table 6. Values for the goodness of fit negative affect.

Goodness of fit indicator	Values
CFI	0.97
GFI	0.90
AGFI	0.90
RFI	0.93
NFI	0.95

4 DISCUSSION

SWB is defined as an individual's cognitive and affective evaluation of his life (Diener 2000). SWB is a general category or phenomenon that includes individual emotional responses, satisfaction in the life domain, and global judgments related to life satisfaction (Diener *et al.* 1999). SWB focuses on how individuals evaluate their lives, especially in relation to the assessment of satisfaction (both with life in general and with specific domains such as marriage and work) and its relation to moods and emotions that reflect an event (Diener *et al.* 2006). The word 'subjective' refers to the definition and limitation of the construct, SWB researchers are interested in seeing an evaluation of a person's quality of life based on the individual's personal view (Diener *et al.* 2018).

As stated by (Diener 2000), two elements of SWB are cognitive element and the affective element. Cunsolo (2017) stated that the cognitive element refers to how individuals perceived their general life satisfaction based on a retrospective evaluation of life (as a whole) and specific domains (specific areas such as work, relationships, health, etc.). General life satisfaction refers to an individual's conscious assessment of his or her life, often based on factors deemed relevant. Some examples of measuring tools that can be used to measure general life satisfaction based on Diener, Lucas, & Oishi (2018) are the satisfaction with life scale consisting of five items (Diener E. *et al.* 1985) and SLSS intended for adolescents (Huebner 1991).

The scale used in this research for global life satisfaction is SLSS which focused on students'/adolescents' global life satisfaction. The data analysis showed that the adapted scale is reliable and valid. From a total of seven items, only one item that found to be problematic. That item is item number 3 which state "Saya ingin mengubah banyak hal dalam hidup saya." It indicated that even though most of the participants were satisfied with their life in general, there are still a lot of things they want to change. It didn't necessarily mean that the item isn't valid, but it showed the possibility that the participant strived to improve themselves. This is in line with previous research that found that an individual's grit was positively related to life satisfaction (Li *et al.* 2018; Liu *et al.* 2022).

In contrast to general life satisfaction, satisfaction in the life domain refers to evaluations in various life domains such as health, work, and relationships with others. Diener, Lucas & Oishi (2018) state that a measurement that can be used to assess satisfaction in the domain of a child's life is a BMSLSS (Huebner *et al.* 2006). The data analysis showed that all of the items in BMSLSS are reliable and valid. The item with the most contribution toward BMSLSS is item number four related to satisfaction with self. On the other hand, the least contributed item is item number two related to satisfaction with friendship.

Mood and emotion, labeled together as affect, represent individuals' evaluations of events in their lives. Bradburn and Capovitz (Diener 2000) proposed that positive affect and negative affect consist of two independent factors and should be measured separately. Positive affect is a feeling or mood that is positive, pleasant, and desirable. On the one hand, negative affect is a negative, unpleasant, and unwanted feeling or mood (Diener *et al.* 2018). Diener & Emmons (1985) found that positive affective experiences had little effect on negative affective experiences. In other words, positive and negative effects can vary independently. Diener & Emmons (1985) found that what affects an individual is not the frequency or affective experience, but the frequency of positive affect experienced. In other words, feeling pleasant emotions from time to time even in simple terms is enough to make a person feel happy (Diener, Subjective well-being: The science of happiness and a proposal for a national index, 2000). Proctor (2013) revealed that the measurement that is often used is the positive affect negative affect schedule (Watson *et al.* 1988) or the affect balance scale (Bradburn 1969).

Although measured on the same scale, positive and negative effect were analyzed separately. This is because positive affect and negative affect are two separate independent factors. Both positive and negative effects that were measured using PANAS showed good reliability and validity. For positive effect, the item that has the most contribution is item number ten that is "Aktif" and the item that has the least contribution is item number six that is "Waspada." Meanwhile, for negative affect, the item that has the most contribution is item number nine that is "Gelisah" and the least contributed item is item number two that is "Kesal."

The findings in this research are in line with previous research that found a good reliability and validity of SWB measurement in Indonesia (Akhtar 2019). The already validated measurement in this research can be used for further research in a larger population or to identify another factor that contributes to SWB in high school students.

5 CONCLUSION AND IMPLICATION

Based on the results of the analysis and discussion in this study, it can be concluded that the measuring instruments used to measure the SWB in Indonesian high school students, namely SLSS, BMSLSS, and PANAS are reliable and valid. Considering that the participants in this study are still limited, future studies can use a scale that is already validated in a larger population.

REFERENCES

Akhtar, H. (2019). Evaluasi Properti Psikometris dan Perbandingan Model Pengukuran Konstruk Subjective Well-being. *Jurnal Psikologi*, *18*(1), 29–40. https://doi.org/10.14710/jp.18.1.29-40

Arslan, G., & Renshaw, T. L. (2018). Student Subjective Welbeing as a Predictor of Adolescent Problem Behaviors: A Comparison of First-order and Second-order Factor Effects. *Child Indicator Research*, *11*(2), 507–521. https://doi.org/10.1007/s12187-017-9444-0

Bolier, L., Haverman, M., WEsterhof, G. J., Riper, H., Fmit, F., & Bohlmeijer, E. (2013). Positive Psychology Interventions: A Meta Analysis of Randomized Controlled Studies. *BMC Public Health*, 199.

Bradburn, N. M. (1969). *The Structure of Psychological Well-being*. Aldine.

Briggs, S. R., & Cheek, J. M. (1986). The Role of Factor Analysis in the Development and Evaluation of Personality Scales. *Journal of Personality*, *54*(1), 106–148. https://doi.org/10.1111/j.1467-6494.1986.tb00391.x.

Bucker, S., Nuraydin, S., Simonsmeier, B., Schneider, M., & Luhmann, M. (2018). Subjective Well-being and Academic Achievement: A Meta-analysis. *Journal of Research in Personality*, *74*, 83–94. https://doi.org/10.1016/j.jrp.2018.02.007.

Compton, W. C., & Hoffman, E. (2013). *Positive Psychology: The Science of Happiness and Flourishing*, 2nd Edition. Belmont: Wadsworth.

Cunsolo, S. (2017). Subjective wellbeing During Adolescence: A Literature Review on Key Factors Relating to Adolescent's Subjective Wellbeing and Educational Outcomes. *Studi sulla Formazione*, 81–94.

Diener. (2000). Subjective well-being: The Science of Happiness and a Proposal for a National Index. *American Psychologist*, *55*(1), 34–43.

Diener, & Ryan. (2009). Subjective Well-being: A General Overview. *South African Journal of Psychology*, 391–406.

Diener, E., & Emmons, R. A. (1985). The Independence of Positive and Negative Affect. *Journal of Personality and Social Psychology*, *47*(5), 1105–1117.

Diener, E., Heintzelman, S., Kushlev, K., & Tay, L. (2016). Findings all Psychologist Should Know From the New Science on Subjective Well-being. *Canadian Psychology*, 1–42.

Diener, E., Lucas, R. E., & Oishi, S. (2002). Subjective well-being: The Science of Happiness and Life Satisfaction. In C. R. Snyders, & J. Lopez, *Handbook of Positive Psychology* (pp. 63–73). New York: Oxford University Press.

Diener, E., Lucas, R. E., & Oishi, S. (2018). Advances and Open Questions in the Science of Subjective Well-being. *Collabra Psychology*, 1–76.

Emmons, R. A., & McCullough, M. E. (2003). Counting Blesings Versus Burdens: An Experimental Investigation Gratitude and Subjective well-being in daily life. *Journal of Personality and Social Psychology*, *84*(2), 377–389.

Froh, J. J., Sefick, W. J., & Emmons, R. A. (2008). Counting Blessings in Early Adolescents: an Experimental Study of Gratitude and Subjective Well-being. *Journal of School Psychology*, *46*, 213–233.

Gandner, F., Proyer, R. T., Ruch, W., & Wyss, T. (2013). Strength-based Positive Intervention: Further Evidence for Their Potential in Enhancing Well-being and Alleviating Depression. *Journal of Happiness Study*, *14*, 1241–1259.

Gliem, J. A., & Gliem, R. R. (2003). Calculating, Interpreting, and Reporting Cronbach's Alpha Reliability Coefficient for Likert-Type Scales. *Midwest Research to Practice Conference in Adult, Continuing, and Community Education*, (pp. 82–88). Columbus.

Goldbeck, L., Schmitz, T. G., Besier, T., Herschbach, P., & Henrich, G. (2007). Life Satisfaction Decreases During Adolescence. *Quality of Life Research*, 969–979.

Gregmini, P. (2020). The Importance of Using Valid and Reliabel Measures in Psychology and Psychiatry. *EC Psyhology Psychiatry*, *9*(4), 24–25.

Hamdan, F., & Alhamdu. (2015). Subjective Well-being dan Prestasi Belajar Siswa Akselerasi MAN 3 Palembang. *PSIKIS-Jurnal Psikologi Islam*, *1*(2), 115–124.

Huebner, E. S. (1991). Initial Development of the Student's Life Satisfaction Scale. *School Psychology International*, *12*(3), 231–240. https://doi.org/10.1177/0143034391123010.

Huebner, E. S., Seligson, J. L., Valois, R. F., & Suldo, S. M. (2006). A review of the Brief Multidimensional Students' Life Satisfaction Scale. *Social Indicator Researd*, *79*, 477–484. https://doi.org/10.1007/s11205-005-5395-9.

Ismail, A. N., & Yudiana, W. (2020). Subjective Well-being Pada Siswa Pesantren Modern Dan Siswa Madrasah Aliyah. *Jurnal Psikologi Islam dan Budaya*, *3*(1), 13–22.

Jose, P., Lim, B. T., & Bryant, F. B. (2012). Does Savoring Increase Happiness? A Daily Diary Study. *The Journal of Positive Psychology*, *7*(3), 176–187.

Kaplan, Y. (2017). School-specific Subjective Well being and Emotional Problems Among High School Adolescents. *Journal of Positive Psychology & Wellbeing*, *I*(I), 1–9.

Li, J., Fang, M., Wang, W., Sun, G., & Cheng, Z. (2018). The Influence of Grit on Life Satisfaction: Self-esteem as a Mediator. *Psychologica Belgica*, *58*(1), 51–66. http://doi.org/10.5334/pb.400.

Liu, H., Yu, Z., Ye, B., & Yang, Q. (2022). Grit and Life Satisfaction Among College Students During the Recurrent Outbreak of COVID-19 in China: The Mediating Role of Depression and the Moderating Role of Stressful Life Events. *Frontier Public Health*, 1–8. http://doi.10.3389/fpubh.2022.895510.

Manzoor, A., Siddique, A., Riaz, F., & Riaz, A. (2014). Determining the Impact of Subjective Well-being on Academic Achievement of Children in District Faisalabad. *Mediterranean Journal of Social Sciences*, *5*, 2673–2678.

Nayana, F. N. (2013). Kefungsian Keluarga dan Subjective Well-being pada Remaja. *Jurnal Ilmiah Psikologi Terapan*, *1*(2), 230–244.

Nee, C. H., Yaacob, S. N., Baharudin, R., & Jo-Per, T. (2016). Adolescents' life Satisfaction in Selangor, Malaysia: The effect of demographic differences. *Asian Social Science*, *12*(2), 1–8.

Nisa, P. A. (2019). *Hubungan Antara Dukung Sosial dengan Subjective Well-Being pada Siswa Genrus Nusantara Boarding School.* Yogyakart: Skripsi tidak dipublikasikan.

Opshaung, E. A. (2013). *Life Satisfaction in Adolescence: A Longitudinal Study of Associations with Body Image, Negative Emotionality, Social Support, and Life Events.* Oslo: University of Oslo.

Park, N. (2004). The Role of Subjective Well-being in Positive Youth Development. *The Annals of the American Academy of Political and Social Science*, *591*(1), 25–39.

Prasetya, P. (2015). *Perbedaan Subjective Well Being dan Hardiness Pada Siswa SMA Program Akselerasi Dengan Program Reguler di Surakarta.* Surakarta: Skripsi, tidak dipublikasikan.

Prasetyawati, W., Rifameutia, T., Gilles, R. M., & Newcombe, P. A. (2021). Adaptasi Brief Adolescent Subjective Well Being in School Scale (BASWSBSS), Skala Kesejahteraan Subjektif Siswa Dalam Konteks Indonesia. *Anima, Indonesian Psychological Journal*, *36*(2), 184–203.

Proctor, C. L. (2013). Subjective well-being. In A. Michalos, *Encyclopedia of Quality of Life and Well-Being Research* (pp. 2735–2743). Springer.

Proctor, C. L. (2014). Subjective well-being. In A. Michalos, *Encyclopedia of Quality of Life and Well-being Research* (pp. 6437–6441). Springer.

Sukmasari, R. N. (2016, Oktober 11). *Dari Skrining 1.300 Siswa SMA di Jaksel, 30 Persen Berpotensi Depresi.* Retrieved Agustus 19, 2019, from Detik Health: https://health.detik.com/berita-detikhealth/d-3317807/dari-skrining-1300-siswa-sma-di-jaksel-30-persen-berpotensi-depresi

Suldo, S., Thalji, A., & Ferron, J. (2011). Longitudinal Academic Outcomes Predicted by Early Adolescents' Subjective Well-being, Psychopathology, and Mental Helath Status Yielded From a Dual Factor Model. *Journal of Positive Psychology*, *6*(1), 17–30.

Thapar, A., Collinshaw, S., Pine, D., & Thapar, A. K. (2012). Depression in Adolescence. *Lancet*, *379*(9820), 1056–1067.

Watson, D., Clark, L. A., & Tellegen, A. (1988). Development and Validation of Brief Measures of Positive and Negative Affect: The PANAS scales. *Journal of Personality and Social Psychology*, *54*(6), 1063–1070. https://doi.org/10.1037/0022-3514.54.6.1063.

Author index

For Product Safety Concerns and Information please contact our EU
representative GPSR@taylorandfrancis.com
Taylor & Francis Verlag GmbH, Kaufingerstraße 24, 80331 München, Germany

www.ingramcontent.com/pod-product-compliance
Ingram Content Group UK Ltd.
Pitfield, Milton Keynes, MK11 3LW, UK
UKHW052049230926
475702UK00025B/979
9781032514420